THE COMPLETELY REVISED AND UPDATED FAST-FOOD GUIDE

2ND EDITION

2ND EDITION

THE COMPLETELY REVISED AND UPDATED FAST-FOOD GUIDE

What's good, what's bad, and how to tell the difference.

MICHAEL F. JACOBSON, PH.D.

Executive Director
Center for Science in the Public Interest

SARAH FRITSCHNER

Workman Publishing, New York

Library of Congress Cataloging-in-Publication Data

Jacobson, Michael F.
    The fast-food guide / by Michael F. Jacobson and Sarah Fritschner.
    —2nd ed., completely rev. and updated.
        p.   cm.
    Includes index.
    ISBN 0-89480-823-0 (pbk.)
    1. Fast food restaurants—United States. 2. Convenience foods.
I. Fritschner, Sarah.   II. Title.
TX945.J3   1991
647.9573—dc20   90-50952
                                                                 CIP

Cover design: Charles Kreloff
Cover illustration: Karen Kluglein
Book illustration: Kenneth Spengler

Workman Publishing Company, Inc.
708 Broadway
New York, NY 10003

Manufactured in the United States of America

10   9   8   7   6   5   4   3   2   1

Acknowledgments

We acknowledge, with gratitude, the efforts of our colleagues who contributed to the first and second editions of this book. Bonnie Liebman, nutrition director of the Center for Science in the Public Interest, wrote several articles in *Nutrition Action Healthletter*, identifying key problems with fast foods; she and Jayne Hurley, CSPI's associate nutritionist, helped ensure the accuracy of the health information in this book. Jill Mathews, Beth Burns, Richard Layman, and Lorraine Jones were extremely helpful in obtaining information from fast-food companies, entering it into the computer, and proofreading endlessly. CSPI attorneys Bruce Silverglade and Charles Mitchell pressed the issue of nutrition and ingredient disclosure. Thanks, also, to Donna Lenhoff who did not, and to Howard Lenhoff who eagerly did, look forward to our tasting forays into fast-food restaurants.

We'd also like to thank numerous officials of fast-food chains for providing us with nutrition information and alerting us to menu changes, sometimes in advance of public release.

Finally, we are grateful to Peter Workman and Suzanne Rafer at Workman Publishing for encouraging us to produce this revised edition; to Shannon Ryan and Beth Pearson for their painstaking editorial assistance; and to Harry Schroder for making the book lay-out so well.

Michael F. Jacobson
Washington, D.C.

Sarah Fritschner
Louisville, Kentucky

October 1991

FAST-FOOD UPDATE

The winds of change have been blowing through the fast-food industry. Twenty years ago, the word "nutrition" would not have been whispered anywhere near a fast-food restaurant. Then, slowly, Burger King, Wendy's, Carl's Jr., and a few other chains began introducing a salad here and a baked potato there. But hamburger joints were still largely hamburger joints, offering you your fill of calories, sugar, grease, and salt.

When we wrote the first edition of *The Fast-Food Guide* in 1986, we were hard-pressed to identify any healthy foods at many of the restaurants. Most chains refused to disclose ingredients, and several chains had never measured the nutritional values of their products.

We're happy to report progress: It is becoming increasingly possible to have a healthy, tasty meal at a fast-food restaurant. Nutrition previously received only lip service from corporate executives; now nutrition is part of the process of developing new foods and menus.

Companies have been both improving existing foods and introducing healthier new ones. Most have switched from beef fat to vegetable shortening for frying, several have replaced fatty mayonnaise sauces with low-calorie sauces, and McDonald's has reduced the fat content of its shakes. Companies have also added such new items as grilled chicken sandwiches, fat-free muffins, and nonfat sorbet. Even dessert-oriented Baskin-Robbins and Dairy Queen have been touched by the nutrition fairy, adding nonfat yogurt to their virtually endless lines of fatty, sugary cones, sundaes, and shakes.

Of course, that's not to say that fast-food restaurants may now be considered health-food restaurants. No way. The companies have not so much been eliminating their unhealthy items as adding new choices. Cholesterol addicts, who comprise the bulk of the chains' customers, will be comforted to know that they can still choose Jack in the Box's Ultimate Cheeseburger—probably the single worst fast food—Burger King's Double Whopper with Cheese, Dairy Queen's Heath Blizzard, McDonald's Biscuit with Sausage & Egg, and giant, 32-ounce soft drinks.

The evolution of fast foods over the past several years provides a wonderful example of how consumer activists can force changes when an industry is marketing harmful products and when government watchdogs slumber. The Center for Science in the Public Interest (CSPI) began focusing public attention on the deplorable nutritional quality of fast foods in 1983. In the years that followed:

- CSPI recruited prominent physicians and health organizations to demand more healthful foods.

- Consumer groups demanded that companies disclose what their products were made of.

- CSPI petitioned the Food and Drug Administration to require ingredient labeling of fast foods.

- Legislators introduced bills calling for nutrition and ingredient labeling.

- State attorneys general ordered several companies to halt deceptive ads and to provide ingredient and nutrition information to consumers.

- CSPI published newspaper articles, a wall chart, and this book on fast foods.

- A wealthy businessman sponsored full-page newspaper ads criticizing companies that fried in beef fat.

- Countless magazine and newspaper articles and television and radio shows discussed the nutritional merits of fast foods.

Alone, each of those actions would have had no effect. Together, they persuaded multi-billion-dollar corporations to improve their products and provide more information to their customers.

Currently, many major fast-food companies are in a no-growth period, partly because millions of people feel they have no reason to ever eat in such places. To lure those people into their restaurants, the large chains will likely be offering a wider array of healthy foods, such as whole-grain buns, fresh fruit or fruit salad, vegetarian sandwiches and entrées, and nonfat yogurt. Such foods are currently available in one or another (often small) chain. Still, it might take the industry years to persuade health-oriented individuals that fast-food restaurants really do have something to offer.

Several chains already provide nutrition brochures or post charts on their walls. Perhaps some chains will actually print information on labels, just like other packaged foods. One excellent measure would be to post the calorie content next to each item listed on menu boards.

We don't urge anyone to eat at a fast-food restaurant. Frankly, it would save you a lot of time and money to eat at home, pack a lunch, or pick up a salad, fruit, or yogurt at a supermarket. But if you do decide to drop into a fast-food outlet, we hope that this book will enable you to choose meals that contribute to your good health.

Michael F. Jacobson
Sarah Fritschner

A FAST SURVEY OF FAST FOOD

When Ray Kroc opened the doors of that first McDonald's restaurant in Des Plaines, Illinois, in 1955, he opened a whole new world for busy parents, fussy eaters, and people who just plain did not like to cook.

It has come to be a world of burgers and quick-fried chicken ready before the kids begin to yank on each other, ready to pick up on the way home from work, ready without dirtying that first pan or laying the first piece of flatware. Kitchen cleanup is just a trash can away.

It started a love affair with the french fry. And it no doubt encouraged the love affair with soft drinks. From 1963 to 1990, soft-drink consumption per capita more than tripled.

These restaurants fueled other dietary changes occurring in America since the turn of the century. In addition to french fries and soft drinks, Americans consume more meat, fish, poultry, and cheese than they did in 1910. If we take a look at fast-food-restaurant menus, we can understand one major reason why.

Changes in the foods we consume have resulted in changes in the nutrients we consume. In the past 80 years, our fat intake has increased about 20 percent. Sugar consumption has increased 40 percent. We eat more processed foods that often come with high levels of salt and fat. Fast-food restaurants take things one step further by making their own distinct modifications to traditional foods. Offering a ham and Swiss cheese sandwich on a croissant, for instance, adds far more fat and calories than we would get by eating the same sandwich made with two slices of bread.

Consuming three out of five calories as fat or refined sugar, as many people do, affects more than our waistlines. It crowds out fresh fruits and vegetables, milk, and other nutrient-packed foods.

In addition, we don't get the fiber we need—fiber that may help protect us against heart attacks and cancers (which are promoted by a diet high in fat and calories).

Some of our dietary changes were a result of increasing affluence. And as we have prospered since World War II, other changes in society have affected the way we eat.

In 1960, Wendy's wasn't even a twinkle in David Thomas's eye and only 250 McDonald's sprinkled the landscape. In those days, Americans chose fast-food restaurants only 1 in 20 times when they ate out, which wasn't that often. Most moms stayed home with the kids and cooked (whether they liked it or not). Dad worked and brought home the only money to support the family. When they went out to eat, they generally chose an independently run restaurant with a variety of foods offered on the menu.

But then many moms joined the work force. Since Dad was working too, time for doing chores at home diminished. When people work outside the home 40 or more hours a week, they make more money, have less time, and need to make more efficient use of the free time they have. Cooking gets short shrift in such a world. Fast foods were made to order. Nearly everyone agrees that the working woman was a phenomenon that contributed greatly to the success of the fast-food industry.

TODAY'S SEDUCERS

In a society known for its time constraints, even eating is measured carefully against other priorities. Fast foods make the desk or the car seem a suitable dining room. We'd rather sleep late in the morning than eat, so many of us skip breakfast, or detour on the way to work through the drive-in window for biscuits and coffee. We spend lunch hours working or shopping or doing the odd errand so we eat standing at a kiosk, on the run, waiting at a stoplight, or back at the office. Dinner hours are diffused by all manner of distractions, from cheerleading practice to working overtime. Fast-food restaurants allow us to eat quickly without planning, without dressing up, without having to make many decisions, and without getting out of the car. For the hurried, harried, and overworked, it's eat and run at reasonable prices.

The presence of children makes fast foods even more appealing. One or two working adults might content themselves with tuna from a can and a salad, but good parents find it difficult to impose such a diet on their children. We've always been taught that children must have something hot and at least a semblance of a vegetable on their plates. So much the better if they actually eat the stuff. Contenting their children with hamburgers, fries, and soft drinks, parents feel they've done their job.

The fast-food restaurant seems remarkably resilient to anything toddlers can do to it. They can spill or smush food—the tables and floors are made for mopping and wiping. Service is quick enough that you don't have to mollify your children with crackers only to find out they aren't hungry when dinner arrives. And as one mother observed, "You don't have children climbing on chairs that are going to fall over."

Restaurants have their allure for single working parents, whose quality time with children can be critically meager anyway. They don't need to ruin it trying to coax Jason to eat his peas. Fast-food restaurants mean hassle-free dinner, oftentimes with a playground thrown in for good measure.

Our mobility accounts for no small portion of fast food's popularity. With two bored children in the backseat, a car pointed toward Disney World, and 15 hours to drive, you'd like to know that mealtime will be a pleasurable experience. While some restaurants build their reputations on style and creativity, fast-food restaurants build theirs on uniformity and dependability. From Tacoma to Tallahassee, one Original Recipe chicken wing tastes exactly like another.

Mobility has become such a way of life it affects our eating habits. Many of us who grew up on fast foods have adopted the practice called "grazing," or eating small amounts of food all through the day rather than eating at three or four designated times—a feeding pattern that doesn't take much time and leaves the impression of having eaten very little.

"Grazers are largely members of the first generation to grow up on fast foods. They eat on impulse rather than three times a day," *The Wall Street Journal* has reported. As a result, people expect to eat when they are ready, not when the food is. Eating on impulse leaves plenty of opportunity for fast-food restaurants that thrive on selling snacks — chicken nuggets, a slice of pizza, and nachos to go. Now.

Fast-food dining fits in with the flexible suburban lifestyle that puts a premium on speed and privacy. The McDonald brothers seized on these concepts as early as 1948 when, in an effort to reduce personnel, they replaced carhops with window clerks. They used chalk to draw a floor plan of the restaurant on their tennis court "where they meticulously worked out the placement of windows and equipment to eliminate any unnecessary movement by the personnel," wrote Philip Langdon in the *Atlantic Monthly*.

These days it takes no more verbal contact than a "Help you ma'am?" and a total-bill tally, with perhaps a "Have a nice day" thrown in by especially animated clerks. The restaurants are clean, the prices moderate and predictable, and we can get through an entire meal without getting to know the quirks and travails of Bob, our waiter.

Many of us still turn to fast food remembering the early days when hamburgers cost 15 cents, fries a dime, and milk shakes less than a quarter. The low-cost image remains, despite the fact that fast food now sometimes costs as much as the fare at a table-service restaurant. Fast-food meals can cost you more than double what it would cost to make them at home — even without the extra dollar or two for the optional toys that you can buy your children.

As society was changing, baby boomers got older and had more money to spend. The 1980s brought an adult population that had never lived without fast foods, and children who couldn't imagine a world without Whoppers. Now, 5 out of every 10 dollars we spend on restaurant food is handed across the counter of a fast-food restaurant, according to the U.S. Department of Agriculture. Each American spends an average of $250 a year on fast foods.

In the '50s, nobody worried much about burger joints and chicken outlets. They offered respite to Mom and a treat for the kids. Mom-and-pop restaurants offered a variety of foods sure to please every taste. The new-style restaurants reduced the pleasure to a common denominator — a hamburger or chicken or pizza — and became known in the trade as "limited-menu" establishments.

And limited they were. In 1955, your McDonald's meal could consist of nothing but a hamburger or cheeseburger, fries, and a

THE BIGGEST RESTAURANT CHAINS*

COMPANY	1990 SALES ($ BILLION)	NO. OF RESTAURANTS
1. McDonald's	$18.759	11,803
2. Burger King	$6.100	6,298
3. KFC (Kentucky Fried Chicken)	$5.800	8,187
4. Pizza Hut	$4.845	8,000
5. Hardee's	$3.400	4,010
6. Wendy's	$2.980	3,721
7. Domino's Pizza	$2.650	5,376
8. Taco Bell	$2.545	3,277
9. Dairy Queen	$2.280	5,207
10. Arby's	$1.535	2,490
11. Little Caesars Pizza	$1.500	3,300
12. Denny's	$1.370	1,358
13. Subway	$1.200	5,170
14. Big Boy	$1.100	970
15. Jack in the Box	$0.953	1,040
16. Dunkin' Donuts	$0.943	2,166
17. Long John Silver's	$0.781	1,446
18. Baskin-Robbins	$0.714	3,318
19. Carl's Jr.	$0.614	572
20. Church's Fried Chicken	$0.575	1,217
21. Popeyes Famous Fried Chicken	$0.557	755
TOTAL	$64.556	82,413

* *Adapted with permission from* Restaurant Business, *March 20, 1991. Figures include overseas outlets.*

beverage. That formula was good enough for Wendy's, too, which opened almost 15 years later, adding only chili—a handy receptacle for unsold burgers.

But one McDonald's has grown to more than 11,000 outlets, at which a half million people work. One new McDonald's goes up every 15 hours somewhere around the world, and sales have topped the $18-billion-a-year mark. Other chains built more restaurants to cash in on the quest for convenience. Now, upward of 160,000 fast-food restaurants compete for the consumer's dollars—about $70 billion in 1990. The Department of Agriculture reported that in 1987 fast-food operations for the first time outnumbered traditional restaurants and in 1988 came out on top in terms of sales.

One out of five Americans eats at a fast-food restaurant on a typical day, according to *Consumer Reports*. Four out of five Americans visit a fast-food restaurant every month, says Opinion Research Corporation.

As for cashing in, the top 50 restaurant chains (not all of them "limited menu") sold more than $66 billion worth of food in the U.S. and abroad in 1989, according to the trade journal *Restaurant Business*. Not surprisingly, McDonald's topped the list, with Burger King, KFC (Kentucky Fried Chicken), Pizza Hut, and Hardee's filling out the top five. And those five fast-food restaurants account for 52 percent of all sales in the top 50.

In the last couple of years, restaurant growth has slowed. Ron Paul, an industry analyst, told *Adweek* that "all the [restaurant] growth will be in quick-service," but even that segment will grow by only 2 or 3 percent per year. An exception to the slowdown is the pizza business, which *Adweek's Marketing Week* says is growing by 10 percent a year.

To boost sales in late 1990 and 1991, chains began competing more on the basis of price than toys and games. Restaurants started offering 29-cent miniature tacos, two-for-one pizzas, and cut-rate burgers. Welcome as this may be to consumers, the price cutting is tough on small- and mid-size companies. When Marriott Corporation announced its intention to sell its Roy Rogers fast-food chain, J. W. Marriott, Jr., said that price cutting by McDonald's and other competition forced them to sell. The price war has cut the industry's profits and had the major chains looking overseas for growth.

The fast-food industry feels that until all tables are full at every time of day and there's a steady stream of cars in the drive-through lanes, restaurants will not be fulfilling their potential. To keep profits growing, fast-food executives will diversify, improve, and advertise their products, luring us in for yet one more cheeseburger. They are building new restaurants, redecorating old ones, and even toting some around on wheels to make that cheeseburger more pleasurable and more accessible than ever before.

IMAGE FACELIFTS

All over America fast-food restaurants are changing the way they look, from the gaudy candy striping of the early days to subtler and more sophisticated designs. These attempts toward tasteful design reflect two significant changes in the fast-food market.

First, fast-food patrons are getting older, and they appreciate the sophisticated look. In pursuit of individual style, "typically, operators are aiming for subtler lighting and warm, earthy environments, reminiscent of the fern bars that overtook urban areas . . . ," reported *The Wall Street Journal*, which added that restaurant volume nearly always goes up after remodeling. Burger King, seeking to reverse a reputation for slowness and sloppiness, has remodeled hundreds of outlets in the past year. Its restaurants feature real plants, brass railings to queue customers, and etched glass to separate dining areas.

Just as important as appealing to the older customers, fast-food restaurants need to attract new ones. In the old days, restaurant operators simply needed to stake out a spot on a corner, put up four walls, and turn on the fryer. These days, competition is tougher, and they need to attract new customers to keep production up.

Upscale designs are one lure that attracts a broader range of people, "especially those people who have traditionally said, 'I don't go into those kinds of restaurants,'" one Wendy's franchisee told *The Wall Street Journal*.

In the financial district of New York, one McDonald's, according to *The Lempert Report* newsletter, "has black marble, mirrored atriums, chandeliers, palm trees, flowers on each table, and lots of (live) plants. There is a doorman in uniform, a pianist on [sic] a baby grand,

FAST-FOOD BUSINESS STATISTICS*

COST OF BUILDING A NEW OUTLET

McDonald's	$1,400,000
Hardee's	$1,100,000
Kentucky Fried Chicken	$700,000 to $1,100,000
Long John Silver's	$600,000

AVERAGE ANNUAL SALES PER OUTLET

McDonald's	$1,645,000
Burger King	$943,000
Kentucky Fried Chicken	$714,000
Pizza Hut	$558,000
Baskin-Robbins	$190,000

FRANCHISE FEES

Burger King	$40,000
McDonald's	$22,500
Kentucky Fried Chicken	$20,000
Long John Silver's	$20,000
Hardee's	$15,000
Subway	$7,500
Domino's Pizza	$6,000

and hostesses." Coffee machines dispense cappuccino and espresso at over two dollars a cup. A McDonald's in Columbus, Ohio, according to *Nation's Restaurant News*, features Italian marble and original artwork. Meanwhile, a Burger King in Honolulu is dressed in chrome and neon and provides free newspapers at breakfast.

In fact, fast-food restaurants may get so comfortable that they

ROYALTIES AND ADVERTISING FEES (% of sales)	
McDonald's	12%
Subway	10.5%
Long John Silver's	9%
Domino's Pizza	8%
Burger King	7.5%
Wendy's	4%
ODDS AND ENDS	
Number of BK Broilers sold per day:	1 million
Pounds of sesame seeds served on Big Mac buns annually:	7,000,000
Pounds of french fries prepared by McDonald's daily:	2,000,000
Number of chickens used by Kentucky Fried Chicken annually:	500 million
Of every $100 Americans spend eating out, McDonald's share is	$7.30
Average price of an Arby's sandwich:	$2.02
Average cost of ingredients:	$0.54

* *Business statistics for this chart taken from* Venture, *December, 1988;* The Wall Street Journal, *September 29, 1989;* Restaurant Business, *March 20, 1990;* Nation's Restaurant News, *August 27, 1990; and* Adweek's Marketing Week, *June 25, 1990.*

will lose the one attribute that Ray Kroc perfected so well—turnover. Kroc didn't allow pay phones or cigarette machines in his restaurants, because he didn't want to have to deal with the bane of the burger joint: loitering teenagers. "Kroc wanted the teenage trade," Philip Langdon wrote in the *Atlantic Monthly*, "but he hoped to keep the transactions quick and limited."

In addition, he kept restaurant seating hard and immobile. "Sitting down at a typical McDonald's," Langdon wrote, "customers felt immediate relief in the cushioning of the seat backs. A few minutes later, however, they became restless because the seat bottom consisted of hard, uncushioned plastic." In an effort to keep the restaurant neat and the customers moving, seats were screwed to the floor. After shifting in his seat once or twice, the typical customer was on his way. Since Wendy's was designed to appeal to an older crowd, restaurant design didn't need to be so austere. Still, according to Wendy's president David Thomas, the tables at his restaurants are fitted with "four not particularly comfortable chairs." Several McDonald's in Washington, D.C., go one step further than hard chairs. They have posted signs warning, "No Loitering. 20 minute time limit consuming food." Next thing you know they'll put parking meters by the tables and give tickets to slow eaters.

MORE CONVENIENT

Fast-food restaurants are being spruced up, but beauty's only skin deep. Convenience is the core of the fast-food business and another way to compete for more customers.

Proximity is an important aspect of convenience — nobody wants to drive far or make dangerous turns against oncoming traffic in order to get a mediocre hamburger. In fact, many of us would rather sacrifice flavor than spend time or effort getting better food, according to a 1988 *Consumer Reports* survey. Ask people what their favorite fast food is and you're likely to get a response similar to that of this Indiana diner, "I can tell you where I eat all the time. I can't tell you I like the food they serve there."

As competition gets tougher, the restaurants strive to make their outlets more convenient. McDonald's, according to *Nation's Restaurant News*, has even resorted to using satellite photos to help situate its restaurants so that hungry travelers won't go too far without bumping into the famous golden arches.

More than half of fast-food business is now done at drive-through windows. To accommodate drivers more quickly, many restaurants have double lines, and one McDonald's in Chicago has five windows plus a home-delivery car fleet. Burger King is testing two-way color monitors so clerk and customer can see one another.

In some cities, Burger King now hauls its Whoppers right up to your door or playground. It's been converting recreational vehicles into mobile restaurants that offer everything a full-size restaurant can except onion rings, shakes, and salad bars. "It's the ultimate evolution of the fast-food industry; bring the restaurant to the people," Burger King vice president Stephen Finn told *The New York Times*. Restaurants on wheels mean additional low-cost, high-volume outlets for fast-food chains. Summer crowds on New Jersey beaches don't scare these guys. If lines get too long, clerks will "work the crowd, punching in sales on a hand-held keyboard that electronically relays the message to a printer in the van. A cook fills the order just as the customer reaches the front of the line. Employees can call in special requests on a radio headset," the newspaper reported. When summer ends and the tourists go home, the restaurant operator puts the transmission in drive and heads for another spot. Burger King even sent a mobile unit to Alaska after the March, 1989, Exxon Valdez oil spill. Mobile restaurant sales have been reported as high as $3,000 per day, about average for a Burger King restaurant without wheels. KFC's $200,000 mobile restaurant has been showing up at jazz festivals in Texas, while Taco Bell's $30,000 cart is in the Minneapolis airport.

The search for more convenient locations inspired Wendy's to set up next to the gorilla cage at the zoo in Columbus, Ohio. There are fast-food restaurants in hospitals, on military bases, on college campuses, at museums, at airports, and floating on the Mississippi River. Burger King has installed shops in Greyhound Bus stations. One analyst said toll roads are "wildly successful" locations for fast foods. State governments "would much rather have a McDonald's than a Howard Johnson's because McDonald's will double the sales," he said, and higher sales mean more money for the state because of rents and sales taxes.

Anywhere there is consumer traffic and a lunch hour, there is restaurant potential. "We finally took our blinders off and realized we had captive markets in places like K Mart," Wendy's spokesman Denny Lynch told *The New York Times*. Captive markets allow chains to install outlets not just on street corners, but everywhere from office buildings to naval ships. K Mart, in fact, will soon house Little Ceasars Pizza outlets in 400 of their discount stores.

Hospitals—symbols of health and healing—are particularly surprising sites for "limited menu" restaurants. But hospital managers are attracted to such restaurants because they can convert money-losing, headache-inducing cafeterias into profit centers. Staff and patients may like the fast-food outlets, but many doctors and dietitians are troubled by the easy availability to patients of junk foods.

When Valley Medical Center in Seattle leased space to McDonald's in 1990, Kathy Hunt, president of the Greater Seattle Dieticians Association, told the *Seattle Times*, "When you invite McDonald's to your hospital campus, you're saying it's OK to eat it all the time." Liz Stanton, chief nutritionist at Children's Hospital in Los Angeles, said that many doctors were upset by the McDonald's in that hospital, which was having a "devastating" effect on the full-service cafeteria. Carole Marcus, who also works at a California children's hospital, decried in *The New England Journal of Medicine* in 1990 "the symbolic implications of a major teaching hospital's not only condoning but actually endorsing the consumption of fast food."

One hospital that considered installing a McDonald's but then declined to do so is the Baptist Medical Center in Oklahoma City. The hospital's food-service director, Bill Barkley, said, "[Having the McDonald's] doesn't really mesh with our goals. It's like having a liquor store in a dependency clinic."

Ironically, while fast-food operations in hospitals infuriate many nutritionists, these outlets may turn out to be perfect sites for test-marketing more healthful foods. For instance, McDonald's first sold carrot and celery sticks in the 18 hospitals in which it has outlets. Those healthful items are now sold in all McDonald's. Burger King's unit in the parking garage of Harris Methodist Hospital in Fort Worth, Texas, paved the way for other Burger Kings by frying potatoes in vegetable shortening instead of beef fat.

Another irony is that fast-food companies are moving onto college campuses, and even into dormitories. Not too many years ago, those restaurants would have been greeted with picket signs condemning "plastic food." Now they're selling thousands of Pizza Hut pizzas, Dunkin' Donuts doughnuts, and Dairy Queen cones a day at such institutions as the University of South Carolina at Columbia, Central Missouri State University, and the University of North Carolina in Chapel Hill. According to *The Wall Street Journal*, some companies have arranged for students to be able to use their dormitory-cafeteria meal ticket to buy the brand-name fast foods.

DIVERSIFIED MENUS

In 1964, a brand-new Arby's opened in Boardman, Ohio, selling roast-beef sandwiches, potato chips, and beverages. Today, Arby's restaurants sell turkey sandwiches, fries, salads, stuffed potatoes, and much, much more.

The menu expansion that took Arby's from roast beef to croissants and baked potatoes is a relatively new concept in fast food. In the 1970s, fast food restaurants had no interest in new products, said Wendy's Denny Lynch. "If you were a hamburger chain, all you had to do was serve a hamburger, fries, and soft drinks; that was your formula for success." It was Wendy's president who told *Business Week* in 1977, "I've always felt that menu diversification was a sign of weakness."

In the 1980s, however, it was a sign of strength. The high price of beef forced executives to look at fish, chicken, and pork possibilities. Competing for the women's market, the restaurants began offering salad options—Wendy's found it could keep costs low by using hearts of lettuce left over from their fresh lettuce-topped hamburgers. In addition, many consumers' taste buds were tiring of burgers.

Restaurants galore have joined the baked-potato bandwagon, stuffing spuds with everything from chicken à la king to taco mix. Salad sales are fiercely competitive. Wendy's salad bar is "one of the industry's most prominent successes," according to *Advertising Age* magazine. And Burger King, McDonald's, Hardee's, and others offer several varieties of prepackaged salads with low-calorie or regular dressings.

Not only are burger joints diversifying their menus, but the fast-food market itself continues to broaden. The Mexican-food business began to realize its potential in the early '80s. Its strong point was a culinary paradox: they offered something more than hamburgers—but not too much more. They serve the same food—beef, cheese, tomato, and lettuce—but throw in a tortilla and just enough spice so that you know you're not at Burger King.

Pizza popularity continues to increase, with home delivery being the big change. Domino's Pizza doubled in sales between 1985 and 1988, while Pizza Hut leapfrogged over Wendy's and Hardee's to become the fourth-largest fast-food chain. McDonald's is now testing pizza and may give the pizza chains a run for their money.

Several factors affect menu changes. If food prices go off the map, as beef prices did in the late '70s, a restaurant will naturally seek alternatives to keep costs low, as Wendy's did with the salad bar. Sometimes menus expand because corporate headquarters wants to test a trend—Pizza Hut tested the regional food trend by selling barbecued pizza in Louisville. If the product succeeds—barbecued pizza did not—it winds up on menus nationwide.

Does diversity mean nutrition? In some cases, yes. The new low-fat hamburgers are a great improvement over the traditional greaseburgers. Plain baked potatoes, salad bars, broiled chicken, and baked fish all potentially improve your diet—especially if you choose them over fried potatoes, fried chicken, and fried fish.

Yet it seems that with every new healthful item comes another unhealthful one. The bacon-cheese-topping syndrome spread from restaurant to restaurant, adding extra fat and sodium to sandwiches. Fatty biscuits and croissants replace bread and rolls. Salads are only as good as their ingredients, and if you pour on the bacon bits and full-fat salad dressing you won't be doing your health a favor; you will be adding extra fat to an otherwise healthy choice.

Many fast-food restaurant executives claim that what people say and what they do are two entirely different things. Fast foods are popular, the restaurateurs say, because although people say they want nutritious foods, they actually buy for taste and convenience. But Americans' attitudes about eating are changing, and eating habits will also change, as day follows night. Fifteen years ago, people would have laughed if someone had suggested that hamburger joints would have handsome salad bars. Could vegetarian burgers be too far away?

Diversity generally does not mean speed. As fast-food restaurants offer more menu choices—McDonald's now offers more than 40 items—they become decidedly less fast. It's not uncommon to stand in line for five minutes and then have to wait, standing, another five minutes for an item to be prepared.

ADVERTISING

If menu diversity, convenient location, and chic decor aren't enough to lure you in, the restaurants also saturate the airwaves — and you — with commercials. In 1989, just eight of the large restaurant chains spent more than $1 billion on television advertising alone, according to *Restaurant Business*. The budget for all forms of advertising and promotions for all chains was probably over $2 billion.

McDonald's, the restaurant with top sales, not coincidentally is the single most advertised brand in the United States and probably the world. In 1990, the company spent $1.2 billion in the United States and overseas on ads and promotions. More than a billion dollars! It boggles the mind. McDonald's films so many commercials that it built a specially equipped "restaurant" to serve as a set for use by its ad agency and filmmakers.

That kind of investment seems to pay off: a survey of 24,000 people found that McDonald's had the most popular television commercials of 1989. The massive advertising campaigns that the big companies mount make life very difficult for small chains, let alone luncheonettes and other traditional, independently owned eateries that have largely disappeared from the American scene.

But some chains have figured out how to grow without spending billions on advertising. Arby's, one of the fastest-growing chains, spends only about $40 million a year on media advertising and has not advertised on network television for several years (of course, that doesn't mean that Arby's doesn't wish it could spend at McDonald's level and grow even faster).

The massive advertising campaigns for fast foods sometimes provide useful information, such as special prices, new products, and nutritional improvements. Other times, though, companies step over the line of honesty or tastefulness. The non-profit Action for Children's Television has criticized Burger King ads featuring the Kids Club, because the "club" is nothing more than a membership card, a poster, some stickers, and periodic promotional mailings, and not a fun social activity for kids. McDonald's ads for Big Macs have depicted fantasy sandwiches: crisp lettuce leaves, a fluffy bun covered generously with sesame seeds, and burgers larger than life. An earlier McDonald's ad claimed that Chicken McNuggets were made of pure chicken, even though at the time they contained ground-up

RESTAURANT
TV ADVERTISING (1989)*

COMPANY	AD SPENDING ($ Millions)
1. McDonald's	$412
2. Burger King	$162
3. Kentucky Fried Chicken	$116
4. Pizza Hut	$91
5. Wendy's	$88
6. Taco Bell	$58
7. Red Lobster	$57
8. Domino's Pizza	$56
9. Hardee's	$38
10. Long John Silver's	$26
11. Jack in the Box	$26
12. Sizzler	$24
13. Arby's	$24
14. Little Caesars Pizza	$20
15. Dairy Queen	$19
TOTAL	$1,217

* *Expenditures in the United States; excerpted with permission from the May 20, 1990, issue of* Restaurant Business *copyright © 1990. Companies devote millions more to other forms of advertising.*

chicken skin and were fried in beef fat. Individual deceptive ads can be blocked by citizen protests or government regulatory agencies. A much bigger problem is that even honest ads cultivate a hamburger-fries-and-soda mentality, especially among children.

THE CHILDREN'S MARKET

Arnold Fege, the national PTA's director of government relations, says, "Fast-food companies are getting rich at the expense of our children's health. Companies that spend millions of dollars on ads aimed at luring children into fast-food outlets have a responsibility to offer much more healthful foods." Sorry, Arnie, the companies are listening to their cash registers, not the PTA. As a Burger King spokeswoman told *The Wall Street Journal*, "We want to increase our share of the belly. And kids are the future."

During the 1989 Christmas season, Wendy's lured kids into its outlets by promoting plastic figurines based on the movie *All Dogs Go to Heaven*. McDonald's fought back with free bath toys based on the Disney film *The Little Mermaid*. And Burger King responded with toys based on *Teenage Mutant Ninja Turtles* and *The Simpsons*. In other years, fast-food chains offered Huggletts, *Star Wars* and *Star Trek* toys, Care Bears, and *Masters of the Universe* premiums. Hardee's sold about 15 million plush Pound Puppies and Pound Purriers in four weeks in 1987. Taco Bell beat that record with 21 million *Batman* cups in '89. In 1990 Domino's Pizza opened a mail-order operation that offered a Noid doll, a Domino's truck, an insignia cap, and many other items. Dennis the Menace has been luring kids to Dairy Queen. And McDonald's was giving miniature Barbie dolls and Hot Wheel cars with the purchase of a Happy Meal.

In 1989, McDonald's spent $20 million on Saturday-morning television ads aimed at kids, according to *Advertising Age*. That level of spending means that the average child saw about 100 commercials per year, plus countless commercials at other times of day. It costs a fast-food company about ten bucks to reach 1,000 kids with one commercial.

Children mean big business for the fast-food industry. The National Restaurant Association reports that:

- The number of meals purchased for children under six jumped by 36 percent between 1982 and 1986, twice the rate of increase as for adults.

- 83 percent of the time that children under 17 eat out, they do so at fast-food outlets.

- When 6- to 17-year-olds eat out, they order soft drinks 43 percent of the time, french fries 30 percent, hamburgers and cheeseburgers 24 percent, pizza 21 percent, fried chicken 10 percent, ice cream 9 percent, and side-dish salads 6 percent of the time.

Once you've hooked the children, getting their parents' business is no problem. Survey after survey shows that parents let their children make restaurant choices. "I'll say, 'What do you want to eat tonight, kids?' and it's a majority vote," says Angela Partee, a single mother whose family eats out about once a week. Her children like the Happy Meal (burger, fries, soft drink or milk, and toys) because "it makes it more like a treat . . . something more than dinner," despite the fact that "I don't think the hamburgers are very good there; it reminds me of economy meat." Advertising research has shown that "parents will pay 20 percent more for an advertised product with child appeal—even when the less expensive, nonadvertised product is no different," writes Stan Luxenberg in *Roadside Empires*, a book about the business of franchising.

"The popularity and success of fast food has been characterized as a marketing phenomenon rather than as a food phenomenon," then Senator George McGovern said when he convened a Senate hearing to look at fast-food labeling in 1976. He referred specifically to a 1972 survey showing that 96 percent of schoolchildren knew of Ronald McDonald. If that were true (apparently the survey was very limited), it would mean that Ronald was better known than any other figure except Santa Claus.

It probably isn't too far from the truth. After all, Santa only visits once a year, but Ronald visits every Saturday morning via television and sometimes in person—50 Ronalds travel to restaurants around the United States spreading the word about Big Macs. The executive of a competing chain called Ronald a "Pied Piper" who's brought millions of children into McDonald's.

In some outlets, parents can arrange a birthday party for their child and at least seven others at McDonald's. The restaurant supplies each child with a Happy Meal, party hats, and favor-type toys. McDonald's hasn't forgotten the cake—decorated not with "Happy Birthday, Jennifer," but with a picture of Ronald himself and sugar-formed edible characters from McDonaldland.

For several dollars a head, "you have a place to have a party and your home isn't wrecked," said one mother. Obviously it's a plus for parents. Not surprisingly, about one fifth of McDonald's sales come from people under 15 years old, according to *Business Week*. McDonald's thousands and thousands of ads on Saturday morning television is a major reason why McDonald's dominates the children's market.

As mentioned earlier, in 1990 Burger King tried to gain a larger share of the children's market by starting a Kids Club. Every month or two, outlets seek to lure kids into the so-called club with new badges and toys linked to popular TV shows and movies. This ploy was so successful that more than one million children added their names to Burger King's mailing list in just the first three months of the program, exciting Burger King's franchisees and executives no end. Not only do clubs lure children, but one registered member is a burger-loving dog named Howard from Washington, D.C. He prefers the local McDonald's Ronald McDonald Club and Birthday Club over Burger King's Kids Club, because at least McDonald's sends occasional coupons for free meals.

Playgrounds are another attraction and so good for business that McDonald's will put them indoors, if need be, even in Manhattan where floor space is at a premium. "Restaurants with playgrounds have higher sales than restaurants without," said Joe Edwards of the *Nation's Restaurant News*. Burger King is planning to add "clubhouses" for kids, as well as more bite-sized finger foods.

While the right hand sells birthday cakes, hamburgers, fries, cookies, and the image of Ronald McDonald to young children, the left hand reinforces the McDonald's logo to elementary schoolchildren—and with the blessing of school systems. McDonald's representatives developed an "Eating Right, Feeling Fit" nutrition and exercise package for schoolchildren with reproducible materials—all bearing the big M logo—and a comic book with pictures of squeaky-clean Olympic gymnast Mary Lou Retton doing exercises with all the McDonaldland characters. McDonald's ploy of "innocence by association" spurred a nutritionist with the Chicago Heart Association to say, "Personally, I find it incongruous. McDonald's has incredible chutzpah." A food company that really cared about fitness would stop selling products that are literally oozing with fat.

Most of the major fast-food chains offer special children's meals, usually a small sandwich, fries, and beverage. For instance, the McDonald's Happy Meal is created with mix-and-match choices that allow some flexibility in ordering. The cost of the total meal, including toy, is more than the sum of its food parts. In Washington, D.C., we paid 44 cents extra for the toy and packaging. Other restaurants don't always allow parents to substitute more nutritious foods. An Arby's clerk in Louisville had no problem substituting milk for a soft drink, but a Wendy's clerk in Olean, New York, refused to make a similar switch. Even if both child and restaurant agree to substitute milk, the meal is generally lacking fiber, vitamin A, vitamin C, and several B vitamins (B6, folic acid, and pantothenic acid).

Fast-food commercialism now extends into every corner of a child's life. Both McDonald's and Burger King have licensed their name to toymakers, so three-year-olds can dress up in caps and aprons and pretend they are serving burgers and fries. McDonald's and Sears have teamed up to create a line of McKids clothing. Play-Doh's "Build a Whopper Playset" lets kids grind out not-so-reasonable facsimiles of everything from burgers and buns to onion rings and pickles. Fisher-Price's Little People McDonald's Restaurant comes complete with a miniature playground. Who can predict what they'll think of next? — tie-ins with the National Education Association enabling teachers to dress like Domino's Pizza delivery people?

Fast-food chains target teens, a prime market, by planting restaurants near high schools and by filling teens' eyes and ears with commercials. Fast-food restaurants are an obvious choice for teens who want to congregate with their friends after school. For the price of a cola and maybe some fries, they have an afternoon of sociability. A 1989 survey conducted by the Center for Science in the Public Interest found that teens' favorite TV commercials were, not surprisingly, for fast foods (especially McDonald's).

But why wait for the kids to get out of school? Pizza Hut is now delivering pizzas to hundreds of schools. And McDonald's has actually opened outlets in one high school in San Antonio, Texas, and another in Boulder, Colorado.

"Growing children shouldn't eat in fast-food restaurants more than once a week because the amount of calories is out of proportion to the amount of nutrients," Dr. Judith Anderson told *The Detroit*

News. A dietitian and nutrition professor at Michigan State University, Anderson cautioned that a high proportion of the calories in typical fast foods comes from fat and sugar. The average 7- to 10-year-old needs about 2,000 calories per day and should consume no more than about 2,000 milligrams of sodium and 67 grams (15 teaspoons) of fat throughout the day. (See page 106, for the appropriate intakes for other age groups.) A McDonald's Quarter Pounder, medium order of fries, and low-fat vanilla shake yield 1,020 calories and 9 teaspoons of fat, plus 970 milligrams of sodium and 8 teaspoons of sugar.

A McNugget here and a fried pie there—pretty soon it adds up to bad nutrition. It's unfortunate that the giant restaurant chains don't work harder to promote good eating habits while they make a buck. It would be so easy to develop special kids' packages containing fruit salad or carrot sticks, juice or milk, instead of soft drinks and french fries.

BREAKFAST

Other battles are fought on the breakfast front, a relatively new territory for the fast-food industry. The typical morning meal is proving to be as lucrative for the restaurants as it is unhealthful for consumers.

In 1972, McDonald's put an egg, cheese, and Canadian bacon in the middle of an English muffin and began selling the product in selected markets. By 1976, Egg McMuffin was a national product and breakfast was a fast-food phenomenon.

"None of us would be serving breakfast if McDonald's hadn't spent tons of money educating the public that you could get a good breakfast at a fast-food restaurant," said Paul Mitchell, of California-based Carl's Jr. restaurant chain. Carl's now sells orange juice and its own "Sunrise sandwich"—an English muffin topped with egg, meat, and cheese.

How "good" these meals are depends on your point of view. Certainly, if you have a vested interest in the fast-food business you would say they are very good. Between 1977 and 1984, restaurant breakfast traffic increased 57 percent according to *The Wall Street Journal*. Breakfast now accounts for about 15 percent of McDonald's $18 billion sales. That's a lot of muffins. Twenty years ago, McDonald's wasn't even open in the morning.

IMPROVEMENTS IN FAST FOODS
*1986–1991**

ARBY'S

- Switched from beef fat to vegetable shortening for all frying.
- Dropped the Country Fried Steak sandwich and added the Grilled Chicken Barbecue sandwich.
- Dropped its fatty 590-calorie Chicken Cashew Salad.
- Halved the standard dollop of mayonnaise/sauce and switched to cholesterol-free mayonnaise.
- Switched from whole milk to 2 percent low-fat milk.
- Removed all sulfites, Yellow No. 5 dye, and tropical oils.
- Added under-300-calorie roast beef, roast chicken, and roast turkey sandwiches.

BASKIN-ROBBINS

- Introduced nonfat yogurt and other low-fat frozen dairy desserts.

BURGER KING

- Downsized its danish pastries by 100 calories and 4 teaspoons of fat.
- Switched from beef fat to vegetable shortening for all frying.
- Cut one-third of the calories and more than half the fat from the BK Broiler Chicken Sandwich.

DAIRY QUEEN

- Switched from beef fat to vegetable shortening for all frying.
- Added a low-fat Grilled Chicken Sandwich at Dairy Queen/ Brazier outlets.
- Cut 158 calories and 3 teaspoons of fat from the Chicken Breast Fillet Sandwich.

DOMINO'S PIZZA

• Reduced the sodium in its 16-inch cheese pizza from 800 to 483 milligrams.

DUNKIN' DONUTS

• Introduced bagels.
• Is switching to low-calorie mayonnaise.
• Introduced nonfat frozen yogurt at some outlets.
• Eliminated egg yolks (and cholesterol) from doughnuts.

HARDEE'S

• Introduced the Real Lean Deluxe burger with one-third less fat than typical ground beef.
• Switched from beef fat to vegetable shortening for all frying.
• Switched to low-calorie mayonnaise.
• Removed sulfites from its fish breading, Yellow No. 5 dye from its reduced-calorie French and Italian dressings, and MSG from its steak, sausage, and Thousand Island dressing.

JACK IN THE BOX

• Switched from beef fat to vegetable shortening for all frying.
• Added a low-fat Chicken Fajita Pita to its menu.
• Stopped salting burgers during the grilling process.

(continued)

---

* *These comparisons are based on company data published between 1986 and 1991. Some apparent improvements may be an accident of sampling (products, cooking times, etc., may vary slightly from restaurant to restaurant) or errors in booklets that do not reflect true changes.*

IMPROVEMENTS IN FAST FOODS, CONTINUED

LONG JOHN SILVER'S

- Introduced baked fish and chicken.
- Dropped chicken nuggets made with ground-up chicken.
- Cut 3 teaspoons of fat and one-third of the sodium from the 3-piece Chicken Planks Dinner.
- Cut 6 teaspoons of fat and half a teaspoon of salt from the 3-piece Fish Dinner.
- Makes Hushpuppies with 87 percent less sodium and one-third less fat.

McDONALD'S

- Switched from beef fat to vegetable shortening for all frying.
- Introduced fat-free muffins for breakfast.
- Reduced the fat content of milk shakes by 80 percent.
- Replaced ice milk with frozen yogurt, cutting the fat and cholesterol by two-thirds and calories by one-fourth.
- Stopped adding ground-up chicken skin to chicken nuggets and reduced their calorie content by 20 percent.

In addition, breakfast attracts customers who don't usually eat fast foods. Traditionally, the older you get, the less fast food appeals to you. But breakfast attracts these older consumers and brings in new dollars for the companies. "Some restaurants are attracting people who used to skip breakfast but now consider it an important part of their diet," reported *The Wall Street Journal*. While it isn't wise to skip breakfast, a look at the nutrient contribution of some fast-food breakfast choices might provoke consumers to seek more healthful choices.

Hardee's, which has long emphasized breakfast, features a plain

- Reduced the fat in its hotcakes by 40 percent.
- Became the first chain to offer 1 percent low-fat milk and breakfast cereals.
- Introduced McLean Deluxe Burgers, which contain half the usual amount of fat and dropped its fatty McD.L.T.

TACO BELL

- Switched from highly saturated coconut oil to a less saturated vegetable shortening for all frying.
- Removed lard from the beans and tortillas.
- Dropped its very fatty Taco Light.
- Added chicken and steak tacos.
- Cut 750 milligrams of sodium from the Taco Salad with shell.

WENDY'S

- Switched from beef fat to vegetable shortening for all its frying.
- Dropped the fatty Triple Cheeseburger from the menu.
- Reduced the sodium in chili by 30 percent.
- Added a Grilled Chicken Sandwich.

biscuit containing 4 teaspoons of fat. Hardee's also tempts customers with chicken-, ham-, sausage-, or bacon-topped biscuits with or without eggs. After experimenting with breakfast for six years, Burger King abandoned biscuits and muffins in favor of croissants that, already high in fat, come topped with fatty cheese and meats. Burger King's bagel sandwiches are just a tad better.

Still, there's more than a glimmer of hope. Hardee's and McDonald's offer pancakes. Dunkin' Donuts has bran muffins and bagels. McDonald's offers by far the best breakfast with its nonfat apple-bran muffins, Wheaties, Cheerios, and 1 percent low-fat milk.

A GLIMPSE INTO THE FUTURE

As the restaurants proliferate, so do concerns about health and about diets built on fried meat, fried potatoes, and soft drinks. Restaurant officials defend the industry by insisting that nobody encourages consumers to eat 21 meals a week at fast-food restaurants, although it's hard to imagine their objecting if consumers suddenly chose to do so. Statistics show that we eat more fast food every year, and the chain restaurants aren't spending almost $2 billion a year suggesting that maybe you should stay home tonight and fix yourself low-fat cottage cheese with sliced fruit.

What does the future hold for fast-food restaurants and fast-food consumers? Will fast food become more nutritious or less so? Are there too many fast-food restaurants now or will the industry continue to grow? Will they deliver foods to our homes? Will they change their menus in order to compete with innovative supermarkets?

Judging from the past, the fast-food industry will become bigger and more diverse. That means fewer independent restaurants and more empty wrappers littering our neighborhoods. While the conventional fast-food market seems as stuffed as the consumer of a double cheeseburger, the future holds promise for a good product under good management, said one analyst. The U.S. market has room for unconventional fast-food outlets, and there is plenty of room on the highways of the Far East, Latin America, and Europe. But you can bet that some large chains will falter, fold, or be taken over by other chains.

One thing is certain. We're all getting older. And as we get older, we worry more about health and the environment and care less about speedy eating and plastic toys.

ENVIRONMENTAL CONCERNS

Environmentalists have made us aware of new reasons to focus on fast-food chains. Until recently, litter, ugly outlets, the smell of grease, and late-night crowds of teens were sufficient reason for people in several well-to-do neighborhoods and resort towns to successfully fend off the fast-food invasion. Recently, though, other problems have given critics new ammunition.

For instance, some critics charged that hamburger chains buy beef from Central and South America, where ranchers chop down primeval forests to make room for beef ranches. Full-page newspaper ads have decried the desecration of rain forests for the sake of restaurant chains' saving slightly on the price of beef. After several years of controversy, Burger King in 1987 stopped buying Central American beef and says all its beef now comes from the United States, Canada, Australia, or New Zealand. Wendy's and most other large fast-food companies claim that none of their beef comes from rain-forest areas.

McDonald's and other chains have taken a public relations beating for packaging foods in polystyrene foam. A typical restaurant generates thousands of pounds of polystyrene waste a year. All that waste for a product whose useful life may be only about 10 seconds! Foam packaging has been banned by several states and cities. Interestingly, McDonald's argued that paperboard packaging, such as the kind used by Burger King, affects the environment just as much as Styrofoam, because the manufacture of paper is a terribly polluting business.

McDonald's (whose 1989 annual report, in focusing on the environment, states, "It is truly time to act!"), began indicating its concern for the environment by using thinner straws and more recycled paper. In April, 1990, McDonald's announced that it would spend $100 million a year on recycled construction supplies, especially those made from plastics, when building new and remodeling existing restaurants. And later in 1990 it responded to public pressure by announcing that it would stop using foam containers. One industry spokesperson estimated that McDonald's restaurants used about 75 million pounds of foam packaging a year. In 1991, McDonald's announced a series of steps, developed with the assistance of the Environmental Defense Fund, that included using smaller napkins, refillable coffee mugs, and unbleached, recyclable bags. All told, McDonald's expected to cut its trash disposal by up to 80 percent, or 1.6 million pounds a day. To further associate itself with environmentalism, in 1991, McDonald's restaurants gave away about 10 million tree seedlings.

Officials at Burger King, Taco Bell, McDonald's, Wendy's, and other companies say that cutting down on solid waste, recycling, and other environmental issues are among their top corporate concerns. Let's hope that these huge wastemakers help the environment by eliminating as much packaging as possible and recycling the rest. That should be especially easy for foods eaten at the restaurant.

FAST, FASTER, FASTEST

A 1988 *Consumer Reports* magazine survey found that the biggest complaint people have about fast-food restaurants is that they're too slow. So companies are turning to new technologies to return speed to their service. Microwave ovens, now a fixture in millions of homes and offices, can prepare an incredible variety of offerings faster than you can walk to McDonald's. A vending machine has been developed to dispense freshly fried potatoes. To compete with such marvels, the fast-food industry knows it will have to become even faster and more efficient to satisfy increasingly impatient customers.

Some companies are developing automatic hamburger flippers. McDonald's has robots cooking the fries and telling its human co-workers when to prepare sandwich buns. Arby's and Taco Bell are testing computers into which customers punch their orders so that no time is wasted exchanging pleasantries with clerks. Carl's Jr. has even offered customers a dollar off if they wait longer than one minute for their meal.

The pizza industry has been busy figuring out how to deliver their product to more people in less time. Domino's, which promises to deliver pizzas within 30 minutes, has its delivery people riding everything from speedboats in Alaska to roller skates in New York to scooters in London and Hong Kong. Not only is Pizza Hut trying to play catch-up with Domino's by offering home delivery, but, according to *The New York Times*, it is introducing "a kind of Fotomat-equivalent that lets [customers] order and pick up a pizza in 30 seconds." Nunzio's Express, a five-store drive-through newcomer in Albuquerque, sells baked-to-order, one-minute-to-bake, 8-inch pizzas at $1.79 (toppings are 49 cents extra).

Perhaps the ultimate in speedy pizzas is an automated coin-operated pizza maker. A trade journal reported that after the customer drops in the coins, "the device oozes out enough crust for a slice of pizza, squirts on some sauce, adds the chosen toppings, and bakes it in short order." The magazine does not report on the product's taste. In addition, several companies are marketing vending machines that store frozen pizzas and, when fed several dollars, reheat them in about one minute with infrared or microwave technology. One brand of machine costs $8,900, according to *The Washington Post*, making it the most expensive vending machine around.

OVERSEAS EXPANSION

Fast-food restaurants are growing like wildfire overseas. In 1990 over one-third of McDonald's sales came from more than 3,000 outlets in 52 countries outside of the United States. Three thousand of KFC's (Kentucky Fried Chicken) 8,187 outlets were abroad. Yet many markets remain virtually untouched and represent literally billions of potential consumers of burgers, fried chicken, franchise-style pizza, and doughnuts.

The Pacific Rim is especially attractive to these expanding chains, some of which are already well established there. KFC has about 800 restaurants in Japan, and its Beijing store has claimed to be the busiest in the world. McDonald's has over 700 restaurants in Japan, more than in Canada and more than one-fourth of all holdings abroad. After touring Japan, one American restaurant executive told *Restaurant Business*, "I think the Japanese are more American than we are."

An average McDonald's overseas does 25 percent more business than the average U.S. restaurant. And they adapt to the needs and preferences of the countries they are in: restaurants serve wine with fast food in France, beer with the burgers in Germany.

Even before the 1989 political revolutions in Eastern Europe, fast-food companies began opening outlets in China and the Soviet Union. In fact, the 700-seat McDonald's in Moscow was expected to capture the title of the busiest in the world, having served 20,469 diners in March 1990, just a month after it opened. Though the industry may be stagnating in the United States, golden arches and Pizza Huts may become as common in Eastern Europe in the coming decade as pictures of Karl Marx were in the past decade.

THE DINNER MARKET

Currently, fast foods appeal to lunch crowds with 40 percent of business typically being done between 11 A.M. and 2 P.M., said Joe Edwards, of the *Nation's Restaurant News*. Industry analyst Ron Paul estimates that McDonald's does only 25 percent to 30 percent of its business during dinner hours.

A special higher-priced dinner menu would have to be different enough to attract customers who have already eaten a fast-food lunch and attract people who presently don't think of fast foods as a real meal. Initial attempts haven't been particularly successful. Burger King's attempts at dinner platters with vegetables failed. Wendy's couldn't move the "dinner" idea out of test markets in Cincinnati. McDonald's, which opened the breakfast hours to fast-food meals, has several outlets in Nashville offering $3.69 platters of fried shrimp and chicken and fish fillets, along with coleslaw and curly fries, according to *Adweek's Marketing Week*. In other cities, McDonald's is testing enchiladas and "a form of sushi" in California.

THE NEW TV DINNERS

High technology, including VCRs, giant-screen televisions, and microwave ovens, may affect fast-food restaurants indirectly. One pizza-chain executive told *Restaurant Business*, "A large part of [discretionary income] is spent on the home environment. The baby-boomer family is beginning to seek an eating experience that can be enjoyed within the luxury of the home." As a result, TV dinners may take on a whole new meaning. The next feeding boom is likely to be in microwavable prepared dinners and supermarket delis and salad bars. Supermarkets and manufacturers of packaged foods are, in effect, waging a major counterattack on the franchised restaurants that have taken billions of dollars a year out of supermarket cash registers.

If those predictions prove accurate, pizza makers may be in better shape than many competitors. Baby boomers became familiar with delivered pizza on their college campuses, and Domino's carries the tradition into every urban and suburban neighborhood.

MORE NUTRITIOUS ALTERNATIVES?

Will the food get more nutritious? Fast-food operators discuss nutrition out of both sides of their mouths. From one side, they say the public doesn't want nutritious food. From the other side, they are touting the nutritious qualities of the foods they serve.

"Our research states that consumers will talk about nutrition, but they buy taste, convenience, and value," said Wendy's spokesman Denny Lynch. This may mean that, in the future, fast-food companies might try a novel approach such as offering nutritious food that tastes good.

New-age restaurateur Clark Heinrich seems to have met the nutrition challenge—with soybean foods, no less. His vegetarian fast-food McDharma opened in Santa Cruz, California, in 1982 to offer an alternative to traditional fast food—double-decker vegetarian sandwiches, beans, and frozen yogurt. The restaurant faced legal as well as gustatorial challenges, when McDonald's successfully sued the new little restaurant to drop the "Mc" from its name. Dharma's, as the restaurant was renamed, offers vegetarian hot dogs and "I'm Not Chicken" patties that are very low in fat. But don't automatically assume that vegetarian equals nutritious. Dharma's Killer Nachos (over 1,400 calories and 23 teaspoons of fat), grilled-cheese sandwich, and other items are loaded with cheese, butter, and other fatty ingredients.

From Muncie, Indiana, comes another upstart. The hypertensive wife of a former KFC franchisee developed a low-fat, low-sodium chicken product that resembles fast-food fried chicken. Carolyn Duncan, who has high blood cholesterol and triglycerides in addition to high blood pressure, said, "You can't find any place out there to go eat" if you're concerned about your health.

Saying, "There's got to be a way to [sell chicken] without the grease," Mrs. Duncan opened C.J. Carryl's in September 1985. The restaurant offers her secret chicken recipe that is cooked with hot air in a method Mrs. Duncan compares to hot-air popcorn. The chicken dinner offers sides of green beans and carrots, in addition to her low-sodium mashed potatoes and gravy. She's lowered the sodium in her biscuits and offers slaw and a salad bar. She's created a fast-food restaurant with alternatives to the high-fat, high-sodium foods offered by restaurants that marinate chicken in a high-salt mixture and then fry it in deep fat. The challenge for the Mrs. Duncans of the world, of course, is to expand from a few outlets into a major chain.

Several companies have also developed hot-air ovens that may let us have our french fries and our nutrition, too. For instance, Trak-Air, Inc., of Colorado rents an oven that quickly cooks french fries, pizza, and other foods in high-velocity hot air. Time will tell whether appliances like this oven will turn up in chain restaurants.

MOVES IN THE WRONG DIRECTION: NOT EVERY FOOD HAS GOTTEN HEALTHIER*

ARBY'S

- Mushroom and Cheese Croissant gained 150 calories and 5 teaspoons of fat.
- Mushroom and Cheese Baked Potato and Deluxe Baked Potato added several teaspoons of fat.
- Soups added 200 to 500 milligrams of sodium.
- Added Polar Swirls—milk shakes mixed with high-calorie cookies and candy bars.

CARL'S JR.

- Two new roast-beef sandwiches (Roast Beef Club Sandwich and Roast Beef Deluxe Sandwich) provide over 200 more calories and 4 to 6 teaspoons more fat than the California Roast Beef'n Swiss sandwich they replaced.
- The Double Western Bacon Cheeseburger gained over 200 calories and 2 teaspoons of fat.
- Stuffed potatoes gained 2 to 3 teaspoons of fat and 80 to 170 calories.

Paul Wenner, who heads Wholesome & Hearty Foods in Portland, Oregon, is an entrepreneur who has another approach to healthful fast-food eating. He invented the Gardenburger. While it is shaped like a hamburger, it is made of mushrooms, onions, rolled oats, and other natural ingredients. It has one-third fewer calories and less than half as much fat as a comparable hamburger. And judging from its growing popularity in dozens of restaurants, hospitals, and even local sports stadiums, the Gardenburger might spread like wildfire. So far it has spawned a Gardentaco, Gardensausage, and Gar-

- Danish pastries gained 220 calories and a teaspoon of fat.
- Regular hamburger gained 100 calories and a teaspoon of fat.

JACK IN THE BOX

- Jumbo Jack has 2 teaspoons more fat.

McDONALD'S

- Filet-O-Fish sodium level jumped by 15 percent to 930 milligrams.

TACO BELL

- Added a teaspoon of fat and about 250 milligrams of sodium to its Bean and Beef Burritos and Burrito Supreme.

WENDYS

- Taco Salad grew up 270 calories and over 4 teaspoons of fat.

These comparisons are based on company data published between 1986 and 1991. Some apparent worsenings may be an accident of sampling (products, cooking times, etc., may vary slightly from restaurant to restaurant) or errors in booklets that do not reflect true changes.

densteak, all relatively low in fat and calories and available at many health-food stores around the country.

The 11-unit Macheesmo Mouse chain in California and Oregon offers a variety of healthful burritos, tacos, enchiladas, and salads. Complete dinners come with one-fifth the fat of a large hamburger sandwich. The Chicken Majita (the same as a fajita) gets only 11 percent of its calories from fat, the company claims.

Another West Coast mouse is also testing the fast-food waters. The Walt Disney Company is opening up Mickey's Kitchen restaurants, which could be irresistible to kids. The still-small chain offers a meatless Mickey Burger that is made from ingredients similar to those in the Gardenburger. Without the sauce it's quite low in fat. Meanwhile, restaurants at Disneyland are offering a meatless burger, spaghetti with meatless sauce, vegetable lasagna, and an avocado and alfalfa-sprout sandwich. Across the country, Walt Disney World's restaurants tossed out its fatty turkey, ham, and cheese sandwich and started offering pasta, eggplant Parmesan, and grapes.

The surge of Mexican-style fast food in the '80s foreshadowed what may be the fast-food craze of the '90s: pasta and Chinese fast food. Analysts say Italian fast food (other than pizza) is virgin territory, and only small inroads have been made by Chinese chains. Both of these cuisines offer many low-fat possibilities, high in complex carbohydrates, that would provide nutritious and delicious alternatives to burgers.

Since the first edition of *The Fast-Food Guide* was published in 1986, the big chains, too, have begun modifying current products and adding new ones (see page 26). Many other products are being tested, ranging from low-fat breakfast sausage to skinless baked chicken to a new cooking process at Arby's that eliminates deep-fat fryers and frying oil.

In the future, fast-food restaurants could do an even better job of responding to their customers' taste desires and health needs if they introduced:

- More baked and broiled foods
- Low-fat pasta and Chinese meals
- Vegetarian burgers
- Fresh fruit or fruit salad
- Hot green vegetables
- Nonfat plain or flavored yogurt
- 1 percent milk as the standard milk
- Buns and pizza crust made with whole-wheat flour

Fast-food restaurants have come a long way since they sold hamburgers and fries for 25 cents, and even since the first edition of this book. For the good of the country's health and economy, they could go a lot further. The public can take an active role in changing fast food by choosing the most nutritious food offered. Highly motivated consumers might write companies with compliments and/or complaints (see Appendix for addresses) and also write their legislators to encourage passage of laws that would make ingredient and nutrition information readily accessible.

For many reasons, fast food has become part of our way of life. But the more we eat out, the more questions we have about the quality of our diets. Many of the questions are basic: Is fast food junk food? Other questions aren't so basic. People with food allergies or other health concerns often ask specific questions about preservatives, artificial colorings, and other additives.

Zealous health advocates might declare that all fast food is poison, but those of us with jobs, children, and/or little time or enthusiasm for cooking know better. We aren't going to die from eating an occasional double cheeseburger and fries.

Still we also know that burgers and fries don't rate up there with broccoli as food of the week. And if one fifth of American adults, on a typical day, go in for "a grease job," as *Washington Post* columnist Colman McCarthy so pithily stated, it might be prudent to consider what food choices will keep us healthier, longer.

Is it possible that we can have convenience and nutrition all wrapped in food that the kids will eat? A little nutrition knowledge goes a long way in helping us make decisions. As we shall see in the next chapter, when we know about common health problems and how food affects them, we can make wiser choices about what we eat.

THE HEALTH YOU SAVE MAY BE YOUR OWN

If food had no more effect on our well-being than socks or wristwatches, Americans would hardly be concerned about it. But what we eat does have a major impact on our health.

Health has long been thought of as freedom from infectious diseases or nutritional deficiencies. For most Americans, that definition is no longer relevant. Over the years, vaccines and medicines have freed us from the fear of infectious diseases like polio and tuberculosis. Federal food programs, such as school lunches, combined with higher incomes and food fortification have all but eliminated scurvy, rickets, and other "deficiency diseases" arising from inadequate diets.

Now we define health a little differently. Since most infectious and deficiency diseases have been eradicated or are easily controlled, our attention has turned to combating chronic, degenerative diseases. Degenerative diseases are the long-term conditions that gradually erode our health before they kill us — insidiously harmful diseases that sometimes lurk unnoticed for years, like cancer or high blood pressure, or that debilitate us, such as heart attack and stroke.

Unfortunately, drugs can't always treat these conditions. Those drugs that do treat don't cure, and many have unpleasant side effects. Diuretics, used to treat hypertension, may also increase blood cholesterol and cause sexual dysfunction, diarrhea, drowsiness, and headaches. One antihypertensive drug, Selacryn, was removed from the market after it was shown to cause liver damage and ultimately death. Even if the drugs didn't have side effects, however, a lifetime of paying for medications would cost a bundle.

Coronary-bypass surgery has been a widely used method of treating heart disease, yet this incredibly expensive procedure increases life expectancy for only a small proportion of patients.

Fortunately, sophisticated research has shown that many chronic diseases can be postponed for many years, or even avoided entirely, simply by taking good care of ourselves—by not smoking, by exercising, and by eating the proper foods.

Fast food, which comprises a growing portion of the national diet, is sometimes sold as "fun food," as much a toy as sustenance for our bodies. The U.S. Department of Agriculture, which analyzes virtually every aspect of the nation's food system, has expressed concern about the nutritional adequacy of a diet that relies on fast foods. Karen Bunch, then at USDA's Economic Research Service, contended,

> The trend toward eating more fast food reduces the variety in our diets and may increase the risk of nutritional deficiency. For example, a typical fast-food meal of a hamburger, french fries, and milk shake contains approximately half the Recommended Dietary Allowances (RDA) of calories and protein for the adult male. Yet that meal gives him only one third of the RDA of vitamin C, thiamin, and niacin, and lesser amounts of iron, calcium, vitamin A, and riboflavin. The meal's large calorie count reflects a high fat level.

Connie Roberts, a registered dietitian at Brigham and Women's Hospital in Boston, has chided burger-based meals for being low in biotin, folic acid, pantothenic acid, vitamin B6, vitamin E, copper, magnesium, and other nutrients.

So, what are the proper foods? It can be difficult to tell in these days of mixed messages, when tricky advertising claims make doughnuts and potato chips sound like health foods. Sometimes even health professionals seem to disagree among themselves.

DIETARY GUIDELINES

A look at the history of nutrition explains the confusion. Some health professionals, who may have studied nutrition several decades ago, still subscribe to the "all food is good food" concept that was popular before research showed that too much of some foods was, in fact, linked to bad health. These people learned World War II nutrition theories—when many men who enlisted in the armed forces were so malnourished that the government started a school-lunch program to make sure the next generation received at least one nutritious meal each day.

You can recognize these old-style nutritionists (some of whom aren't very old) when they insist that even empty-calorie foods, such as soft drinks, have a place in the diet, and that fat is an important source of vitamin E and calories, makes us feel full, and makes our food taste good. That philosophy comes in part from a Depression-era food-shortage mentality. It is still appropriate today in many developing countries where starvation is more threatening than chronic disease, and where all food may indeed be good food. But overfed Americans cannot afford to live by the same rules by which rural Ethiopians live. Even Americans who aren't overweight still need a diet that reduces their risk of heart disease and cancer.

Degenerative diseases are difficult to study because, unlike infectious diseases, there's no one "bug" that causes them. Decades must pass before their ravages become apparent, and reversibility is difficult to measure.

But the U.S. Departments of Agriculture and Health and Human Services recognize that a massive amount of evidence links the typical American diet to many of our most common diseases. These two government agencies have developed a national nutrition policy known as the "Dietary Guidelines for Americans." The current guidelines, first issued in 1980 and revised in 1990, are:

- Eat a variety of foods.

- Maintain healthy weight.

- Choose a diet low in fat, saturated fat, and cholesterol.

- Choose a diet with plenty of vegetables, fruits, and grain products.

- Use sugars only in moderation.

- Use salt and sodium only in moderation.

- If you drink alcoholic beverages, do so in moderation.

Those are the modern recommendations in their mildest form. Other health organizations—including the National Cancer Institute and the American Heart Association—make somewhat stronger and more specific recommendations about the quantities of fat, cholesterol, sodium, and dietary fiber that people should consume. These rules imply that in our affluent society not all food is necessarily good food. Some foods *are* better than others.

Where do fast foods fit in? Chosen with care, fast foods can be included in a healthful diet. But there is a wide world of high-fat, high-sodium, low-fiber fast foods, and it takes some expertise to pick your way through a menu for healthful choices.

The fast-food companies' ad writers, food technologists, and consumer psychologists devote their careers to developing ways to persuade you to buy their products. While even a Ph.D. nutritionist can be tricked by some of the industry's carefully worded claims, understanding basic nutrition principles can shift the odds in your favor when you sally forth into the fast-food-restaurant jungle.

CLARIFYING THE OLD RULES

Back in the days when all food was considered good food, health professionals were guided by a basic tenet: Eating a variety of foods would provide you with all the necessary nutrients. Times changed and so did the concepts. When obesity and chronic diseases became more troublesome, health professionals added "moderation" to the concept of variety. But the health status of Americans shows that this advice wasn't nearly specific enough to clear up questions about healthful eating. Obviously, if you think that variety means choosing from among soft drinks—Coke today, 7-Up tomorrow, and so on—you won't be the healthier for it. Likewise, if you believe that moderation means limiting your dinner to one double cheeseburger, a chocolate shake, and a large order of fries, you might have a misdirected sense of nutrition. The old Basic Four could easily mean a hot dog (protein) on a white bun (grain), fries (vegetable), and ice cream (dairy).

Variety and moderation make sense, but only if the diet makes sense in other ways. On a daily basis, a good diet should get no more than about 20 to 25 percent of its calories from fat, and should contain no more than about 2,000 milligrams of sodium, and 150 to 250 milligrams of cholesterol, the less the better in each case. That diet would also include about 30 grams of fiber from vegetables, whole grains, and fruits and would be low in refined sugar.

Without the help of a computer, it's difficult to monitor your diet with any accuracy. It's only a little easier if you remember that 1 teaspoon of salt contains about 2,000 milligrams of sodium, and that 25 percent of calories from fat would be roughly 15 teaspoons fat or oil per day, if you eat a 2,400-calorie diet.

Obviously, working with those numbers is much more difficult than simply identifying and cutting down on foods high in fat and sodium.

A health-promoting diet contains many foods close to their natural state. Eating 3 or 4 servings of vegetables, the same of fruits, plus a variety of whole grains, beans, lean meats and chicken, fish, and low-fat milk products will help you attain your nutritional goals. To be healthier, Americans need to readjust the proportions of food they eat, depending less on red meat, fried foods, fatty rich desserts, and soft drinks. All too often the narrow variety of fast-food options excludes foods high in vitamins A and C and fiber, while giving us foods loaded with fat.

If fast foods were a limited phenomenon, there would be little cause for alarm. But fast foods have gone beyond the drive-through window. An occasional fast-food meal no longer merely interrupts a prudent pattern of food choices. Food from well over 100,000 fast-food restaurants feeds 50 million people every day. In addition, fast foods survive where meat loaf fails: In school lunchrooms, military dining rooms, and even hospitals, fast foods or their facsimiles are making money where other foods didn't. Hamburgers, pizza, and chicken nuggets—all foods from the notorious limited menus—are a way of life. Variety cannot compete with the bottom line.

"Wherever they go, young people now demand menus based on fast food," writes Stan Luxenberg in *Roadside Empires*. He recounts the story of the men aboard the U.S.S. *Saratoga* who suffered low morale and, when surveyed, said they wanted fast food served at meals. So the Navy redecorated the dining room in bright colors, set up a food-production method similar to a fast-food restaurant, and replaced roast beef, turkey, and tuna with cheeseburgers, fries, and shakes. "The Navy has since expanded the program to cover all its aircraft carriers," writes Luxenberg.

In their oversize burgers the fast-food restaurants dish out several portions of fatty, salty food, stick it inside a bun, and call it a sandwich. For instance, Burger King's Double Whopper with Cheese is one blatant example of excess, because its 935 calories come drenched in sodium and fat. That sandwich is hardly the sole example, however. Some people consider a plain hamburger or a melted-cheese sandwich with bacon two good choices for lunch. Hardee's combines them into one sandwich and calls it a Bacon Cheeseburger. It is a 610-calorie choice. A single fast-food sandwich like that provides a hefty share of a person's daily calorie and fat allowance.

QUEST FOR PROTEIN

When fast-food proponents begin to talk nutrition, protein leads the list of attributes. It's an effective selling point. Our bodies *do* need protein to build new tissues and to repair old ones. Many hormones and all enzymes consist of protein, and hair, skin, nails, blood, and bones all contain protein. Protein helps us fight disease and carries oxygen through the blood.

Millions of Americans buy the protein pitch. Perhaps because muscle is made of protein, we associate it with strength and assume that our body would never convert excess protein into fat. While we need less than two ounces of pure protein a day to function normally, most of us consume much more than that. One large hamburger sandwich contains more than half the protein we need in a day. And meat isn't the only food that contains protein. If we lived on nothing but beans and macaroni we'd more than meet our protein needs. Many foods, from bread to broccoli, contain protein; it isn't uncommon for Americans to eat twice as much as they need.

Protein requirements vary with age, sex, and ideal weight. Young children's diets should have a higher percentage of protein than those of adults (but less actual protein because children weigh less). Adults should consume about 0.34 grams of protein per pound of ideal body weight (an ideal weight is used because your body doesn't require extra protein if you put on extra fat tissue). So if you're a woman whose ideal weight is 120 pounds, your protein requirement is $0.34 \times 120 = 41$ grams, or the amount of protein found in 4 ounces of tuna fish (a bit more than half a can).

A meal of a Burger King Whopper with Cheese, a large chocolate

FAST FOODS HIGHEST IN CALORIES

COMPANY/PRODUCT	CALORIES
Carl's Jr. Double Western Bacon Cheeseburger	1030
Jack in the Box Ultimate Cheeseburger	942
Burger King Double Whopper with Cheese	935
Taco Bell Taco Salad, with Shell	905
Burger King Double Whopper	844
Carl's Jr. Super Star Hamburger	820
Wendy's Double Big Classic, with Cheese	820
Dairy Queen Heath Blizzard, regular	820
Pizza Hut Supreme Pan Pizza, large, 2 slices (estimate)	766
Jack in the Box Egg Rolls, 5 pieces	753
Wendy's Double Big Classic	750
Carl's Jr. Western Bacon Cheeseburger	730
Carl's Jr. Fiesta Potato	720
Jack in the Box Grilled Sourdough Burger	712
Dairy Queen Peanut Buster Parfait	710
Burger King Whopper with Cheese	706
Jack in the Box Bacon Cheeseburger	705
Dairy Queen DQ Homestyle Ultimate Burger	700
Burger King Chicken Sandwich	685
Wendy's Frosty Dairy Dessert, large	680
Pizza Hut Personal Pan, Pepperoni, whole pizza	675
Arby's Chicken Cordon Bleu	658
Taco Bell Nachos Bellgrande	649

To put these figures in context, consider that an average woman consumes about 2,200 calories per day; an average man about 2,900 calories per day.

shake, and fries contains all the protein that a 120-pound woman needs in a day, but along with it comes 1,779 calories. For her, that's almost enough calories for an entire day, much less for one meal. If she opts for a diet soft drink instead of the shake she'll still be downing over 1,000 calories and need just 4 more grams of protein to get her requirement. A meal of a burger, shake, and fries has lots of protein, but it's also loaded with fat and is weak in other substances such as fiber and vitamins A and C.

If some protein is good, more must be better. Right? Wrong. More is not better in the case of protein. The protein you consume each day goes to producing tissues and enzymes, and for other metabolic functions. When you don't need a protein molecule for building, the nitrogen component is removed and discarded into urine. The remaining portion is used by the body for energy. No matter where they come from, excess calories all end up in the same place: on our hips, thighs, and bellies.

Too much protein may cause other problems. Excess protein promotes calcium excretion, increasing the risk of osteoporosis (the brittle-bone disease). And protein accelerates kidney damage in people whose kidneys are just beginning to deteriorate.

The most misleading aspect of the protein argument is that proponents rarely mention fat. Unless your favorite fast-food restaurant serves lentil loaf, you're most likely getting a lot of fat along with your protein. Ground beef and many varieties of cheese get more calories from fat than protein. Chicken and fish are lower in fat, but not when they're batter-dipped and fried.

Obviously, protein is a valuable nutrient and no one would recommend that you try living without it. On the other hand, Americans get more than enough, so regard with skepticism any sales pitch featuring protein.

Eschewing the Fat: A Look at Heart Disease, Cancer, and Diabetes

If any one smell characterizes a fast-food restaurant, it is the smell of grease. A fast-food restaurant without the heavy fragrance of grease is like a bicycle without wheels. Now, if all the grease did was smell up the neighborhood, that would be one

FAST FOODS HIGHEST IN FAT

COMPANY/PRODUCT	FAT (tsp.)
Jack in the Box Ultimate Cheeseburger	16
Carl's Jr. Double Western Bacon Cheeseburger	14
Taco Bell Taco Salad with Shell	14
Burger King Double Whopper with Cheese	14
Burger King Double Whopper	12
Carl's Jr. Super Star Hamburger	12
Wendy's Double Big Classic with Cheese	12
Jack in the Box Grilled Sourdough Burger	11
KFC Extra Tasty Crispy Thigh and Wing	11
Carl's Jr. Classic Double Cheeseburger	11
Dairy Queen DQ Homestyle Ultimate Burger	11
Wendy's Double Big Classic	10
Burger King Whopper with Cheese	10
Jack in the Box Sausage Crescent	10
Carl's Jr. Bacon & Cheese Potato	10
Carl's Jr. Country Fried Steak Sandwich	10
Dairy Queen Quarter Pound Super Dog	9
Pizza Hut Supreme Pan Pizza, large, 2 slices (estimate)	9
McDonald's Thousand Island Salad Dressing, 2.5 fl. oz.	9
Burger King Chicken Sandwich	9
Arby's Sausage and Egg Croissant	9
Hardee's Bacon Cheeseburger	9
Wendy's Taco Salad	8

A high fat content is the number-one problem with most diets. Adults should try to limit their fat intake to 15 teaspoons a day, active children 18 to 20. Typical fast-food meals can easily put you over the top.

thing. But medical researchers have proven that much worse than smelling too much grease is *eating* too much grease. Of course, a few greasy meals won't cause anything worse than a little heartburn. However, someone who eats rich, fatty foods from childhood through adulthood stands a high risk of developing the most serious and sometimes fatal diseases.

Many an old-fashioned nutritionist will staunchly defend fat's role in the diet, saying that fat adds an essential fatty acid (linoleic), and contributes fat-soluble vitamins (A, D, E, and K). Again, this mentality may be appropriate for subsistence-level societies, where people get barely enough to eat, but not for well-fed Americans who are never more than a Big Mac away from satisfaction. Natural foods— particularly grains, seeds, and nuts—contain all the essential fatty acids we need were we never to fry the first potato. As it is, Americans eat too much fat, and no one we've ever heard of in this country suffers from nutritional deficiencies because he or she eats too little fat.

Plenty of people suffer from overconsumption, however. High-fat diets complicate diabetes and are implicated in several forms of cancer. Fat in general, and saturated fat in particular, plays a significant part in heart and cardiovascular diseases, as does the fat-like substance cholesterol.

The American Heart Association, the National Cancer Institute, and many other health authorities recommend that everyone two years of age or older consume no more than 30 percent of his or her calories from fat. Our bodies actually need no more than about 10 percent to function properly. The typical American consumes 35 to 40 percent.

HIDDEN SOURCES OF FAT ... AND CALORIES

Many of our dietary fat sources are obvious: we butter our bread, dress our salads, and fry our potatoes. Other fat sources aren't so obvious, and sometimes fat shows up where we least expect it. Red meat contributes the largest amount of fat to the American diet. Bologna, hot dogs, hamburgers, whole milk, cheese, and ice cream are major sources of fat. But there are others. Snack foods such as chips, chocolate, and nuts are high in fat, as are sweets such as pies, doughnuts, danish pastries, and most cakes and cookies.

Even creamed soups, nondairy creamers, and whipped-cream substitutes contain hidden fat.

Rather than counting every single gram of fat in your favorite foods, you will probably find it easier to identify and cut back on the major sources of fat in your diet. That allows some flexibility and doesn't force you to give up food you "can't live without." If you can't live without your nightly bowl of rich ice cream, substitute fresh fruit and yogurt for your morning doughnut, or skinless chicken breast for your midday bologna.

You'll be hard pressed to find low-fat foods on fast-food menus. A glance at the charts later in this book (see Chapter Four) shows that not many fast foods get less than 30 percent of their calories from fat. Indeed, many fall into the 40- and 50-percent-fat range. Some go even higher. That's because the food that isn't high in fat to begin with gets dunked in fat before it's sold.

Fortunately, the grease and burger monopoly has been eroding. Wendy's started salad bars to draw new customers into its restaurants. Several other chains have followed its lead. Salad bars provide terrific low-fat alternatives, provided that the eater avoids such fatty foods as regular salad dressings, bacon, potato salad, cheese, seeds, and high-cholesterol eggs. You can also get plain baked potatoes. They're nourishing, filling, and virtually fat- and sodium-free. (If you must add a topping, use a minimum of sour cream, margarine, or butter.) Burger King's BK Broiler Chicken Sandwich and Hardee's Grilled Chicken Sandwich (Hold the mayo!) are also good choices, as is McDonald's McLean Deluxe.

Cutting down on fatty foods will have pleasant side effects for many people—it may help them lose weight without a rigorous diet. Fatty foods are generally high-calorie foods as well, because fat has more than twice as many calories per gram as do carbohydrates and protein. A plain, large baked potato weighing about eight ounces provides 250 calories, while a medium order of McDonald's french fries weighing under 4 ounces provides 320 calories.

"Staying trim" is as basic to health as not smoking. Maintaining an ideal weight is a major defense against diabetes, hypertension, and several types of cancer. Even if we never gained a pound, however, eating too much fat is our number-one nutrition problem, playing a leading role in chronic disorders, from cancer to heart disease.

FAST FOODS HIGHEST IN SATURATED FATS

COMPANY/PRODUCT	SATURATED FAT (tsp.)
Carl's Jr. Double Western Bacon Cheeseburger	7
Jack in the Box Ultimate Cheeseburger	6
Burger King Double Whopper with Cheese	5
Carl's Jr. Super Star Hamburger	5
Dairy Queen DQ Homestyle Ultimate Burger	5
Carl's Jr. Western Bacon Cheeseburger	5
Taco Bell Taco Salad	4
Burger King Double Whopper	4
Arby's Sausage and Egg Croissant	4
Arby's Deluxe Baked Potato	4
Dairy Queen Hamburger, Double with Cheese	4
Burger King Whopper with Cheese	4
Pizza Hut Pan Pizza, Supreme, large, 2 slices (estimate)	4
Hardee's Bacon Cheeseburger	4
Burger King Bacon Double Cheeseburger Deluxe	4
Jack in the Box French Fries, Jumbo	4
Jack in the Box Grilled Sourdough Burger	4
Jack in the Box Sausage Crescent	4
Arby's Mushroom and Cheese Croissant	4
Jack in the Box Bacon Cheeseburger	3
Carl's Jr. Bacon & Cheese Potato	3
Carl's Jr. Cheese Potato	3
Wendy's Double Big Classic with Cheese	3

Saturated fat is your heart's number-one dietary enemy. An adult's healthy diet should include a maximum of 5 to 6 teaspoons of saturated fat in an entire day. Several individual fast foods reach or exceed this limit.

HEART DISEASE AND A HIGH-FAT DIET

Coronary heart disease, stroke, and related disorders account for almost half of all deaths in this country. In 1988, heart and blood-vessel diseases killed nearly a million Americans, according to the American Heart Association. They cost our nation an estimated $100 billion a year to pay for medical care and for time lost from work.

Heart disease and high-fat diets are inextricably linked. When you eat foods that are high in certain fats or cholesterol, the amount of cholesterol in your blood is likely to rise. The more cholesterol you have in your blood, the more likely your arteries will become clogged. And then comes the heart attack.

A LINEUP OF FATS

The fatty acids contained in fats come in three types: saturated, monounsaturated, and polyunsaturated (they are named for the type of bonds that link the atoms). Most saturated fatty acids raise blood cholesterol levels, while the monounsaturated and polyunsaturated varieties either reduce or have no effect on blood cholesterol.

Foods contain a combination of those fatty acids, but one of the three types often predominates. Animal fats, for instance, are high in saturated fatty acids. So the fat found in meat, milk, butter, cream, and lard is no friend of your arteries. In addition, two vegetable fats, coconut and palm kernel oils, are even more highly saturated, but they are rarely used by restaurants.

Until the late 1980s most fast-food restaurants fried all their foods in shortening made with 95 percent beef fat or coconut oil. In 1984 the companies started shifting to vegetable shortenings after the Center for Science in the Public Interest began its campaign to convince the companies to use shortenings that were more heart-healthy. In the late 1980s, Hardee's, Carl's Jr., Arby's, and Taco Bell all stopped using highly saturated fats. McDonald's and Burger King used beef fat only for frying potatoes. In 1990 the National Heart Savers Association, which is run by Omaha businessman Philip Sokolof, ran powerful full-page newspaper ads throughout the country urging McDonald's to stop frying potatoes in beef fat. Within weeks McDonald's, Burger King, Dairy Queen, Jack in the Box, and Wendy's all said they would switch to vegetable shortening by 1991.

The shortenings to which the companies switched are made of

partially hydrogenated vegetable oils. Hydrogenation makes the oils harder. Margarine, for example, is almost as solid as butter, because it is partially hydrogenated. These shortenings contain less saturated fat than beef fat, but more than pure vegetable oils. Though partially hydrogenated shortenings have been used for decades and were always thought to be safe, one careful study published in 1990 suggests that they might raise blood cholesterol levels more than previously thought. That study needs to be confirmed, but it suggests another reason why it's best to avoid fried foods as much as possible.

Adults eating a healthy diet should consume no more than about 15 to 20 teaspoons of fat a day, including no more than 5 to 6 teaspoons of saturated fat. The table on page 51 lists some of the fast foods highest in saturated fat. Note that several large burgers provide an entire day's worth of saturated fat.

THE RESEARCH

Studies of large population groups reveal that diets high in saturated fat tend to increase the risk of heart disease. The Japanese, who traditionally consume very little saturated fat, have a low incidence of heart disease (even though they smoke more cigarettes and have more high blood pressure than Americans). Japanese immigrants to Hawaii tripled their fat intake and increased their risk of heart disease as well. Dr. Ancel Keys and his colleagues, heart researchers at the University of Minnesota, found that Americanized Japanese between 50 and 69 years old were twice as likely to have severely clogged arteries as their counterparts living in Japan.

Keys and his coworkers also found that Scandinavians got as much as 22 percent of their calories from *saturated* fat. Americans average about 15 percent. A healthy diet would have one-third that much. Scandinavians had much higher rates of cardiovascular disease than people in Mediterranean countries, such as Greece and Italy, where much more of the fat consumed is unsaturated.

In 1984, the National Heart, Lung and Blood Institute announced the results of a landmark study on heart disease. The 10-year study involving 3,806 men showed that a cholesterol-lowering drug significantly reduced the risk of heart attack. For every 1-percent decrease in blood cholesterol, the participants experienced a 2-percent decrease in heart-attack risk. That is, a 10-percent reduction in blood

Fast Foods Highest in Cholesterol

COMPANY/PRODUCT	CHOLESTEROL (mg.)
McDonald's Scrambled Eggs	399
Jack in the Box Scrambled Egg Pocket	354
Carl's Jr. Breakfast Burrito	285
McDonald's Biscuit with Sausage & Egg	270
Arby's Sausage and Egg Croissant	271
Burger King Croissan'wich with Sausage, Egg and Cheese	268
McDonald's Sausage with Egg McMuffin	256
McDonald's Biscuit with Bacon, Egg & Cheese	248
Carl's Jr. Scrambled Eggs	245
Burger King Croissan'wich with Ham, Egg and Cheese	241
McDonald's Egg McMuffin	224
Arby's Bacon and Egg Croissant	221
Jack in the Box Breakfast Jack	203
Carl's Jr. Jr. Crisp Burritos, 3	195
Burger King Double Whopper with Cheese	194
Carl's Jr. Sunrise Sandwich with Sausage	190
Jack in the Box Sausage Crescent	187
Hardee's Canadian Rise 'N' Shine Biscuit	180
Hardee's Country Ham & Egg Biscuit	175
Wendy's Double Big Classic with Cheese	170
Burger King Double Whopper	169
Hardee's Fried Chicken Breast and Wing	165

The cholesterol in food increases the cholesterol level in blood and increases the risk of heart disease. You should try to limit your cholesterol intake to 250 milligrams a day—the less the better. Avoid breakfast sandwiches and large burgers.

cholesterol (e.g., from a count of 250 to 225) reduced the risk by 20 percent. Dr. Basil Rifkind, director of the study, said it was the "first study to demonstrate conclusively that the risk of coronary heart disease can be reduced by lowering blood cholesterol." While the study employed a drug, as well as diet, to reduce cholesterol levels, the researchers said that diet alone should have exactly the same benefit.

Just five years later a new study showed that diets containing only 10 percent of calories from fat, together with exercise and stress reduction, can actually reverse coronary heart disease. Dr. Dean Ornish and his colleagues at the University of California at San Francisco School of Medicine put 28 people with clogged arteries on a low-fat vegetarian diet that included unlimited amounts of vegetables, grains, legumes, and soy products. The only animal products allowed were egg white and one cup of nonfat milk or yogurt a day. The patients were also urged to exercise for at least three hours a week and to do breathing exercises and meditation for an hour a day. Another 20 patients served as a "control" group that was not instructed to make those lifestyle changes.

After one year, the openings in the vegetarian group's arteries were significantly larger, while the control group's openings had shrunk further. What's more, the patients who most closely followed the diet, exercise, and stress-management advice were most successful in unclogging their arteries. Ornish's enormously important study settled once and for all the enormous potential for treating serious heart disease through diet, exercise, and stress reduction, rather than drugs and surgery.

CHOLESTEROL

Animal-derived foods generally contain cholesterol, which, like saturated fats, can lead to higher levels of cholesterol in your blood. The cholesterol in American diets comes mostly from egg yolks, red meat, fish, and poultry. Dairy products, some baked goods, and cooking fats make up the rest. The chart on the facing page lists some of the greatest sources of cholesterol at fast-food restaurants. Several products provide enough cholesterol for two days' worth of eating!

Our bodies need cholesterol to function properly. But because our livers are dependable manufacturers of cholesterol, that substance is not a critical nutrient. Unlike vitamin C, for example, cholesterol needn't be included in our diets.

Dr. William Castelli, director of the Framingham Heart Study in Massachusetts, points out that for every 100 milligrams of cholesterol you remove from your daily diet, you can decrease your blood cholesterol by five points within several weeks. People who cut down from 450 milligrams to 250–300 milligrams a day would lower their blood cholesterol by 10 points. Pair that with decreased intake of saturated fat and the changes are indeed significant. "The point is that the change comes from the diet," says Castelli.

The great body of biomedical research into the causes of heart disease has led health organizations from New Zealand to Great Britain, and from Canada to Scandinavia, to call for decreased consumption of fat and cholesterol. In the United States, the surgeon general, the American Heart Association, the Department of Health and Human Services, the Department of Agriculture, and other distinguished private and authoritative governmental agencies have all said, "Eat less fat—especially saturated fat—and cholesterol." *That is the single most important dietary change that most Americans could make.*

HOW MUCH CHOLESTEROL IN YOUR BLOOD?

So, what should your blood cholesterol be? Generally speaking, the lower the better. In countries where people rarely get cardiovascular disease, cholesterol levels don't go much above 200, according to Castelli. A "consensus conference" convened at the National Institutes of Health in 1984 concluded that people under the age of 30 should have blood cholesterol levels under 180. Those over 30 should aim for a cholesterol level under 200. According to a 1988 report from the National Cholesterol Education Program, a level between 200 and 239 is considered "borderline high." Levels above 240 are definitely high. Anyone in either group should have their cholesterol rechecked (because measurements are not always accurate) and should find out how much of their total cholesterol is the "good" kind (HDL) and how much the "bad" kind (LDL). To achieve their desirable level, millions of Americans will have to make major changes in their eating habits. If you haven't had your cholesterol level checked recently, get it done very soon at your doctor's office or local clinic.

Even children aged two and above should cut back on fatty foods, because heart disease begins early in life. Autopsies of young American soldiers killed in the Korean and Vietnam wars showed significant narrowing of the arteries. By contrast, the arteries of Asian soldiers, whose diets were low in fat and cholesterol, were wide open and healthy. In Bogalusa, Louisiana, researchers discovered that children as young as seven years old had fatty streaks on the walls of their blood vessels. Also, in general, their blood-cholesterol levels were high. Some of those streaks represent the first stage of coronary heart disease.

When children start chomping on cheeseburgers, they initiate a process that will gradually clog their arteries and increase the odds of their having heart attacks or strokes several decades later. They're also learning to love the greasy, salty food that could ultimately cause their deaths. Childhood is clearly not too early a time to develop healthful eating habits.

CANCER

Cancer—long dreaded as a disease that reels out of control with no regard for science or medicine—may in some cases be affected by the amount of fat in our diets. Population studies "have repeatedly shown an association between dietary fat and the occurrence of cancer at several parts [of the body], especially the breast, prostate, and large bowel," a special panel of the National Academy of Sciences stated in its landmark 1982 report, *Diet, Nutrition and Cancer.*

A look at the same body of research convinced the government's National Cancer Institute to begin a Cancer Prevention Awareness Program that publicizes, among other things, the role of diet in cancer. The prevention project stresses three points: Americans should "keep trim, eat a variety of foods that are low in fat, and eat foods high in fiber," according to Dr. Peter Greenwald of the cancer institute. Unlike heart disease, which is promoted specifically by saturated fat, breast and colon cancer incidences appear to increase when diets are high in any type of fat.

Throughout the world, high-fat diets correlate with a high risk of breast cancer, which killed 44,000 American women in 1990 and is tied with lung cancer as the leading cause of cancer deaths in women. Consider some of the evidence:

- The traditional Japanese diet contains only about one-fourth as much fat as the average American diet. Breast cancer is about one-fifth as common in Japan as in the United States.

- The rate of breast cancer among Japanese and Italian immigrants increased when they settled in the United States and began to consume the typical high-fat American diet.

- A high-fat diet increased the number of tumors in rats and mice that were simultaneously fed cancer-causing chemicals.

"If you pick countries that have half the amount of calories from fat that we do, they have half the incidence of breast cancer," said Greenwald.

A recent study by Walter Willett of the Harvard School of Public Health found that women who ate fatty diets did not have an increased risk of breast cancer. However, a 1991 Canadian study published in the *Journal of the National Cancer Institute* found that women whose diets were high in fat, but not necessarily high in calories, did have a higher risk. Until the discrepancies between these (and other) studies are ironed out, eating a lower fat diet is prudent, especially because fatty foods are linked to colon cancer as well.

In addition to eating less fat, maintaining ideal weight is another crucial step to reducing your risk of breast cancer. Women who reach middle age more than 10 percent heavier than their ideal weight have an increased rate of breast cancer, according to Dr. John Spratt, a cancer researcher and surgeon at the University of Louisville School of Medicine. And once overweight women get it, he said, the cancer will be "more virulent, more lethal." Because fat is such a concentrated form of calories, and because it is easy for the body to turn food fat into body fat, diets high in fat promote obesity.

But the cancer-diet link doesn't stop with breast cancer. Eating less fat and more fiber may reduce the chances of a man or woman developing colon cancer — which, at about 61,000 deaths per year, is the second most common cancer killer (after lung cancer) in America. Population studies, again, show that Japanese immigrants got colon cancer at a higher rate when they settled on American soil and

FAST FOODS LOWEST IN FATS

COMPANY/PRODUCT	FAT (tsp.)
Side Salad (several companies)	0
McDonald's Apple Bran Muffin	0
McDonald's Wheaties	0
Long John Silver's Green Beans	0
TCBY Nonfat Frozen Yogurt	0
Wendy's Three Bean Salad, ¼ cup	0
Arby's Old Fashioned Chicken Noodle Soup	<1
Long John Silver's Light Portion Baked Fish dinner	<1
McDonald's Low-fat Milk Shake	<1
Carl's Jr. Lite Potato	<1
Dunkin' Donuts Plain Bagel	<1
KFC Mashed Potatoes and Gravy	<1
Hardee's Pancakes, 3 (without margarine)	<1
Long John Silver's Rice Pilaf	1
KFC Corn on the Cob	1
McDonald's Chunky Chicken Salad	1
Carl's Jr. Charbroiler BBQ Chicken Sandwich	1
Wendy's Chili, 9 oz.	2
Wendy's Seafood Salad	2
Jack in the Box Chicken Fajita Pita	2
Carl's Jr. Bran Muffin	2
McDonald's McLean Deluxe	2

It is increasingly possible to get low-fat foods in a fast-food restaurant. Each of the foods in this chart contains two teaspoons of fat or less. Omitting the mayonnaise or tartar sauce will reduce the fat content of many other sandwiches.

started eating an Americanized diet. If people "increase [their] fiber intake and reduce the fat [they] eat," cancer experts predict a "30-percent reduction in colon cancer," said then Secretary of Health and Human Services Margaret Heckler in 1984. Unfortunately, Secretary Heckler's words had little effect on our diets.

In 1990, researchers at Stanford University compared the diets of Chinese (those living in both the United States and China) who had colon cancer to those who did not. Men and women whose diets contained as little as 10 grams of saturated fat had an increased risk of colon cancer. (That's about one-third as much saturated fat as most Americans consume.) Harvard's Walter Willett also found that saturated fat—and red meat in particular—were correlated with a higher risk of colon cancer. Whether the culprit was the fat or some other component of the meat is not known. In Willett's study, fat in dairy products or vegetable oils did not seem to be a problem.

While stopping short of saying that a high-fat diet all by itself causes cancer, Dr. Greenwald says fat "probably acts as a promoter" and that researchers are "pretty confident" that even if you cut your fat intake late in life, you can decrease your cancer risk. Cancer experts specifically recommend that we lower our fat consumption to a maximum of 30 percent of our calorie intake (from the current average of about 40 percent), and preferably considerably less. That's equivalent to about 15 teaspoons of fat a day for someone on a 2,000-calorie diet.

In terms of fat, all restaurants have good and bad choices, but often it's difficult for the average consumer to know what the best choices are. Who would guess that a stuffed baked potato would contain as much as 10 teaspoons of fat . . . or that a "lean" chicken or fish meal would provide upward of 12 teaspoons of fat . . . or that a taco salad would have 14 teaspoons of fat? You don't have to be a genius, though, to know to avoid foods that contain fatty items piled on top of fatty items and foods that are deep-fried.

DIABETES

The common wisdom about diabetes turned out to be all wrong. Because people with diabetes have difficulty moving sugar from the blood into the cells, diabetics were encouraged to eat less sugar.

"Sugar's been looked at as the problem," says diabetes researcher Dr. James Anderson, chief of endocrinology at the Veterans Administration hospital in Lexington, Kentucky. "Sugar's not the big problem. Fat's the big problem. The major problem with fast foods is that they are high in fat. They are so packed with calories from fat that they should be a cause of concern for anybody with a metabolic disorder. People need to cut down on all fat."

There are two types of diabetes: In Type I (commonly known as childhood diabetes), the pancreas fails to produce insulin. Insulin is the hormone that blood sugar must hook up with in order to enter the cells where it is used for energy. The Type-I diabetic is insulin-dependent — since he or she produces no insulin and must inject some at every meal.

Type-II diabetics (who usually develop the condition as adults) actually produce plenty of insulin, but the cells either have too few insulin-receptor sites (entrances to the cell) or the receptor sites don't function properly. Type-II diabetics often take insulin or other medications, but if they change their diet or lose weight, they may not need any at all.

Being overweight is a major cause of Type-II diabetes. Eight out of 10 Type-II diabetics are obese. But most obese people are not diabetics. Only one in four obese people becomes diabetic because of his or her faulty insulin receptors. It appears that the functioning of the receptors is linked to a number of factors, including genetics, weight, exercise, and the fat content of the diet — the more overweight a diabetic is, the less he or she exercises, and the higher the fat content of his or her diet, the less efficiently the receptor sites function.

But even Type-I diabetics, who do need some insulin, may be able to use lower doses with a low-fat diet. That's a benefit because the infamous side effects of diabetes — blindness, kidney failure, nerve damage in the feet and legs, and clogged arteries — may be heightened by the amount of insulin injected. "The general consensus is the less insulin [injected] the better," says Janet Tietyen, a dietitian, who worked with Anderson.

Insulin aside, however, diabetics should avoid high-fat diets because they come with lots of calories. Excess calories lead to obesity, and obesity can exacerbate diabetes. Also, diets high in saturated fat tend to raise blood cholesterol, which is linked to heart disease and stroke. Heart disease is the number-one cause of death for diabetics.

FATS OF THE FUTURE

Fat foods of today may be slimmed down dramatically in just a few years through ingenious chemistry. New fat substitutes promise to give the "mouth feel" of fat without the calories. If these chemicals are safe, they could lead to a wide variety of interesting new foods.

NutraSweet Company's Simplesse is made primarily from egg protein or milk protein. It can replace much of the fat in cold foods, such as ice cream, sour cream, and cream cheese. Protein being protein, though, Simplesse cannot be used as shortening, because it would coagulate just like an egg white does when it is heated. Simplesse does not pose any health problem, though people allergic to milk or eggs have to avoid it. Kraft and other companies have developed similar products made from cellulose or vegetable gums, enabling them to make fat-free "ice creams" and "mayonnaise."

More versatile are fat substitutes like olestra, which looks like Crisco and is also produced by Crisco's maker, Procter & Gamble. Olestra, a chemical combination of sucrose (table sugar) and fatty acids, passes through the body largely unabsorbed. It can be used to fry foods. Procter & Gamble originally petitioned the Food and Drug Administration (FDA) to approve olestra as a substitute for one-third to two-thirds of the oil in home cooking oils and commercial shortenings. In 1990, though, it limited its request to replacing the fat used in making potato chips, corn chips, and similar snack foods.

The fly in olestra's ointment is that Procter & Gamble's safety tests indicate problems. Olestra caused a kind of liver damage in rats that suggests it might cause cancer in humans. It also appeared to cause pituitary-gland cancer in two studies. Tests on mice were still in progress in 1991. The Center for Science in the Public Interest has urged the FDA not to approve olestra until the safety questions have been resolved.

While olestra was the first shortening substitute at the starting gate, many other companies are testing their own versions. It's probably only a matter of time before one or more of these substances is approved. You can bet that image-conscious fast-food chains will try to introduce them quickly to help dispel the "fat-food" image.

DIETARY FIBER

Dietary fiber is a type of carbohydrate that passes through the body largely undigested and unabsorbed. Nutrition lessons used to dismiss fiber as a nonnutrient. Teachers called it "bulk" or "roughage" and said the only thing it was good for was proper bowel movement. These days, the lessons have changed. It may just be that fiber is one of the best things to be discovered since penicillin.

It seems that for every ill that you can attribute to fat, you can attribute good to fiber-rich foods. Dr. Denis Burkitt is the English physician-researcher who has waged a worldwide publicity campaign to popularize the benefits of high-fiber diets. Burkitt believes that such diets can help us avoid such troublesome, though rarely fatal, diseases as diverticulosis, hemorrhoids, hiatus hernia, appendicitis, varicose veins, and gallstones.

Recalling his earliest thoughts on fiber, Burkitt told one interviewer that he became "acutely conscious that a high proportion of the beds in any Western hospital are filled with patients suffering from diseases which are rare or unknown in the rest of the world." These diseases, he thought, might well come from the Western diet—a diet high in refined foods from which fiber is removed and fat and sodium added. "When you take out the fiber," Burkitt told a medical conference, "you pass small stools and you need big hospitals." Burkitt's observations provoked a great deal of research, and some of his theories are being borne out. The evidence is mounting that diets high in fiber-rich foods help protect against diabetes, cancer, and heart disease.

The term *dietary fiber* includes a large family of substances that occur in foods of plant origin but never of animal origin. Beans, bran, fruits, and vegetables are among the best sources of dietary fiber.

Researchers are now trying to identify the benefits of different types of fiber. They have identified at least two general categories of fiber—soluble and insoluble. Within these two categories are numerous fibers that seem to affect specific diseases. For example, whole wheat is rich in fiber that is insoluble in water. It seems to protect against diverticulosis, constipation, and possibly colon cancer. Beans and oat bran, on the other hand, contain not only insoluble, but also water-soluble forms of fiber that appear to help control diabetes and may lower blood-cholesterol levels, thereby decreasing the chances of heart disease.

FIBER AND COLON CANCER

In the pamphlet "Good News: Better News: Best News," the National Cancer Institute explains that "every day you can do something to help protect yourself from cancer." Eating a high-fiber diet is one of those things.

High-fiber diets appear to decrease the risk of colon (large intestine) cancer, which Burkitt calls a "deficiency disease of Western civilization." The way in which a high-fiber diet protects against colon cancer is not absolutely clear, but researchers think that it goes something like this: Bile acids are secreted into the intestines to help digest fat. The fattier the diet, the more bile acids in the colon. A high concentration of bile acids in the colon increases the risk of colon cancer. High-fiber diets increase the size and water content of stools as they are forming in the large intestine, thereby diluting the bile acids and reducing the risk of cancer. Another theory holds that diets high in fiber lead to faster passage of stools. That shortens the time during which carcinogens in the gut could attack the colon.

Studies on fiber and cancer should encourage consumers to head for the salad bar and whole-wheat bread:

- International comparisons show that countries consuming more fiber in their diets, especially from grains, have lower rates of colon cancer.

- People in Finland consume high-fat, high-fiber diets. When compared with people in the United States, whose diets are high in fat but low in fiber, the Finns showed half the incidence of colon cancer.

- In 1982, scientists from the International Agency for Cancer Research performed a carefully controlled study on 30 randomly selected residents from each of four populations: low-risk Parikkala, Finland; intermediate-risk Helsinki, Finland; intermediate-risk Them, Denmark; and high-risk Copenhagen, Denmark. For each group of 30, the higher the cancer rate in the population they represented, the lower the group's fiber intake and the higher the concentration of bile acids in their feces.

All that said, it's possible that something other than the fiber in fiber-rich foods protects against cancer. In addition to the fiber that they contain, fruits, beans, and vegetables are loaded with vitamin C, carotenes, indoles, phenolic acids, saponins, terpenes, and many other unfamiliar constituents that sometimes prevent cancer in animal studies. Furthermore, studies suggest that people who eat more fruits and vegetables have lower risks not just of colon, but lung and oral cancers as well. That's why it is smart to replace the fat in your diet with at least 3 servings of fruit and 4 servings of vegetables. According to a 1976–1980 government study of more than 10,000 people, the average adult consumed only one serving of fruit and 2 servings of vegetables in a day.

Finally, high-fiber diets may help control weight. Foods rich in fiber can make you feel full. For instance, eating a 150-calorie apple is more filling than eating the same number of calories in the form of apple juice. And keeping weight close to your ideal seems to reduce the risks of cancers of the breast and endometrium (lining of the uterus).

FIBER AND HEART DISEASE

High-fiber diets help prevent heart disease in two ways. First, if you eat foods high in fiber—such as fruits, vegetables, grains, and beans—chances are you will simultaneously eat fewer fatty foods, because you have less room for them. That alone will help prevent heart disease.

In addition, specific types of fiber may lower blood-cholesterol levels, thereby lowering the risk of heart attack. The soluble fiber in prunes, peas, corn, apples, many types of beans, and possibly oat bran lowers blood cholesterol.

FIBER AND DIABETES

A high-fiber diet helps prevent the onset of, and helps curb the symptoms of, diabetes. First, it can help people maintain suitable weight, which reduces the risk of diabetes. Second, dietary fiber seems to improve the way we use other carbohydrates. "Starches

FIBER CONTENT OF FOODS

FOOD	SERVING SIZE	FIBER (*gm.*)
Nabisco 100% Bran	½ cup	10
Kidney beans	½ cup cooked	9
Pizza Hut Supreme Personal Pan Pizza	whole pizza	9
Pizza Hut Traditional Hand-Tossed Pizza, Cheese	2 slices, medium	7
Ralston Bran Chex	⅔ cup	6
Wendy's Chili	9 ounces	6
Pizza Hut Traditional Hand-Tossed Pizza, Pepperoni	2 slices, medium	6
Kellogg's Raisin Bran	¾ cup	5
Apple	1 large	5
Baked potato	1 medium with skin	4
Whole-wheat bread	2 slices	4
Lentils	½ cup cooked	4
Corn	½ cup cooked	3
Orange	1	3
Nabisco Shredded Wheat	1 biscuit	3
Peas	½ cup canned	3
Popcorn	2 cups air popped	3
Carrot	1 raw	2
Broccoli	½ cup cooked	2
Strawberries	½ cup	2
Brown rice	½ cup cooked	2
Peach	1	1

Dietary fiber is present in whole grains, dried beans, fruits, and vegetables. For optimal health, try to consume 20 to 30 grams of fiber daily.

and sugars are more slowly absorbed when they are eaten as part of a high-fiber diet," according to Anderson. Therefore, blood sugar does not rise as much after a high-fiber meal as it would after a low-fiber meal, and does not require a rush of insulin to empty sugar from the blood.

By increasing the fiber in their diets, Anderson reduced the insulin needs of Type-II diabetics. "We have been able to discontinue insulin therapy in more than half of the patients who have followed [a high-fiber, low-fat] program." Overall, hospital studies involving more than 1,000 patients have shown that a low-fat, high-fiber, high-carbohydrate diet allowed insulin doses to be lowered by an average of 70 percent. Blood cholesterol levels dropped by an average of 30 percent. Among outpatients, insulin doses decreased by 72 percent and blood cholesterol levels dropped by 11 percent over four years.

High-fiber diets are a double benefit to diabetics, who run twice the risk of heart disease as the nondiabetic population. Because a high-fiber, low-fat diet not only alleviates the symptoms of diabetes, but also can lower blood cholesterol, such diets reduce a diabetic's risk of heart disease and stroke.

GETTING ENOUGH FIBER

How much dietary fiber is enough? Right now, Americans consume an average of about 10 grams per day. The National Cancer Institute recommends that we consume two to three times that much. Back-to-basics nutrition that includes more fresh fruits, vegetables, whole grains, and beans should provide adequate fiber. Those unrefined plant foods not only contain lots of fiber and little fat, but they generally contribute more vitamins and minerals to the diet. A baked potato has more vitamin C than french fries; whole grains have more vitamins and minerals than refined grains. By including more high-fiber foods in your diet, you will naturally include a better balance of nutrients in a lower-calorie package.

While a bowl of wheat-bran or oat-bran cereal or oatmeal—which has not yet appeared on the fast-food scene—gets you off to a good fiber-rich start in the morning, it doesn't mean you can forget about fiber for the rest of the day. As we have seen, it takes a variety of fibers—found in a variety of foods—to promote optimal health.

Beans are one of our best sources of fiber. Unfortunately, legumes are not easy to find, that is with the exception of a few fried-chicken restaurants that offer baked beans; Wendy's, which serves chili with beans; and the Mexican restaurants that serve refried beans.

High-fiber foods show up at salad bars, especially those bolstered with kidney or garbanzo beans. Also, baked potatoes eaten with their skins are a good source of fiber. Ordering them plain (and perhaps adding one pat of sour cream or margarine) helps you avoid excess salt and fat. (While baked potato skins are an excellent fiber source, ordering skins only is a risky proposition—they are usually deep-fat fried and often come with salty, fatty toppings.)

Whole-grain bread or buns, good sources of fiber, have not yet shown up in restaurants. Subway offers what it calls wheat buns, but white flour is the main ingredient. Likewise, the "multi-grain" buns that some Wendy's outlets offer contain only 10 percent whole grain, not enough to warrant much fiber excitement. In many "wheat" bread products, such as Subway's buns, caramel coloring is added to make them look dark, as if they were rich in whole grains. The same applies to Burger King's oat-bran bun, which contains mostly white flour. Except for Pizza Hut, Dunkin' Donuts, and Domino's, the nutrition brochures that companies offer do not list dietary fiber.

SODIUM

Salty and greasy are two of the primary taste sensations associated with fast foods. They are simple tastes, inexpensive ones to include in foods, and, unfortunately, linked to serious health problems.

Most of the salt in fast foods (and in other processed foods) comes from table salt, good old sodium chloride. Like protein and fat, sodium is essential to a properly functioning body. It keeps cells from bursting with excess fluid and helps transmit electrical messages from nerves to muscles. But like protein and fat, a little dab'll do ya. The average person needs only about 200 milligrams a day—the amount found in $\frac{1}{10}$ teaspoon of salt or a couple of slices of bread—to function properly. "Our body's physiology is designed to hold on to sodium," says Dr. Norman Kaplan, of the University of Texas Health Science Center in Dallas. Therefore, our daily need is small. Knowing that it is virtually impossible to eat so little, The National Academy of Sciences recommends that we try not to consume more than 2,400 milligrams a day, which is equivalent to just over 1 teaspoon of salt. Ideally, says the NAS, we should consume no more than 1,800 milligrams in a day. Most Americans consume 4,000 to 6,000 milligrams per day (some estimates are even higher). Those high levels have been linked to high blood pressure.

SODIUM AND HYPERTENSION

An estimated 60 million Americans — one out of three adults — have high blood pressure (hypertension). Among people over 65, half have high blood pressure. African-Americans have significantly higher rates than whites. Dr. Elijah Saunders, president of the Association of Black Cardiologists, says "The sodium content of some fast foods is astonishingly and irresponsibly high. I hope that someday people will be able to eat at fast-food outlets without harming themselves."

Hypertension is a silent disease that has no outward symptoms, but can be detected easily by a simple, painless procedure. Compared to people with normal blood pressure, people with high blood pressure have eight times the risk of a stroke, three times the risk of a heart attack, and five times the risk of congestive heart failure.

No one can predict exactly who will develop the disease. Family history and obesity certainly influence the onset of hypertension, but they aren't the only risk factors. A diet high in sodium is an important factor in many cases.

A great deal of research supports the connection between high-sodium diets and hypertension. In animal experiments, researchers raised the blood pressure of rats, dogs, pigs, chickens, and monkeys by feeding them high levels of salt. In other studies, rats bred to be especially prone to hypertension did not develop high blood pressure when fed a diet low in sodium; blood pressure rose only when dietary sodium did.

The Japanese, traditionally protected from heart attacks and from many cancers by their low-fat diets, do suffer high rates of hypertension and stroke. Their diet, high in pickled and fermented foods, and high in soy sauce and other high-sodium foods, makes their sodium consumption and their incidence of high blood pressure greater than ours. Moreover, populations in northern Japan who eat particularly large amounts of sodium (more than 20,000 milligrams a day) have a higher incidence of high blood pressure than their southern neighbors.

An important study done in 1982 in England showed quite clearly the effect of sodium on blood pressure. In this study, the participants were put on a diet containing 1,900 milligrams of sodium, as compared to their customary intake of 4,400 milligrams of sodium per day. Their blood pressure dropped significantly. Then the participants were given placebo pills containing no active ingredient. Their blood pressure remained stable. Finally, they were given time-release sodium tablets to bring their sodium intake back to their normal levels. The higher-sodium diet returned their blood pressure to what it had been at the start of the study. Not only did these experiments show that blood pressure is affected by sodium, but also they showed that even mild sodium restriction can often lower blood pressure.

A few skeptical scientists believe that the link between hypertension and sodium is unproven. "Perhaps," says University of Texas' Kaplan, but "the circumstantial evidence is strong." So strong, in fact, that he and thousands of other physicians confidently place hypertensive patients on low-sodium diets. So strong that other scientific and government authorities advise limits on sodium intake. The U.S. surgeon general, the Food and Drug Administration, the U.S. Department of Agriculture, and the National Academy of Sciences all advocate lower-sodium diets.

In 1972, hypertension was so rampant, disabling, and life-threatening that the National Institutes of Health began a massive project to educate both the public and physicians about the hazards of the disease. The two slogans associated with the program were "See Your Doctor" and "Treat It for Life." Partly as a result of the program, deaths from stroke have been cut in half, and heart-disease rates dropped by 40 percent between 1970 and 1985.

But the program had at least one drawback: It implied that drugs were the primary answer to the problem. Drugs are certainly useful, but they are not necessarily the best answer. The drugs that are used to treat hypertension can have side effects ranging from

Fast Foods Highest in Sodium

COMPANY/PRODUCT	SODIUM (mg.)
Subway Seafood and Crab Sandwich, 12 in. (estimate)	2612
Pizza Hut Traditional Hand-Tossed Pizza, Super Supreme, large, 2 slices (estimate)	2197
Carl's Jr. Roast Beef Club Sandwich	1950
Arby's Chicken Cordon Bleu	1824
Carl's Jr. Double Western Bacon Cheeseburger	1810
Arby's Bac 'N Cheddar Deluxe	1672
Carl's Jr. Bacon & Cheese Potato	1670
Pizza Hut Traditional Hand-Tossed Pizza, Super Supreme, medium, 2 slices	1648
Jack in the Box Egg Rolls, 5 pieces	1640
Hardee's Country Ham & Egg Biscuit	1600
Jack in the Box Taco Salad	1600
Wendy's Double Big Classic with Cheese	1555
Hardee's Canadian Rise 'N' Shine Biscuit	1550
Arby's Sub Deluxe	1530
Carl's Jr. Western Bacon Cheeseburger	1490
Jack in the Box Chicken Supreme	1470
Wendy's Baked Potato, Bacon & Cheese	1460
Arby's Roast Chicken Club	1423
Hardee's Hot Ham 'N' Cheese	1420
Burger King Chicken Sandwich	1417
KFC Colonel's Deluxe Chicken Sandwich	1362
Dairy Queen Quarter Pound Super Dog	1360

Salt is a cheap seasoning and abundant in fast foods. One teaspoon of salt contains 2,000 milligrams of sodium. Each of the foods listed contains two-thirds teaspoon of salt or more. A reasonable target is 2,000 milligrams a day.

headaches and insomnia to depression and sexual dysfunction—not a pleasant prospect for 60 million Americans. And the cost of putting one-third of American adults on drugs for the rest of their lives was estimated at $30 billion a year.

Dr. Robert Levy, then director of the National Heart, Lung and Blood Institute, which organized the education project, recommended in 1979, that the 35 million "borderline hypertensives" deal with their condition by reducing their weight and their dietary salt. Drugs are necessary for many of the remaining 25 million "definite" hypertensives. (Both borderline and definite hypertensives have an increased risk of heart attack and stroke.)

SALT, SALT, EVERYWHERE

Most of our dietary sodium comes from table salt (sodium chloride). Though all foods contain some naturally occurring sodium, 90 percent of the sodium we eat comes from salt, but generally not from the saltshaker. Estimates differ, but it's clear that much of the salt we eat is added during manufacturing. Natural Cheddar cheese, for instance, contains less than 200 milligrams of sodium per ounce, while processed cheese contains more than twice that. Two ounces of broiled pork contain 40 milligrams of sodium, but the same amount of bologna contains over 500 milligrams.

Nearly all fast food is riddled with salt. A Burger King Chicken Sandwich contains 1,417 milligrams of sodium—well over half of the maximum recommended daily intake. Hardee's Big Country Breakfast with Sausage contains 1,980 milligrams—equivalent to almost a whole teaspoon of salt and as much as a person should get from a whole day's worth of eating. Several companies have reduced sodium levels in some of their products since the first edition of this book was published; it is to be hoped that other companies will make similar or larger reductions.

It's difficult to know by tasting just how much sodium is in a food—processed cheese doesn't seem to taste saltier than natural cheese and we probably don't think of biscuits as being saltier than English muffins. Without lists like the ones in this book (see Chapter Four), it's difficult (if not impossible) to cope with fast foods as an informed consumer.

Surprisingly, french fries tend to be one of the lowest-sodium foods offered. That's because the salt is all on the outside of the product, making them taste particularly salty. Ask the clerk to hold the salt and you'll save yourself one hundred or more milligrams. Other safe bets, as usual, are the baked potato and salad bar (without added bacon, cheese, croutons, and Chinese noodles).

Cyril Brickfield, executive director of the American Association of Retired Persons, spoke for many older Americans with high blood pressure when he urged fast-food restaurants "to show more concern for the health of their elderly customers by expanding the number of nutritious items they sell."

SUGAR

The alluring sweetness of sugar tempts us so predictably that food processors put it in all manner of foods they make, from shakes and colas to ketchup and peanut butter. Many foods that don't contain sucrose (table sugar made from sugar cane or sugar beets) contain other caloric sweeteners—principally those from corn, which are increasingly popular as the cost of table sugar rises. Corn syrup is often used to sweeten soft drinks, for instance.

Executives of fast-food companies know their food will sell better if they add sugar for taste or appearance. The french fries, for instance, have a sugar coating that browns when it hits the hot grease. The batter coatings on many foods contain sugar. Colas and shakes contain a great deal of sugar. Unfortunately, most fast-food (and other) companies do not reveal how much sugar their foods contain. Instead, their charts show the total amount of carbohydrates, which includes sugar, starch, and fiber. For foods such as malts, sundaes, and pastries there is no way to calculate the precise sugar content without knowing the recipes or having the foods tested (which we did in several cases). The figures in this book represent our best estimates.

With sugar being added to so many processed foods, it's no wonder that our consumption of it has climbed steadily. In 1951, average yearly sugar consumption—including sugar, corn syrup, and other refined sweeteners—was 110 pounds per person. By 1990, consumption had climbed to 135 pounds per person. In 1870, the figure was only about 40 pounds. Yet refined sugar is not a required

Fast Foods Highest in Sugar

COMPANY/PRODUCT	SUGAR (tsp.)
McDonald's Coca-Cola Classic, 32 fl. oz.	25
Wendy's Coca-Cola, Biggie, (28 fl. oz.)	22
Dairy Queen Mr. Misty, large	21
Burger King Strawberry Shake, large	17
Dairy Queen Mr. Misty, regular	16
Carl's Jr. Carbonated Beverage, regular (20 fl. oz.)	16
Dairy Queen Mr. Misty Freeze	15
Arby's Peanut Butter Cup Polar Swirl	14
Burger King Sprite, medium (18 fl. oz.)	14
Dairy Queen Heath Blizzard, regular	14
Hardee's Strawberry Shake	14
Carl's Jr. Shake, large	13
Dairy Queen Vanilla Shake, large	13
Dairy Queen Chocolate Shake, regular	12
Arby's Chocolate Shake	12
Arby's Butterfinger Polar Swirl	10
Baskin-Robbins Frozen Low-fat Yogurt, large (9 oz.)	10
Wendy's Lemonade, 8 fl. oz.	10
Dairy Queen Vanilla Malt, regular	9
McDonald's Raspberry Danish	9
McDonald's Strawberry Low-fat Milk Shake	9
Wendy's Frosty Dairy Dessert, large (14.5 oz.)	9

Companies do not disclose sugar contents of their foods, so most of these figures are estimates. Cutting back on sugary desserts is a simple way of cutting back on calories.

dietary nutrient and there is no Recommended Dietary Allowance for it.

The single greatest contributor of sugar to our diet is soft drinks, for which Americans spent $43 billion in 1989. Each year our consumption climbs, aided no doubt by our more frequent trips to fast-food restaurants. In 1950, the average person drank the equivalent of 106 twelve-ounce cans of soft drinks a year. By 1989, that figure had climbed to 490, or one and one-third per day. Almost three-fourths of the sodas Americans consume are sugar-sweetened and contain from 8 to 10 teaspoons per 12-ounce can.

Compare the nutritional contribution of soft drinks (calories) with that of orange juice (vitamin C, potassium, folic acid, and thiamin) or milk (protein, calcium, and vitamins A, B2, and D) and you may order something other than a soft drink the next time you eat out. And if you're thinking "milk shake," be aware that they offer as much sugar as a soft drink and nearly three times the calories of a glass of 2-percent low-fat milk. Though they're as high in sugar as other companies' shakes, McDonald's low-fat shakes have one-fifth as much fat.

FAST-FOOD THIRST QUENCHERS

NUTRIENTS	COCA-COLA 12 fl. oz.	ORANGE JUICE 6 fl. oz.	2% MILK 8 fl. oz.	McDONALD'S LOW-FAT SHAKE 10 oz.
Calories	144	85	121	290
Added sugar (tsp.)	9	0	0	9
Vitamin A (% U.S. RDA)	0	8	10	6
Vitamin C (% U.S. RDA)	0	140	4	0
Calcium (% U.S. RDA)	0	0	44	35
Fat (tsp.)	0	0	1	⅓
Protein (gm.)	0	1	8	11

Sugar can have one of two effects on a diet. It either adds extra calories, contributing to obesity, or it replaces more nutritious foods. The first problem is obvious; the second needs some explaining. Refined sugar can easily constitute 20 to 25 percent of a diet. That means the eater must obtain 100 percent of the necessary nutrients from only 75 to 80 percent of the remaining food. While theoretically possible, the typical sugary diet is likely to fall short in the vitamin and mineral department. Many people, especially women and children, get less than 70 percent of their daily requirements for vitamins A and C. Even more people get too little calcium and vitamin B6.

Also, one shouldn't ignore sugar's common companion: fat. There's a whole realm of food we call "dessert" where sugar is cooked up with fat. A batch of homemade sugar cookies, for instance, gets 800 calories from fat, only 240 from sugar. One-eighth of a double-crust apple pie gets 130 calories from fat, only 30 from sugar. Many doughnut recipes call for a minimum of sugar, but deep-frying gives them many calories from fat. Pies and fruit turnovers, like the rest of their genre, taste sweet but get as many as half of their calories from fat. For example, McDonald's Apple Pie and Arby's Apple Turnover rack up more than 50 percent of their calories from fat.

Of course, regardless of calories, sugar promotes tooth decay. It does this by providing food for oral bacteria that, in turn, produce the acid that decays teeth. In 1973, Dr. Abraham Nizel, associate professor of oral health services at Tufts University, told a Senate nutrition committee that he and his students had "never found a single patient whose [cavity] problem could not, in part, be traced to the patient's inordinate consumption of sugar."

Another black mark against sugar has been registered by many parents who claim that sugar alters their kids' behavior. In fact, until recently, the only studies done suggested that sugar did exactly the opposite—that it might slow kids down a bit. But in early 1990, Yale University medical researchers discovered that when healthy children were fed a breakfast containing as much sugar as is found in a couple of cupcakes, the sugar caused their adrenalin levels to increase ten-fold. Adults' adrenalin levels changed relatively little. The doctors suggested that the increased adrenalin could lead to anxiety, difficulty concentrating, and crankiness. This whole area clearly needs more research to determine if sugar has adverse behavioral effects on significant numbers of children.

Though they may not be quite the villains they're sometimes made out to be, sugary foods can easily take over too large a part of our diets. They replace nutrient-laden foods in our diet, rot our teeth, contribute to obesity, and make such high-fat foods as pies and cookies taste so delicious we'll eat them even if it kills us. The U.S. Department of Health and Human Services and the U.S. Department of Agriculture recommend that Americans cut back on sugar.

VITAMINS AND MINERALS

Although we need them in much smaller amounts, vitamins and minerals are no less vital to our health than are protein, fat, and carbohydrates. And while there is no lack of protein, fat, and carbohydrate in the American diet, the same is not necessarily the case for vitamins and minerals.

Our old-fashioned ideas of health relate severe deficiencies of nutrients to such diseases as scurvy, beriberi, and rickets. "Good nutrition" traditionally meant freedom from those diseases. But a modern evaluation of vitamins and minerals shows us that their health benefits are more extensive. A growing volume of evidence indicates, but has not yet proved, that the consumption of adequate quantities of vitamins and minerals helps fend off chronic diseases.

Numerous surveys have found that millions of Americans are not consuming adequate amounts of vitamins A and C, calcium, and iron. While no nutrient is more important than the others, these are commonly deficient in American diets.

VITAMIN A AND BETA-CAROTENE

Vitamin A is well known for its role in the light-detection mechanism in the retina of the eye. It also plays a part in the formation of teeth, bones, and the mucous membranes. Lack of vitamin A can result in reproductive problems and dry, scaly skin — even blindness when deficiencies are severe. Furthermore, vitamin A and beta-carotene (a yellow-orange pigment — found in carrots and many other foods — that the body converts to vitamin A) may help prevent cancer.

Fast Foods Highest in Vitamin A

COMPANY/PRODUCT	VITAMIN A % U.S. RDA
McDonald's Chunky Chicken Salad	170
Carl's Jr. Charbroiler Chicken Salad	150
Wendy's Garden Salad	110
Burger King Garden Salad	100
Burger King Chunky Chicken Salad	92
McDonald's Garden Salad	90
Wendy's Taco Salad	80
Wendy's Carrots, fresh, ¼ cup	80
Jack in the Box Chef Salad	73
Long John Silver's Ocean Chef Salad	70
Burger King Thousand Island Salad Dressing, 2 fl. oz.	64
Long John Silver's Seafood Salad	60
Arby's Lumberjack Mixed Vegetable Soup, 6 fl. oz.	50
Wendy's Baked Potato with Sour Cream & Chives	50
Carl's Jr. Fiesta Potato	45
Arby's Pilgrim's Corn Chowder Soup, 6 fl. oz.	35
Taco Bell Taco Salad	33
Carl's Jr. Breakfast Burrito	30
Carl's Jr. Scrambled Eggs	25
Taco Bell Nachos Bellgrande	23
McDonald's Wheaties (fortified)	20
Wendy's Cantaloupe, 2 oz.	20
Jack in the Box Scrambled Egg Pocket	21
Carl's Jr. Zucchini (fried)	20
Arby's Tomato Florentine Soup, 6 fl. oz.	20

Studies on human populations indicate that beta-carotene may help the body ward off lung cancer. In one study of 2,000 men, the 488 who reported eating diets lowest in beta-carotene had seven times the rate of lung cancer as the 488 who reported eating the highest amounts of beta-carotene. Promising studies on animals suggest that various forms of vitamin A may help prevent cancers of the skin, breast, and urinary bladder as well. Those and other findings prompted the National Academy of Sciences and the government's National Cancer Institute to recommend that people eat more fruits and vegetables high in beta-carotene.

Other research suggests that beta-carotene has a wider variety of benefits than anyone suspected a few years ago. For instance, a preliminary 1990 study found that large doses of beta-carotene helped prevent heart attacks in people who already had clogged arteries. In fact, beta-carotene slashed the rate of heart attacks and strokes in half. Eye researchers have noted that people who eat more fruits and vegetables rich in carotenes have lower rates of cataracts and macular degeneration, the most common cause of blindness in older Americans. In related research, animal studies have shown that beta-carotene can prevent light-induced injury to the retina.

Beta-carotene can be found in abundance in dark green leafy vegetables, such as collards, broccoli, and spinach, and in bright orange fruits and vegetables, such as carrots, sweet potatoes, cantaloupes, and peaches. The salad bar is the best place to load up on beta-carotene. By contrast, a hamburger, cola, fries, and apple pie provides 1,000 calories but only 4 percent of the United States Recommended Daily Allowance (U.S. RDA) for vitamin A and virtually none of it is beta-carotene. That means that 96 percent of your vitamin A — and beta-carotene — allowance must come at other meals.

Although it is possible to get too much vitamin A, it is unlikely you will do so from food alone. Prompted by stories that vitamin A could diminish acne, some teenagers began taking massive doses of the vitamin and, as a result, suffered abdominal discomfort, fatigue, headaches, and painful areas over bones. The U.S. RDA for vitamin A is 5,000 International Units (I.U.); rare overdoses are usually caused by 50,000 to 100,000 I.U. taken every day for many months, even years. Overdoses of beta-carotene, while possible in carrot lovers, result only in harmless yellowing of the skin.

VITAMIN C

Vitamin C made its reputation more than 200 years ago when the British doctor James Lind identified a factor in certain foods that could control symptoms of scurvy. This deadly disease was rampant among sailors who went to sea for long periods. Ship captains had fought scurvy by carrying citrus fruits, including limes, to sea long before Lind's controlled experiment proved the benefit of the fruit and before the vitamin itself was isolated.

Vitamin C is a sensitive vitamin—heat and exposure to air (oxidation) can destroy it. Since it is water soluble, it tends to leach out of foods and into cooking water. Steaming or microwaving, therefore, are the best ways to cook vegetables. Stir-frying also minimizes losses since it is done quickly and no broth is discarded.

Vitamin C hit the big time in 1970 when two-time Nobel laureate Linus Pauling claimed that massive doses could prevent the common cold or diminish its symptoms. Subsequent scientific studies did not substantiate Pauling's sweeping theories but did indicate that high doses of vitamin C could sometimes shorten the duration of colds.

Beyond scurvy, and the common cold, vitamin C may play an important role in maintaining health, including reducing the risk of cancer. Several epidemiological studies link low stomach-cancer rates to diets high in vitamin C-rich foods. Animal studies demonstrate that vitamin C can inhibit the formation of cancer-causing nitrosamines and sometimes reduce the number of cancers caused by nitrosamines as well. Both the National Academy of Sciences' Committee on Diet, Nutrition, and Cancer and the American Cancer Society have recommended that people increase their consumption of fruits and vegetables rich in vitamin C.

Good sources include all citrus fruits—oranges, grapefruits, limes, lemons, and tangerines. But other sources shouldn't be overlooked—baked potatoes, red and green peppers, strawberries, tomatoes, dark leafy greens, papaya, Brussels sprouts, and broccoli all contain generous amounts of vitamin C. The U.S. RDA for vitamin C is 60 milligrams, an amount found in three ounces of orange juice.

Orange juice is the most common vitamin C–rich food served in fast-food restaurants. The oversized baked potatoes at Wendy's contain more than half of the daily vitamin C allowance—even more if the potato is topped with broccoli. Otherwise, vitamin C–rich foods

FAST FOODS HIGHEST IN VITAMIN C

COMPANY/PRODUCT	VITAMIN C % U.S. RDA
Taco Bell Taco Salad, with shell	125
Orange Juice, 6 fl. oz.	120
Wendy's Chef Salad	110
Grapefruit Juice, 6 fl. oz.	100
Carl's Jr. Fiesta Potato	100
Taco Bell Bean Burrito with Red Sauce	88
Subway Garden Salad, large	84
KFC Apple Shortcake Parfait	83
Wendy's Baked Potato, Sour Cream & Chives	75
Arby's Baked Potato, Broccoli & Cheddar	75
Taco Bell Beef Tostada with Red Sauce	75
Wendy's Cauliflower, ½ cup	70
Wendy's Baked Potato, Broccoli & Cheese	60
Wendy's Green Peppers, ¼ cup	60
Burger King Garden Salad	58
Taco Bell Mexican Pizza	51
Jack in the Box Chef Salad	46
Taco Bell Combination Burrito	45
Subway Chef Salad or Subway Sandwiches, 6 in.	42
KFC Cole Slaw	36
Long John Silver's Seafood Salad	30
Wendy's Honeydew Melon, 2 oz.	25
Pizza Hut Traditional Hand-Tossed Pizza, Supreme, medium, 2 slices	20
McDonald's Wheaties (fortified)	20

FAST FOODS HIGHEST IN CALCIUM

COMPANY/PRODUCT	CALCIUM % U.S. RDA
Pizza Hut Traditional Hand-Tossed Pizza, Cheese, large, 2 slices (estimate)	100
Wendy's Taco Salad	80
Pizza Hut Personal Pan, Pepperoni, whole pizza	73
Pizza Hut Thin n' Crispy Pizza, Cheese, medium, 2 slices	66
Jack in the Box Ultimate Cheeseburger	60
Carl's Jr. Shake, large	59
Carl's Jr. Double Western Bacon Cheeseburger	50
Hardee's Shakes	50
Dairy Queen Heath Breeze, regular	50
Carl's Jr. Cheese Potato	45
Domino's Pizza Double Cheese/ Pepperoni, 2 slices, 16 in.	45
Dairy Queen Shake, large	45
Jack in the Box Double Cheeseburger	40
Domino's Pizza Veggie Pizza, 2 slices, 16 in.	39
Wendy's Frosty Dairy Dessert, medium, 11 oz.	39
Carl's Jr. Breakfast Burrito	35
McDonald's Low-fat Milk Shakes	35
Baskin-Robbins Low-fat Frozen Yogurt, large, 9 oz.	34
Taco Bell Taco Salad, with shell	32
Milk, 1% or 2% low-fat, 8 fl. oz.	30

Calcium is a vital nutrient, especially for women and children. Each of the foods shown provides one-third or more of the U.S. RDA for adults. But note that many of the foods are loaded with sugar, calories, or fat. Skim milk or yogurt made from skim milk, which are unlikely menu items at fast-food restaurants, are among the best sources of calcium.

are found primarily at the salad bar, discounting the sliver of tomato that might garnish your hamburger.

CALCIUM

If vitamin C is the most famous vitamin, then calcium probably holds top celebrity honors in the mineral category. Its prominence has been boosted by large advertising campaigns from both the dairy and the nutrient-supplement industries. Our bones and teeth, of course, contain large amounts of calcium. In addition, calcium is necessary for muscle contraction, for blood coagulation, and for the "cement" that holds cells together.

American diets are frequently deficient in calcium. Kids don't do too badly, despite their voluminous consumption of soft drinks, because most of them also down two or three glasses of milk a day. But adults tend to drink just coffee, soft drinks, or booze. According to a major USDA dietary survey in 1986, 50 percent of 19- to 50-year-old women consumed less than 70 percent of the recommended intake of calcium, which was 800 milligrams a day (a glass of milk contains about 300 milligrams of calcium). But even 800 milligrams daily may be too little for some people. In 1984 a Consensus Panel of the National Institutes of Health (NIH) urged premenopausal women to consume 1,000 milligrams daily and postmenopausal women 1,500 milligrams. More recently, a committee of the National Academy of Sciences recommended that everyone between the ages of 11 and 24 consume 1,200 milligrams of calcium daily to promote maximal bone mass. If food consumption were measured against those higher standards, many more people would be considered deficient.

Health officials are recommending higher calcium intakes primarily to combat osteoporosis, a condition of porous, brittle bones that is common in older Americans, usually women. It shows up most often in the form of broken hips and wrists. "Osteoporosis is a common condition affecting as many as 15 to 20 million persons in the United States," according to a 1984 report from NIH. When the diet is deficient in calcium, the body will steal a little from the bones to keep muscles, such as the heart, contracting properly. Sustained deficiency could result in osteoporosis, which results in approximately 1.3 million fractures and costs Americans as much as $10 billion annually, according to the National Osteoporosis Foundation.

In the land where the soft drink is king, sustained deficiency is a distinct possibility. According to a USDA study, teens who were "high" consumers of soft drinks were more likely than "low" consumers to run short of calcium. Milk and milk products are the most concentrated sources of calcium, yet milk consumption has been on a steady decline, while consumption of soft drinks keeps increasing. Milkophobes are at high risk of calcium deficiency unless they obtain calcium from other foods, such as green vegetables, salmon bones and sardine bones, cheese, yogurt, and cottage cheese.

Fast-food restaurants make a handsome profit off soft drinks and show no signs of promoting milk instead, even in their "kids' meals." Virtually all high-calcium products in fast-food restaurants are also high in fat and/or calories. As a starter, it's impossible to get skim or, except at McDonald's and Carl's Jr., 1-percent low-fat milk at major chains. The "shakes" contain about one-third of the adult U.S. RDA of calcium, but the price tag is between 300 and 400 calories— three times higher than in low-fat milk. The processed cheese slapped onto burgers or melted onto potatoes is a good source of calcium, but it is also high in saturated fat, sodium, and calories.

IRON

Iron helps carry oxygen from our lungs to other parts of our body. Adults low in iron are likely to feel tired and "run down." Children with iron deficiencies may suffer decreased attention spans and, if severely deficient, may have learning disabilities. An iron supplement quickly cures the problem.

Blood loss is the one major way the body loses significant amounts of iron, so adequate intake is especially important for women between puberty and menopause. Since iron is essential for growth, an adequate intake is imperative for children and pregnant or nursing women. Yet these are the very groups who often don't eat enough iron-rich foods to maintain their iron stores.

Iron is one vital nutrient that abounds in meat-oriented fast-food restaurants. A large hamburger or roast-beef sandwich, such as McDonald's Quarter Pounder, Burger King's Whopper, or Hardee's Big Roast Beef, contains about one-fourth of the U.S. RDA for iron. Some double-hamburger or steak sandwiches may have twice as

FAST FOODS HIGHEST IN IRON

COMPANY/PRODUCT	IRON % U.S. RDA
Pizza Hut Traditional Hand-Tossed Pizza, Supreme, large, 2 slices (estimate)	60
Wendy's Double Big Classic	55
Long John Silver's Homestyle Fish, 1 piece	45
Dairy Queen DQ Homestyle Ultimate Burger	40
Burger King Double Whopper	40
Pizza Hut Personal Pan Pizza, Supreme, whole pizza	37
Arby's Giant Roast Beef	35
Carl's Jr. Double Western Bacon Cheeseburger	35
Jack in the Box Ultimate Cheeseburger	35
Wendy's Big Classic	35
Wendy's Chili, 9 oz.	35
Pizza Hut Pan Pizza, Pepperoni, medium, 2 slices	35
Taco Bell Taco Salad, with shell	33
Carl's Jr. Fiesta Potato	30
Hardee's Big Roast Beef	30
McDonald's Cheerios (fortified)	30
Hardee's Mushroom 'N' Swiss Burger	30
Carl's Jr. Potato, Broccoli & Cheese	30
Hardee's Real Lean Deluxe burger	25

Women, especially, need iron. Be careful, though, most of the iron-rich foods are also high in fat and calories. (The Food and Drug Administration is expected to announce a lower U.S. RDA for iron, meaning that the percent of the daily requirement that these foods provide will probably increase by about one-fifth.)

ARE SUPPLEMENTS FOR YOU?

The first rule of taking nutrient supplements is that they should *supplement*, not replace, a good diet. Interestingly, surveys have shown that supplement users tend to have better diets than non-users.

A standard multi-vitamin-and-mineral tablet, which typically provides about 100 percent of the U.S. RDA of many nutrients, offers an inexpensive and safe insurance policy against deficiencies. Such a tablet ought to include vitamin C, beta-carotene (as a source of vitamin A), chromium, magnesium, and zinc. If you want to be more aggressive in taking advantage of possible benefits without risking harmful overdoses, consider taking daily supplements providing perhaps 40,000 IU (25 milligrams) of beta-carotene, 750 milligrams of vitamin C, and 400 IU of vitamin E. Women may want to add 500 milligrams of calcium before menopause and 800 milligrams after menopause.

We can't say that any given dose will actually be beneficial, but doses approximating those suggested above have been used safely in studies. Also, it's unclear whether the price of supplements has anything to do with quality, so don't think the most expensive brand is necessarily the best.

Finally, don't automatically assume that if 100 milligrams of a nutrient is good, 500 milligrams would be five times better. Large amounts of certain nutrients — including vitamins A, D, and B6 and selenium and zinc — can be toxic and could possibly cause imbalances of other nutrients.

much iron. Remember, though, that you may pay quite a price in terms of fat and calories for all that iron. Burger King's Whopper with Cheese gives you more than one-fourth of your daily iron need, but you'll also be downing 10 teaspoons of fat. You'll find additional iron in the skin of a baked potato and in the garbanzo beans, broccoli, cauliflower, and spinach at the salad bar.

The exact amount of iron absorbed from food depends on the form of the iron in the food, the overall makeup of the meal, and whether the person is iron deficient. Iron is absorbed much better from meat, poultry, and fish than from eggs, beans, grains, and green vegetables. However, if these latter foods are eaten along with meat, poultry, fish, or with a good source of vitamin C, more iron becomes available. High-vitamin-C fast foods that enhance iron absorption include citrus juices, baked potatoes, and tomatoes. Because of the body's wondrous adaptive mechanisms, someone with adequate iron stores may absorb only 10 percent of the iron in his or her diet while someone who needs iron will absorb up to 20 percent. Interestingly, vegetarians normally have excellent levels of iron in their blood.

IODINE

Iodine is usually thought of as being a valuable mineral that prevents goiter, a thyroid condition. But *excess* iodine may also cause thyroid problems. Many foods in addition to iodized salt contain iodine, not intentionally, but as a result of contamination. Iodine-containing compounds are added to cattle feed and bread dough, and used to sanitize milking equipment at dairies and to clean restaurants.

Dr. Harvey Arbesman, of the State University of New York at Buffalo, has suggested—on the basis of very limited data—that ingested iodine can cause or exacerbate acne. A 1975 *Consumer Reports* article reported high levels of iodine in typical fast-food meals. New research needs to be done to determine whether iodine really promotes acne and to determine which (if any) foods are high in iodine. Depending on the research results, Dr. Arbesman may have pinpointed one cause of the problem many teenagers dread most of all.

ADDING IT ALL UP

Evaluating the nutritional value of a food, as you have probably concluded, is complicated. A certain food might be low in cholesterol, but high in sodium. Or it might be high in vitamins and minerals, but literally oozing with fat. How does one weigh all these different factors and decide whether to buy or not buy?

To help we have devised the Gloom factor. This number, which we calculated with the aid of a computer for hundreds of fast-food products, rates the overall nutritional value of foods or meals. Excesses of fat, cholesterol, sodium, and sugar are the biggest dietary bugaboos for most Americans, so our formula gives the most weight to them. The Gloom factor also considers the calorie, vitamin A, vitamin C, iron, and calcium content of a food. (A more detailed explanation of the Gloom factor is provided in Chapter Four.)

The more saturated fat, cholesterol, sodium, refined sugar, and calories a food contains, the greater the Gloom score. A high content of vitamins A and C and the minerals iron and calcium, on the other hand, will lower the Gloom rating. For instance, Arby's plain baked potato has a Gloom rating of 2. But Arby's baked potato with margarine and sour cream has a rating of 31.

You can add up the Gloom ratings of individual foods to get approximate Gloom ratings of a meal or even a whole day's diet. To place the Gloom scale in context, an excellent diet for males between 15 and 50 years old would rate about 120 points on the Gloom scale. Children 7 to 10, females up to the age of 50, and men over 50 should shoot for about 90 points. Women over 50 should shoot for about 80 points. If you're calculating your total daily Gloom scores, estimate the Gloom ratings of homemade foods by looking for similar items listed in this book.

CHEMICAL CUISINE:
COLORINGS, PRESERVATIVES, AND OTHER ADDITIVES

Like other manufactured foods, many fast foods contain a wide variety of chemical additives. The additives preserve, extend, emulsify, color, and otherwise "enhance" the products to ensure uniformity, taste, and profitability. However, KFC, Dairy Queen, and several other chains refuse to disclose what additives are in which foods. This secrecy makes it impossible to discuss the potential consequences of all the fast foods. The profiles on individual companies (see Chapter Four) include as much specific information about additives as we have been able to learn.

A few additives deserve special attention. Aspartame (Nutra-Sweet or Equal) or saccharin, for instance, is used in most diet

FAST FOODS WITH HIGHEST GLOOM RATINGS

COMPANY/PRODUCT	GLOOM
Carl's Jr. Double Western Bacon Cheeseburger	91
Jack in the Box Ultimate Cheeseburger	88
Burger King Double Whopper with Cheese	83
KFC Extra Tasty Crispy Thigh and Wing	78
Wendy's Double Big Classic with Cheese	72
Burger King Double Whopper	72
Arby's Sausage and Egg Croissant	69
Burger King Croissan'wich with Sausage, Egg and Cheese	69
Jack in the Box Sausage Crescent	68
Taco Bell Taco Salad, with shell	68
Carl's Jr. Country Fried Steak Sandwich	66
Dairy Queen Quarter Pound Super Dog	64
Wendy's Double Big Classic	64
McDonald's Biscuit with Sausage & Egg	63
Wendy's Blue Cheese Salad Dressing, 4 tbsp.	63
Dairy Queen DQ Homestyle Ultimate Burger	63
Burger King Whopper with Cheese	61
Arby's Chicken Cordon Bleu	60
Hardee's Steak & Egg Biscuit	57
Carl's Jr. Bacon & Cheese Potato	57
Hardee's Bacon Cheeseburger	54
Arby's Bac 'N Cheddar Deluxe	52

An excellent diet for teenage girls and women up to 50 should contain no more than about 90 Gloom points a day. Active teenage boys and men could handle up to 120 points. See page 106 for more details. The highest Gloom foods are loaded with fat and calories.

products, particularly soft drinks. Nitrite is a preservative, coloring, and flavoring in hot dogs, bacon, ham, and cured sausage. Monosodium glutamate (MSG) shows up as a flavor enhancer. These and others may pose health risks.

As you read about the hazardous additives, don't forget that the fat, cholesterol, sodium, or sugar content of a food generally poses a greater health risk than the additives. Additives may contribute to several thousands of deaths a year. But fat, sodium, and cholesterol excesses, and fiber deficiences, contribute to hundreds of thousands.

ARTIFICIAL COLORINGS

A rainbow of food dyes accounts for pink strawberry shakes, green and purple candy, and lemon-yellow soft drinks. For decades dyes have been suspected of being toxic or carcinogenic, and many have been banned. Safety questions swirl around the few that remain.

Yellow No. 5 dye, America's second most popular food color (behind Red No. 40), is associated with allergic reactions in some people. The dye causes hives, runny or stuffy noses, and occasionally severe breathing difficulties. For some reason, most of those who have been found to be sensitive to the dye are also sensitive to aspirin. The FDA has estimated that between 47,000 and 94,000 Americans are sensitive to the dye.

The FDA requires food manufacturers to list the presence of Yellow No. 5 dye specifically on ingredient labels rather than putting it under the general heading "artificial coloring." Once these manufactured foods make it to the fast-food restaurant, however, the consumer is cast back into ignorance—not many would presume that an apparently white milkshake contains a yellow food dye.

Several artificial colorings, including the dyes Blue No. 2 and Red No. 40, need to be better tested to determine whether or not they promote cancer. Whenever possible, choose foods without dyes. After all, natural ingredients should provide all the color that is needed.

ASPARTAME

This sugar substitute, sold commercially as NutraSweet and Equal, was hailed as the savior of dieters who for decades had put up with saccharin's nasty aftertaste. Aspartame would save the taste buds of diet-cola drinkers, sweet-iced-tea fans, and waist watchers across the country. It tasted like sugar and was made primarily of two apparently harmless amino acids, the building blocks of protein. Proponents touted it as the closest you could come to natural without the calories. Discovered in 1965, aspartame was approved for limited use in 1981 and for soft drinks in 1983. Aspartame is not suitable for use in foods that are cooked or baked for any length of time, because it breaks down into chemicals that are no longer sweet.

As loudly as it was hailed by the industry, aspartame was assailed in other quarters. The first problem is phenylketonuria (PKU). One out of 20,000 babies is born without the ability to metabolize phenylalanine, one of the two amino acids that make up aspartame (and also found in many proteins). Toxic levels of this substance in their blood can result in mental retardation. As a result, the FDA requires all packaged goods that contain aspartame to bear a warning notice for the benefit of people with PKU. Foods served at restaurants or cafeterias are not covered by this regulation.

Some scientists believe that high aspartame intakes could pose some risk to the fetuses of pregnant women who carry the trait for PKU but who do not themselves have the disease. If such women consume large amounts of aspartame—more than a few artificially sweetened foods a day—those scientists say, their babies could be born mentally retarded. The FDA disputes this claim and, to date, no problems have been identified.

Beyond PKU, aspartame safety has been bedeviled by other questions. Several scientists are concerned that aspartame might cause altered brain function and behavior changes in consumers. And many people (though still a minuscule fraction of those who have consumed the additive) have reported dizziness, headaches, epileptic-like seizures, and menstrual problems after ingesting aspartame. Those allegations, however, have not yet been confirmed by careful studies. On another point, before aspartame was approved, a rat study indicated that it increased the incidence of brain tumors. A second study did not show an increase, but it was conducted on a different strain of rat and cannot be considered definitive. A new cancer test should certainly be conducted. Lawsuits have been filed to block aspartame's use. The courts, however, upheld FDA's approval process.

If you have PKU or believe aspartame caused you to suffer behavioral problems, avoid it. Pregnant women, just to be on the safe side, should not consume large amounts of aspartame, but need not worry about small amounts. Incidentally, there is little evidence that aspartame helps people lose weight.

BHA AND BHT

These two closely related chemicals are added to oil-containing foods to prevent oxidation and to retard rancidity.

BHA (butylated hydroxyanisole) was once firmly ensconced on the FDA's list of additives that are Generally Recognized As Safe (GRAS). But in 1982, a Japanese researcher found that BHA induced tumors in the forestomachs of rats. While humans don't have forestomachs, any kind of animal tumor is cause for concern. The Japanese government proposed a ban on BHA, but American trade associations and the FDA persuaded the Japanese to wait until the data could be analyzed. A review committee, assembled in 1983 with scientists from Japan, Canada, Britain, and the United States, concluded that the experiment was well conducted, but when it came to banning BHA, the committee balked and, with only the Japanese dissenting, recommended that BHA be used until further studies were done. In 1991, shortly after California's health department declared BHA to be a carcinogen, the FDA was still reviewing its approval of BHA. Meanwhile Americans continue to eat it in processed foods such as potato chips, presweetened cereals, and bouillon cubes. It is often added to the frying shortenings in fast-food restaurants (see Chapter Four for further details).

BHT (butylated hydroxytoluene) has been the subject of numerous tests that have yielded divergent, and sometimes bizarre, results. A 1959 study showed that large doses of BHT caused rats' hair to fall out, increased cholesterol in the rats' blood, and caused birth defects. That study prompted a wave of new studies, none of which could duplicate the 1959 results. Other studies, though, showed that high-BHT diets caused the livers of lab animals to enlarge and develop inappropriately high levels of certain enzymes, provoking at least one researcher to recommend that the government ban BHT.

BHT's cancer-inducing ability is also the subject of conflicting reports. Some studies show reduced incidence of tumors with the addition of BHT; some show increased incidence. The laboratory studies cannot prove whether BHT causes, prevents, or has no effect on cancer in humans, but because of the possibility that it might increase the risk of cancer, BHT should be eliminated from our food supply, including fast-food shortening. The occasional person who is allergic to BHT would also be helped by such a ban.

CAFFEINE

Caffeine is a stimulant occurring naturally in tea, coffee, cocoa, and kola nuts. It is used in no small measure by truck drivers, office workers, students, and others who want to increase alertness and productive time. But a stimulant drug, even if natural, is usually not without its adverse side effects. As caffeine fights fatigue, it also promotes gastric acid secretion (possibly increasing symptoms of peptic ulcers), temporarily raises blood pressure, and dilates some blood vessels while constricting others.

Excess caffeine intake results in a condition called "caffeinism." Its symptoms include nervousness, anxiety, irritability, jitteriness, muscle twitching, and insomnia, according to Dr. John F. Greden, a psychiatrist at the University of Michigan. And while those who habitually consume caffeine may be unaffected by caffeine-containing foods, occasional consumers may be greatly affected. A study at the National Institute of Mental Health showed that 8- to 13-year-old boys who normally did not consume caffeine experienced restlessness, nervousness, nausea, and insomnia after consuming the caffeine equivalent of two or more cans of soft drinks. Because of differences in body weight, the amount of caffeine in a 12-ounce soft drink has about the same impact on a child as the caffeine in a six-ounce cup of coffee has on an adult.

Pregnant women should be aware that caffeine may affect their developing babies. Experiments with laboratory animals link caffeine to birth defects such as cleft palates, missing fingers and toes, and skull malformations. Some of these problems result after hefty intakes of 20 or more cups of coffee per day, but one study showed that the equivalent of three cups per day delayed bone growth. In addition, caffeine consumption seemed to enhance the ability of other ingested chemicals to cause birth defects. Missing fingers and toes have been seen in several babies whose mothers consumed about ten cups of coffee a day. However, other studies found that if caffeine-related birth defects occur, they are not common. The FDA has advised pregnant women to avoid caffeine.

Caffeine is also linked to fibrocystic breast disease (benign breast lumps). Breast lumps are sometimes painful and can be dangerous, because they could hide more serious, malignant lumps. Several studies have found greater caffeine consumption in women with fibrocystic disease. Many women claim that a caffeine-free diet causes lumps to diminish or disappear. For women who experience discomfort from lumps they know to be benign, it's certainly worth avoiding caffeine for a few months to see if the lumps disappear.

While small amounts of caffeine do not pose a problem for everyone, many people, including children, pregnant women, and those inclined to develop benign breast lumps, may be adversely affected. These people should certainly consider cutting down on or cutting out caffeine-containing drinks, such as tea, coffee, and many soft drinks. At fast-food restaurants, water, juice, and milk are the best alternatives.

MONOSODIUM GLUTAMATE (MSG); HYDROLYZED VEGETABLE PROTEIN (HVP)

Early in this century, a Japanese chemist identified monosodium glutamate (MSG) as the substance in certain seasonings that enhanced the flavor of protein-containing foods. Soon Japanese firms produced it commercially, and people around the world began flavoring everything from soup to nuts with a dash of MSG.

Unfortunately, a heavy hand with MSG can lead to headaches, a tightness in the chest, and a burning sensation in the forearms and the back of the neck. These symptoms, also referred to as "Chinese Restaurant Syndrome," were identified in 1968 and linked to soup served in Chinese restaurants. Soup is a particular problem, because it often contains a large dose of MSG and is often consumed on an empty stomach, which means MSG will be readily absorbed into the blood.

If you believe you are sensitive to MSG, you should peruse ingredient listings carefully (see Appendix). You should also avoid hydrolyzed vegetable protein, or HVP, which may contain MSG. Many fast-food giants have been eliminating MSG from their products.

NITRITE AND NITRATE

Sodium nitrite and sodium nitrate are two closely related chemicals that have been used for centuries to preserve meat. They maintain a red color, contribute to the flavor, and prevent the growth of potentially dangerous bacteria. While nitrate is harmless, it eventually breaks down to form nitrite. When nitrite combines with compounds called secondary amines, it forms nitrosamines, extremely powerful cancer-causing molecules. The chemical reaction occurs most readily at the high temperatures of frying, but it may also occur to a lesser extent in the stomach.

Over the past decade, meat processors have gradually reduced the amount of nitrite in food, and have almost totally dropped nitrate, but low levels of nitrosamine contamination still persist, especially in bacon. (Additionally spinach and other vegetables contain natural nitrate that the body may convert to nitrite and nitrosamines.) Nitrite and nitrosamines aside, the high-fat, high-sodium content of most processed meats should be enough to discourage health-conscious people from eating these foods.

Unfortunately, the breakfast menus at fast-food restaurants rely heavily on sausage, bacon, and ham, and many hamburger sandwiches are being made with bacon.

SACCHARIN

A classic combination of the bitter and the sweet, this sugar substitute had been used for nearly a century as a noncaloric sweetener whose one drawback was its unpleasant, bitter aftertaste. But the taste got more bitter when studies associated saccharin with cancer in laboratory animals. In the early '70s several studies linking saccharin with bladder cancer in animals provoked the federal government to remove saccharin from the list of chemicals that are Generally Recognized As Safe (GRAS). Further studies resulted in a 1977 decision by the FDA to ban saccharin. Predictably, public reaction was immense — consumers wanted calorie control and sweet taste too. Food manufacturers saw profits going down the drain — saccharin is cheaper than sugar; often more money can be made from artificially sweetened products.

The resounding objection to the ban caused Congress to intervene and pass a law exempting saccharin from normal food-safety laws. By 1991, the exemption had been extended several times and saccharin had assumed a comfortable place in our diet, despite the cancer warnings on sweetener packets and cans of saccharin-containing diet beverages, and despite the studies linking the sugar substitute to cancer. For better or for worse, though, aspartame has replaced saccharin in many foods.

SULFITES

Sulfites are a class of chemicals famous among restaurateurs for keeping produce looking fresh far beyond the time it actually is fresh. While fresh-cut lettuce, avocados, potatoes, and other foods turn brown rather quickly, sulfite treatment allows them to retain a fresh appearance. Sulfites also prevent discoloration in apricots, raisins, and other dried fruits; control "black spot" in freshly caught shrimp; and prevent discoloration, bacterial growth, and fermentation of wine.

Though sulfites have been used for centuries, only in the past 15 years have doctors discovered that this preservative can cause allergic reactions in many asthmatics and occasionally non-asthmatics —reactions severe enough to have killed at least a dozen people in the mid-1980s. Most of those people unwittingly ate sulfite-treated food in restaurants. In the early 1980s, while the FDA was paralyzed by regulatory inertia, the National Restaurant Association urged its members to abandon the use of sulfiting agents. Subsequently, several local and state governments banned the use of sulfites by restaurants themselves, though those laws do not stop restaurants from using foods that processors have treated with sulfites. In 1986 the federal government finally banned sulfites in most fresh fruits and vegetables. The ban did not cover potatoes and other foods so individuals who are sulfite-sensitive must be careful when eating out. None of the large fast-food chains adds sulfites at its outlets (however, a few products, such as Hardee's and Jack in the Box's Apple Turnovers, contain sulfites previously added by processors).

Now that you're fortified with all this information about the health impact of fast-food ingredients, we bet you're going to run down to the nearest outlet and decipher the information on the wrappers.

Slow down. Before you rush out, please take a few moments to read the next chapter and learn why most restaurant chains long considered ingredient information to be a deep, dark secret.

WHAT'S IN THIS STUFF, ANYWAY?

A t the grocery store, we can wheel our carts down the aisle and when little Billy grabs a flexible carton of "hula punch" we can look at the label and see that it contains sugar, artificial coloring, and other substances we don't think Billy ought to have and that it lacks the vitamins we *do* think Billy ought to have. We can read the labels of similar-looking packages until we run across a natural fruit juice he likes just as well that satisfies our quest for good nutrition. Even if we don't take the time to compare products at the supermarket, we can study the label at the kitchen table and, if necessary, vow "never again."

Unfortunately, when you take Billy into a fast-food restaurant, it's been difficult to evaluate his requests, because ingredient or nutrition information may be completely unavailable. Still, you assume that chicken is better for him because it's lower in calories and fat, and you encourage him to choose the nuggets. You might choose the fish for the very same reason. But in a fast-food restaurant, those products very often aren't better for you. In fact, they may be a worse choice than beef, because many chicken and fish products contain more fat than a hamburger or roast-beef sandwich. Sometimes, the chicken is even fried in beef fat.

The makers of fruit punch and fruit juice are required to put a list of ingredients on the container to comply with federal laws governing packaged foods. Also, a law passed in 1990 will require nutrition information on that label. (Nutrition labeling reports on vitamins, minerals, protein, fat, carbohydrates, and calories, and is essential for people who are watching their calories or are on special diets.)

Fast-food restaurants — though they sell food in boxes, pouches, and other standardized packages — have not been required to comply with federal labeling laws and not one has complied voluntarily. Supermarket shoppers can read the label of frozen french fries to find out what oil and preservatives they contain, but there's nothing in a KFC restaurant that will divulge the ingredients of those french fries; no wrapper on a Dairy Queen Fish Fillet Sandwich that discloses the ingredients in the batter. But more and more information is becoming available, partly as a result of the first edition of this book.

INGREDIENT AND NUTRITION LABELING

Traditional restaurants have never been required to reveal what's in their food, principally because ingredients could change daily and there is really no place for a label, as such, to appear. It's difficult to imagine Emil and Flora down at the Koffee Kup Kafé figuring out ingredient labeling for Thursday's bean soup when they don't know yet how much ham is going to be left over from Wednesday's blue-plate special.

Fast-food restaurants are a whole different ball game, however, one that federal regulators knew little about when they came up with their labeling rules in 1938. Fast-food restaurants can be compared to mini food factories, but instead of turning out cans of chicken and tomato soup, they are producing boxes of cheeseburgers and pouches of fries. Instead of selling via supermarkets, they sell directly to consumers. Most of these foods come in wrappers or containers that could accommodate ingredient labeling.

Government agencies acknowledged in the late 1970s that the restaurant industry was changing. At that time the Food and Drug Administration, the U.S. Department of Agriculture, and the Federal Trade Commission were studying food labeling and wrote,

> . . . in the growing number of fast-food restaurants, food is generally served in individually-wrapped portions, such as foil-encompassed sandwiches, potatoes in bags, and pies in boxes. In these circumstances, the application of labels, and therefore the requirement of ingredient listing, is a realistic possibility.

The government acknowledged that "present legal authority would be adequate to extend ingredient labeling to those restaurant foods that come in 'containers' or that have 'wrappers.'"

As fast foods have become an important part of our lives, numerous health and other problems associated with the ingredients in fast food make it important for consumers to know what they are getting when they order chicken nuggets and a shake. Many people concerned about heart disease would want to know that the nuggets may be fried in saturated beef fat. People sensitive to Yellow No. 5 dye need to know when to avoid milk shakes, and those with dietary and religious concerns would like to know how to avoid beef, pork, or dairy products.

Mark Howat, the former restaurant critic of *The Record* newspaper in Hackensack, New Jersey, described in *Nation's Restaurant News* a litany of horrors that could have been prevented if fast-food and other restaurants provided ingredient information. He told of a woman who couldn't breathe after eating pasta that contained lobster bits. Another woman sensitive to artichoke almost died when she started eating a salad. Howat's grandson, who is sensitive to wheat, became ill when a food the restaurant said did not contain wheat really did. Another boy almost died after eating chili that contained peanut butter.

Indicative of the consumer viewpoint on labeling was an informal opinion survey conducted by the *Detroit Free Press*: 89 percent were in favor; only 11 percent were opposed. Readers said such things as, "We've a right to know what they're selling our children" and "If we did, they'd lose all their sales." A more scientific *Washington Post* poll conducted in 1990 found that 73 percent of the 1,002 adults surveyed felt "that fast food restaurants should be required to provide nutrition information on the wrappers of containers of the food they sell." (That poll also found that 54 percent of respondents believed that fast foods were "not too good for you" or "not good at all for you.")

BENEFITS OF LABELING

By reading labels, consumers who are allergic to specific ingredients can find out instantly if they should avoid a certain product. The National Institute for Allergy and Infectious Diseases says that reading labels may be the most effective way to identify substances that cause allergic reactions. The FDA acknowledged this in 1979 when it stated, "Sufferers of allergies and persons following special diets would benefit from ingredient labeling of restaurant foods." Today, however, millions of allergy-prone Americans must either take chances or avoid fast foods altogether because the ingredients are often kept secret.

Most consumers don't know the trauma of food allergy. Sulfites, corn, soybeans, and milk cause reactions in only a small percentage of the population. But chronic diseases have reached epidemic proportions—heart disease, hypertension, diabetes, and cancer touch all of our lives and often are related to the food we eat. As a result, many people choose to avoid foods that promote those diseases. Labels disclosing the ingredients of foods, including the fats in which they are fried, would benefit millions of people.

In addition, consumers might not want to pay for what they consider inferior products. If a restaurant selling nuggets made of processed chicken meat, skin, and a bunch of additives sits next to a restaurant selling nuggets made from chicken breast fillets, consumers might opt for the latter.

Labeling would be even more useful if it included information about calories and specific nutrients, such as protein, fat, and vitamins. As *Designing Foods*, a 1988 report by the National Academy of Sciences–National Research Council, maintained:

> Point-of-purchase nutrition information at these outlets could have a tremendous impact on the quality of the American diet for two reasons. First, fast-food restaurants provide an ever-increasing share of the calories in the average consumer's diet. Second, they tend to use large amounts of fat in their food.

In *Nutrition Labeling*, a 1990 report, a committee of the National Academy of Sciences specifically urged that "all restaurants be required to have standard menu items evaluated for their nutritional profile and provide this information to patrons upon request." This report said that large chains should place the information either on package wrappers or other location.

Many people would be surprised to learn that the stuffed baked potato they ordered contained 500 calories, the shake 600, and the burger 700. Wouldn't it be nice (and, sometimes, depressing) to have a big, bold calorie-listing on menu boards and on every package as a reminder? In fact, Macheezmo Mouse, a small West Coast chain, already lists the calorie count of every item on its menu board, and Carl's Jr. features the calorie count of its "lite menu" items. While calories aren't everything, that information is certainly important to millions of consumers. And in Germany, McDonald's and all other restaurants disclose on menu boards or menus the presence of artificial colorings and preservatives in each food.

In 1984 the Center for Science in the Public Interest and state attorneys general began pressuring fast-food companies to print ingredient and nutrition information on product labels. Senator John H. Chafee (Republican–RI) and Representative Stephen Solarz (Democrat–NY) introduced legislation on the subject. Senator Chafee said that ingredient labeling "would be a tremendous boon to the public health. It would allow consumers to vote with their fast-food dollars, prompting fast-food chains to compete on the basis of nutrition. Imagine how our diets would improve if the full force of the fast-food giants were put behind a race to offer the most wholesome food!"

To press their case, CSPI, the New York State Consumer Protection Board, and the American College of Allergists in 1985 petitioned the FDA and USDA to order nutrition labeling. Both agencies, however, accepted the industry's argument that labeling is impractical, because the same packages are used for several foods, and because ingredients are always changing. They rejected the petitions. But the publicity began throwing the spotlight on industry's secrecy. *The Washington Post* editorialized that, "Any industry that can exercise such ingenuity in quality control over what goes into our hamburgers can surely find a way, somehow, to let us know what goes into them."

Still, the National Restaurant Association vehemently opposes ingredient labeling but does support "the consumer's right to know" through other approaches. William Fisher, president of the restaurant association, suggested that people "vote with their feet" if a restaurant isn't providing the information they want. Currently, some chains are quite forthright, but many still refuse to provide ingredient information. Those chains include Baskin-Robbins, Church's, Dairy Queen, Hardee's, KFC (Kentucky Fried Chicken), and Popeyes.

LABELING BREAKTHROUGHS

Despite apathy (or worse) at the FDA and USDA, and resistance to labeling from most restaurant chains, public pressure broke the logjam of corporate secrecy. In late 1985, Joliet, Illinois, located just 35 miles southwest of McDonald's headquarters, became the first community to require restaurants to disclose the types of fat they use for frying. That action came at the initiative of Mayor John Bourg, who, after undergoing heart-bypass surgery, learned the importance of cutting back on saturated fat. The Joliet law applies to all restaurants, not just the fast-food variety. Restaurateurs found that the signs stimulated great interest among patrons. In 1986 San Francisco became the first city to require fast-food restaurants to disclose both ingredients and nutrients, though the law is poorly enforced.

THE PRESSURE GETS TO BIG MAC

The first big breakthrough came in April, 1986, and ironically enough, it came from McDonald's. That chain had long refused to disclose ingredients. In fact, just a few weeks earlier McDonald's president Michael Quinlan had sneered at ingredient labeling, telling a group of Wall Street money managers, "I couldn't care less."

But McDonald's became the first chain to consent to a government agency's request to make ingredient and nutrition information available to patrons in its outlets, at least in company-owned ones (about half of all McDonald's) in New York State. What happened was that, at CSPI's request, New York State Attorney General Robert Abrams was investigating the accuracy of a television ad that implied that Chicken McNuggets were made from pure chicken. Negotiations between New York and McDonald's evolved from an examination of the specific ad to the more general problem of consumer information about products. The outcome of the negotiations was that McDonald's agreed to initiate a one-year experiment in which outlets in New York would provide consumers with a pamphlet listing both nutrients and ingredients.

The New York–McDonald's agreement had three immediate benefits for consumers: First, consumers in New York could avoid or choose foods based on the ingredients. Second, the mere fact of knowing that its ingredients would be widely publicized contributed to a corporate decision to reformulate some of its products, including frying fewer foods in beef fat and removing Yellow No. 5 dye from all its products. Third, Wendy's and Burger King said that they would go McDonald's one better by reiterating an earlier vow to provide ingredient and nutrition pamphlets in their company-owned outlets nationwide.

In July, 1986, the attorneys general of Texas and California got into the act and reached additional agreements with fast-food companies. Their pressure spurred McDonald's to expand its information program nationally, and they also got Jack in the Box and KFC (Kentucky Fried Chicken) to disclose certain product information. While Taco Bell refused to cooperate, it was clear that the days of arrogant secrecy were drawing to a close.

But the notion of providing nutrition information in the form of a brochure ran into problems. Clerks weren't aware of the brochures. Shops ran out of them. And interest in complying with the agreements faded with time.

In early 1990, a CSPI survey of 65 restaurants whose headquarters had promised brochures found that only 23 had the brochures. Not one of 14 Wendy's had the information, and only three of 13 KFC outlets did. Only Jack in the Box seemed to be adhering to the agreement. CSPI told the Texas, New York, and California attorneys general that Burger King, KFC, McDonald's, and Wendy's had broken the disclosure agreement, and that many other large chains were not providing consumers with essential information, even by phone or mail. Weeks later, New York City Consumer Affairs Commissioner Mark Green found that few KFC, Burger King, or Wendy's outlets provided nutrition information, though most McDonald's outlets did have the brochure.

CSPI urged that the chains be required not just to provide brochures, but also to list calories, fat content, and certain additives right on the menu board. In August, 1990, the New York City consumer office proposed legislation that would mandate posters and tray liners with key nutrition information, as well as brochures with complete information. Burger King simultaneously said that it would provide all the information in its New York restaurants without waiting for the city council to act. Other cities and states should enact similar legislation.

FILLING THE INFORMATION GAP

In October, 1990, Michael McGinnis, the deputy assistant secretary for health promotion and disease prevention of the U.S. Department of Health and Human Services, told industry officials that "the provision of nutrition information at the point of purchase in fast-food restaurants is an inevitable development."

Until all fast-food chains reveal all the ingredients and nutrients in their foods, consumers will not always be able to make intelligent purchasing decisions. With that in mind, the authors have collected all the available information and paired it with tools to help you make educated decisions. Because many chains refuse to disclose information, our listings are incomplete. Also, be aware that companies may modify existing products or add new ones. If you're a frequent eater of fast foods, you can keep this book updated by asking your favorite chains for their latest information. Addresses and phone numbers are in the Appendix, in case your local outlet does not have a brochure.

Now, let's take a closer look at major companies' products and figure out how to make the best of a sometimes grim situation.

CHOOSING A FAST-FOOD MEAL

S hopping for healthful fast food isn't as easy as shopping for groceries. In a supermarket, virtually all processed foods have labels revealing their ingredients (in descending order by weight). In addition, many labels will tell you how much fat, sodium, protein, and other nutrients the food contains.

On the other hand, fast-food purveyors have not been required to reveal the details of food preparation. Most people don't realize that some companies' french fries get their flavor from beef fat. Tasty as they may be, if you were concerned about heart disease, you might not order them.

Ingredients aren't the only information missing from fast food. Nutrients—protein, fat, minerals, and all the rest—are also important. Without nutrition labeling, there's no way to know that each Wendy's Bacon & Cheese Hot Stuffed Baked Potato is loaded with more than 1,400 milligrams of sodium and 4 teaspoons of fat. A 1990 federal law requires most packaged foods, fresh fruits and vegetables, and seafood to provide nutrition information, but Congress specifically exempted fast-food companies.

Until fast-food restaurants disclose the ingredients and nutrients in their food at the point of purchase or, less likely, stop selling junky, greasy products, we must rely on various other means in order to pick our way healthfully through a fast-food menu.

COPING IN THE FAST-FOOD JUNGLE

Before we examine the products offered by the major fast-food companies, we should fortify ourselves with some basic principles of selecting foods.

Eating out, for some reason, encourages us to abandon our

nutrition principles and throw caution to the wind. Disregarding for a time what we know to be good for us, we often eat differently from the way we do at home. Our tongue plays a dirty trick on the rest of our body.

People see eating out as "an episode and not as a continuous pattern," says Doris Derelian, a consulting dietitian in Santa Monica. When people go into a restaurant they tend to abandon their usual rules for eating. "It's like a birthday party or Thanksgiving dinner," she says. She sees it as a special event when we excuse ourselves for eating something that isn't good for us. But since fast foods have become such an integral part of our lives, it's crucial that people who want to stay healthy use all the meal-planning techniques in these restaurants that they do at home.

To help you fight your way through the fast-food jungle, we have devised the Gloom rating. Gloom ratings emphasize the fat, choles-

DAILY GOALS FOR HEALTHY EATING

	CALORIES*	FAT (tsp.)**	SODIUM (mg.)	GLOOM QUOTA
CHILDREN				
4–6	1,800	14	1,800	87
7–10	2,000	15	2,000	94
FEMALES				
11–50	2,200	14	2,200	91
51+	1,900	12	1,900	79
MALES				
11–14	2,500	16	2,400	104
15–18	3,000	19	2,400	122
19–50	2,900	18	2,400	118
51+	2,300	15	2,300	96

1989 National Academy of Sciences' recommendations; individual needs depend greatly upon activity level.

**Represents 25 percent of calories, except 30 percent for children*

terol, sodium, and sugar content of foods, four of the biggest problems in the American diet. The higher the Gloom rating, the worse the food. For example, a regular McDonald's hamburger, which contains 490 milligrams of sodium and 9 grams of fat, rates 16. The Quarter Pounder with Cheese, which dishes up 1,090 milligrams of sodium and 28 grams of fat, hits 43 on the Gloom scale. (See page 112 for details on how Gloom scores are calculated.)

Boys and men between 15 and 50 years old should limit their intake of Gloom points to about 120. Women should consume no more than about 90 Gloom points a day. In general, the worse your diet, the more Gloom points it will provide. Remember, too, that it's not just fast foods, but all foods, that give you Gloom points.

The tables in this book will help guide you through specific menus. But a few general rules apply everywhere. Until fast-food restaurants label their foods with nutrition data or ingredient information, use these rules to avoid unwanted fat, sodium, and calories.

CUT DOWN ON FAT; HOLD THE SALT; GO FOR THE FIBER

To avoid fat, forgo sauces such as mayonnaise and tartar sauce on your food. Ordering Burger King's BK Broiler Chicken Sandwich without sauce saves 37 calories; getting a Whopper without mayonnaise saves about 150 calories. Having a regular McDonald's Quarter Pounder instead of the cheese version will save you 100 calories. As a rule, toppings add fat to potatoes, so it is wise to order them plain, with vegetables only, or with margarine on the side. Dressing adds fat and calories to salads, typically about 300 calories per 2-ounce packet. Choose a low-calorie dressing or use just a little oil and vinegar.

Processed meats are usually high in fat and sodium, so avoid bacon on your burger, pepperoni on your pizza, sausage on your biscuit.

Deep-fried foods tend to be fattier than grilled or broiled foods. And since some restaurants use saturated fats for frying, the fish and the chicken may be no better for you than a hamburger. When you do order deep-fried foods, discard the breading and batter and avoid the "extra crispy" versions.

To cut your salt intake, hold the pickles, mustard, ketchup, and special sauce. Do without cheese and avoid ordering processed meats such as bacon, ham, sausage, and hot dogs. Order pizza without sausage, pepperoni, salami, ham, or anchovies. Stick with green pepper, mushroom, spinach, and onion toppings. And to really cut down on fat and sodium, try a cheeseless or half-cheese pizza topped with tomato sauce and vegetables. It sounds spartan but is really delicious, and you won't leave the restaurant feeling bloated. (Tell the clerk you're allergic to cheese, if you want to avoid discussions of cholesterol and atherosclerosis.)

To get fiber, choose from the salad bar. Fresh fruits and fruits canned in their own juice, not in syrup, are good sources of fiber without added calories. Load up on vegetables—at least the ones not coated with added mayonnaise, sour cream, and other dressings. Kidney, garbanzo, and green beans are other good salad-bar choices.

BREAKFAST

Probably more than any other meal, breakfast at a fast-food restaurant is fraught with sodium and fat. Biscuits, croissants, bacon, ham, sausage—there's very little to choose from that's good for you.

The Sausage and Egg Croissant at Arby's contains 271 milligrams of cholesterol—roughly the total daily dose recommended by the American Heart Association. An Egg McMuffin at McDonald's contains almost 3 teaspoonsful of fat, plus 710 milligrams of sodium (about one-third teaspoon). Even scrambled eggs get scrambled up with so much grease they tend to have a higher percentage of fat than dinner's hamburger and roast-beef sandwich. A Scrambled Egg Platter at Jack in the Box is eased down your throat by 7 teaspoons of fat and 378 milligrams of cholesterol. More than half of this loser's 559 calories come from fat.

In 1990 McDonald's made a healthy breakfast much easier by offering fat-free apple bran muffins and 1 percent low-fat milk. Most outlets offer Wheaties and Cheerios. A breakfast of cereal and milk, orange juice, and bran muffin will "cost" you about 455 calories and a remarkably low 7 Gloom points. Pancakes without butter are another good low-fat option. On the other hand, a Hardee's Chicken Biscuit with Hash Rounds and 2 percent low-fat milk adds up to 781 calories, a teaspoon of salt, and 65 Gloom points.

KIDS' MEALS

Beware of the special kids' meal packs. You often pay extra for fancy packaging that will end up in the trash bin in 15 minutes — or, even worse, will come home with you and remind your child to pester you to return to the restaurant again and again.

Making nutrition-oriented modifications in kids' meals is certainly worth a try. Swap milk for the soft drink in package deals. Soft drinks have nothing but calories or artificial sweeteners and a flock of additives. Although whole milk has fat, it also has lots of calcium, protein, and B vitamins. Ask for low-fat milk; nonfat milk is as rare as apples and oranges.

BURGERS TO ROAST BEEF TO BACON BITS

If you're thinking beef, think roast beef, not just hamburger. Roy Rogers and Hardee's outlets in the New York–Washington region offer by far the leanest roast beef. In fact, it is the leanest meat product we have found in any fast-food restaurant. McDonald's standard burgers are slightly leaner than the competition's, even Burger King's broiled burgers. (See the chart on page 110.)

In November, 1990, McDonald's announced that it began testing in Harrisburg, Pennsylvania, hamburgers made with meat containing 9 percent fat. That is by far the leanest hamburger meat offered by any restaurant and represents a real breakthrough in meat technology. The meat contains added water and natural flavorings, plus one-half percent carrageenan, a harmless additive, to help retain moisture. The McLean Deluxe sold so well (despite mixed reviews of its taste) that it was marketed nationally within months, displacing the fatty McD.L.T. sandwich — with its much-criticized foam-box packaging — from the menu. And in 1991, Hardee's joined the low-fat league by introducing its Real Lean Deluxe, which contains slightly more fat than McDonald's offering.

Avoid anything that sounds big — "whopper," "double-decker," and "super" all add up to at least twice the calories of the more modest versions and nearly always come with a 100-calorie wallop of special fat-based sauce. A Burger King Double Whopper contains more than three times as many calories (844) as a regular hamburger (272).

Small burgers or the new low-fat burgers are your best burger bet. Order a side salad or a baked potato (if you have calories to spare) and you have a reasonably nutritious meal. Drinking a glass of milk rather than a shake saves about 200 calories.

In pizza and Mexican restaurants, "deep-dish" and "grande" usually promise more fat and calories. Opt for thin crusts and standard size offerings. Get beans instead of cheese when there's a choice, and if you're trying to decide about extras at pizza parlors, get vegetable toppings rather than more cheese or meat.

FAT AND PROTEIN CONTENT OF ROAST BEEF AND HAMBURGER MEAT

COMPANY/PRODUCT	%FAT	%PROTEIN
Roy Rogers/Hardee's** Regular Roast Beef, 3.4 oz.	2	19
McDonald's McLean Deluxe*	—	—
Hardee's Real Lean Deluxe*	—	—
Arby's Regular Roast Beef, 3.2 oz.	13	20
Hardee's** Big Roast Beef, 3.8 oz.	15	18
McDonald's Quarter Pounder, 2.8 oz.	19	25
Hardee's Big Deluxe Hamburger, 3 oz.	20	23
Wendy's Big Classic Hamburger, 2.4 oz.	20	27
Burger King Whopper, 2.7 oz.	21	24
Carl's Jr. Famous Star Hamburger, 2.5 oz.	22	24
Jack in the Box Jumbo Jack, 2.6 oz.	23	24

*A dash means that data not available.
**Hardee's restaurants in the New York–Washington region use the Roy Rogers/Hardee's roast beef. Other Hardee's most likely use the fattier Hardee's roast beef.

The figures are from a 1990 test conducted by Lancaster Laboratories for the Center for Science in the Public Interest; meat was obtained from six restaurants per chain, except five for Carl's Jr. Some of the small differences between products may reflect store-to-store variations.

For the dieter venturing into a fast-food restaurant, salad bars are a godsend. Low-calorie and vitamin-rich choices include vegetables, fruits, and garbanzo and kidney beans. Dress the salad with vinegar, lemon juice, or low-calorie dressings. Avoid coleslaw, potato salads, and pasta salads that are bound with mayonnaise, which is virtually pure fat. Be sure to skip the bacon, cheese, and egg toppings.

FISH AND CHICKEN

You have to take special care to escape with your health at one of the thousands of chicken or seafood restaurants, because the menus are loaded with deep-fried foods. You can strike a major blow for nutrition, if you have the willpower, by removing and discarding the breading and skin from your chicken or fish. Stay away from "extra crispy" coatings — they generally contain more calories, fat, and salt. Skip sweet sauces. And fill out your meal with corn on the cob, a baked potato, mashed potatoes, or a salad. Biscuits, which often come with chicken, get their flakiness from layers of fat and rise with the aid of sodium-rich leavening.

Arby's, Burger King, El Pollo Loco, Long John Silver's, Taco Bell, and several other chains have baked or broiled chicken and fish items — these offer substantial fat and calorie savings. Avoid tartar sauce in favor of the less fattening ketchup or cocktail sauce. Even better, try a squeeze of lemon or a shake of vinegar. When possible, order multigrain buns or rolls, even though they contain just a little whole-grain flour.

The Gloom ratings tell it all. A Grilled Chicken Sandwich at Wendy's has a Gloom score of 16. An order of Chicken McNuggets at McDonald's scores 24. But a Burger King fried Chicken Sandwich, with its 9 teaspoons of fat, has a Gloom rating of 61.

If you like chicken nuggets made from processed chicken, go to McDonald's or KFC. If you prefer nuggets made with whole pieces of chicken, visit Burger King, Hardee's, or Long John Silver's.

OTHER TACTICS

The drive-in window can be a friend to the nutrition-conscious. Order the basics from the window and take them home to serve with skim milk, fruit, and fresh vegetables.

AN EXPLANATION OF THE GLOOM RATING

The Gloom rating is designed to give a quick summary of a food's or a meal's overall nutritional value. This rating emphasizes fat, cholesterol, sodium, and added sugar content. These four substances contribute to heart disease, high blood pressure, diabetes, tooth decay, and certain cancers. The Gloom rating also reflects the "nutrient density" of a food. A food has a high nutrient density if it is rich in vitamins, minerals, and protein compared to calorie content. In general, the lower the Gloom rating of a food or meal, the better it is for your health, because of a low fat, cholesterol, sodium, and sugar content, and a high nutrient density.

For the technically minded, the formula for the Gloom rating first adds between 0.9 point per gram of polyunsaturated oil and 1.1 points per gram for a highly saturated animal fat. It adds 1 point for every 20 milligrams of cholesterol, 1 point for every 133 milligrams of sodium, and 1 point for every 10 grams of refined sugar or corn syrup. This sum is then multiplied by a number ranging from 0.5 to 1.5 depending on the food's nutrient density, which is the ratio of nutrients per calorie (based on protein, calcium, iron, vitamin A, and vitamin C). For instance, the multiplier would be 1 if a food contained 100 percent of the RDA of each of the five nutrients. The higher the nutrient density, the lower the multiplier, thereby lowering the Gloom rating.

We have calculated exact Gloom ratings for several meals of the various fast-food chains. You can calculate approximate Gloom ratings for meals simply by adding up the scores for the components of the meals.

The information provided by Church's and Dunkin' Donuts was too old or scanty to make any Gloom calculations for most of their foods. We have also estimated sugar levels, which companies never disclose; we had the sugar levels of a number of foods analyzed at a testing laboratory. Only Domino's and Pizza Hut provide data on dietary fiber, so that important substance could not be included in the Gloom formula.

If you feel assertive and want to do your bit for society, ask the personnel (preferably the manager) for items you want but don't see listed on the menu. Ask for nonfat milk even if you know whole milk is all they serve. Ask if they have bran muffins at breakfast and whole-grain buns for your burgers. While the person at the counter will inevitably say no, the companies want to make money. If they thought there was a demand for fresh fruit and fruit salad, they just might try offering them.

Of course, one obvious way of coping is to avoid the restaurants altogether since it actually takes less time to wash an apple and grab a carton of yogurt than it does to go out for fast food. Even if it weren't more convenient, food from home can be more healthful, tastier, and cheaper.

If convenience is the key factor in your lunch choice, think about shopping at the supermarket for products that make for low-hassle brown bagging. Small cans and boxes of fruit juices, cups of yogurt, low-fat presliced meats, whole-grain bread, whole fruits, small boxes of raisins, and bags of nuts and seeds are just some of the foods that make taking your lunch easy and nutritious. And more and more grocery stores are installing salad bars to regain customers captured by fast-food and other restaurants.

If your office has a microwave oven, you might want to carry a raw (or precooked) potato to work with you. It doesn't drip or spill and requires no special container. Just stick a washed potato in your briefcase or purse and heat it up at lunchtime. Even better, choose a sweet potato, which is rich in vitamins A and C. Microwaves also offer more exotic options for people who cook. Studies show that many foods have less "leftover" flavor when reheated in a microwave. This bodes well for extras from last night's dinner—including your own homemade, healthful pizza, fish, chicken, and hamburgers.

If you are traveling with children, packing a picnic lunch not only saves money, but also sets a good example. If you pack tuna-fish sandwiches on whole-grain bread, and plums and bananas and crunchy carrots, kids will learn to enjoy healthful foods and are likely to develop good eating habits.

But knowing that we'll sometimes want to head for fast-food land, let's take a closer look at the fare offered by the major fast-food companies.

The lists provided later in this chapter reveal all the information that is now available regarding key fast-food nutrients: calories, fat, sodium, sugar, cholesterol, vitamins A and C, iron, calcium, protein, and carbohydrates (including added sugar, when present). The percentage of calories from fat is another key indicator of nutritional value. We also show the Gloom score, a capsule indication of a food's (or a meal's) overall nutritional rating. The greater the Gloom score, the more you should avoid the product. With each restaurant we point out some of the best choices available, though for some restaurants the "best" is not particularly good. "Best" choices, designated by a ✔, generally have no more than 35 percent of calories from fat, 800 milligrams of sodium, and 30 Gloom points.

Popeyes does not offer any nutrition information, and Church's data has not been updated since the mid 1980s. Baskin-Robbins, Hardee's, Dunkin' Donuts, and Subway have not analyzed the complete vitamin and mineral content of their products. No company provides information on sugar content, so we have calculated our own estimates or, in some cases, actually analyzed products. This lack of information may affect their Gloom ratings somewhat, though we have made adjustments as best we could. Nutrients present in trace amounts (usually under 2 percent of the U.S. RDA) are shown in the charts as 0.

SALADS

COMPANY/PRODUCT	CALORIES	FAT (tsp.)	SODIUM (mg.)	GLOOM
Arby's Side Salad	25	0	30	0
Wendy's Garden Salad	70	<1	60	1
Burger King Chunky Chicken Salad	142	1	443	6
McDonald's Chunky Chicken Salad	150	1	230	6
Wendy's Chef Salad	180	2	140	10
Subway Turkey Salad, small	167	2	479	10
Long John Silver's Seafood Salad	230	2	580	10
Subway Roast Beef Salad, small	185	2	479	11
McDonald's Chef Salad	170	3	400	12
Subway Tuna Salad, small	212	3	545	12
Burger King Chef Salad	178	2	568	13
Carl's Jr. Charbroiler Chicken Salad	200	2	300	15
Hardee's Garden Salad	210	3	270	15
Long John Silver's Ocean Chef Salad	250	2	1340	17
Hardee's Chicken Fiesta Salad	280	3	640	19
Hardee's Chef Salad	240	3	930	20
Jack in the Box Chef Salad	325	4	900	25
Taco Bell Taco Salad, no shell	484	7	680	34
Wendy's Taco Salad	660	8	1110	39
Taco Bell Taco Salad, with shell	905	14	910	68

Most salads, other than taco salads, are low in calories and fat. Be sure to ask for a low-calorie dressing or you risk ruining a good thing with several hundred additional calories.

CHICKEN AND TURKEY

COMPANY/PRODUCT	CALORIES	FAT *(tsp.)*	SODIUM *(mg.)*	GLOOM
Carl's Jr. Charbroiler BBQ Chicken Sandwich	310	1	680	12
Arby's Light Roast Chicken Deluxe	253	1	874	13
Long John Silver's Chicken Plank, 1 piece	130	1	490	13
Jack in the Box Chicken Fajita Pita	292	2	703	14
KFC Original Recipe Drumstick	146	2	275	14
Hardee's Chicken Stix, 6 pieces	210	2	680	15
Burger King BK Broiler Chicken Sandwich	267	2	728	15
Dairy Queen Grilled Chicken Fillet Sandwich	300	2	800	16
Subway Turkey Sandwich, 6 in.	357	2	839	16
Wendy's Grilled Chicken Sandwich	320	2	715	16
Long John Silver's Baked Chicken Sandwich (no sauce)	320	2	900	17
KFC Original Recipe Wing	178	3	372	18
Burger King Chicken Tenders, 6 pieces	236	3	541	21
KFC Extra Tasty Crispy Drumstick	204	3	324	22
Hardee's Chicken Fillet	370	3	1060	24
KFC Original Recipe Center Breast	283	3	672	25
Arby's Grilled Chicken Barbecue	378	3	1059	25

COMPANY/PRODUCT	CALORIES	FAT (tsp.)	SODIUM (mg.)	GLOOM
Wendy's Chicken Sandwich, fried	430	4	725	25
KFC Original Recipe Side Breast	267	4	735	27
Arby's Turkey Deluxe	399	5	1047	28
Jack in the Box Grilled Chicken Fillet	408	4	1130	28
KFC Extra Tasty Crispy Wing	254	4	422	30
McDonald's McChicken	415	7	770	30
Wendy's Crispy Chicken Nuggets, 6 pieces	280	5	600	31
Dairy Queen Breaded Chicken Breast Fillet Sandwich	430	5	760	32
KFC Original Recipe Thigh	294	4	619	33
KFC Hot Wings, 6 pieces	376	5	677	40
Arby's Roast Chicken Club	513	7	1423	43
KFC Colonel's Chicken Sandwich	482	6	1060	45
KFC Extra Tasty Crispy Thigh	406	7	688	48
Arby's Chicken Cordon Bleu	658	8	1824	60
Burger King Chicken Sandwich	685	9	1417	61

Chicken starts out lean and wholesome, but once it is battered, breaded, fried, and smothered with a mayonnaise sauce, it will be loaded with fat and calories. Look for baked or broiled chicken, hold the sauces, and discard the grease-soaked breading.

FISH

COMPANY/PRODUCT	CALORIES	FAT (tsp.)	SODIUM (mg.)	GLOOM
Long John Silver's Shrimp, battered, 1 piece	60	1	180	7
Long John Silver's Fish, Homestyle, 1 piece	125	2	200	8
Wendy's Seafood Salad	110	2	455	9
Subway Tuna Sandwich, 6 in.	402	3	905	17
Long John Silver's Clams, breaded	240	3	410	19
Subway Seafood and Crab Sandwich, 6 in.	388	3	1306	20
Long John Silver's Fish, battered, 1 piece	210	2	570	21
Dairy Queen Fish Fillet Sandwich	370	4	630	26
McDonald's Filet-O-Fish	370	6	930	32
Dairy Queen Fish Fillet Sandwich with Cheese	440	5	880	33
Burger King Ocean Catch Fish Filet	495	6	879	37
Wendy's Fish Fillet Sandwich	460	6	780	38
Hardee's Fisherman's Fillet	500	5	1030	38
Jack in the Box Fish Supreme	510	6	1040	41
Arby's Fish Fillet Sandwich	537	7	994	46
Carl's Jr. Carl's Catch Fish Sandwich	560	7	1220	47

Most fresh fish is low in fat and quite healthful. But most fast-food fish is deep-fried and as fatty as many hamburgers. Seek broiled or baked fish, and season with lemon juice instead of butter and tartar sauce.

FRENCH FRIES

COMPANY/PRODUCT	CALORIES	FAT *(tsp.)*	SODIUM *(mg.)*	GLOOM
Long John Silver's Fryes (3 oz.)	170	2	55	7
Dairy Queen, small (2.5 oz.)	210	2	115	13
Hardee's, regular (2.5 oz.)	230	3	85	14
Jack in the Box, small (2.5 oz.)	219	3	121	14
McDonald's, small (2.4 oz.)	220	3	110	16
KFC (2.5 oz.)	244	3	139	16
Arby's, small (2.5 oz.)	246	3	114	18
Hardee's, large (4 oz.)	360	4	135	21
Wendy's, large (4.2 oz.)	312	4	189	22
Jack in the Box, regular (4 oz.)	351	4	194	22
McDonald's, medium (3.4 oz.)	320	4	150	23
Dairy Queen, large (4.5 oz.)	390	4	200	24
Jack in the Box, Jumbo (5 oz.)	396	5	219	24
Arby's Curly Fries (3.5 oz.)	337	4	167	27
Hardee's Crispy Curls (3 oz.)	300	4	840	27
Carl's Jr., regular (4.5 oz.)	420	5	200	29
McDonald's, large (4.3 oz.)	400	5	200	29
Arby's, medium (4 oz.)	394	5	182	29
Burger King, medium (4 oz.)	372	5	238	30
Hardee's, "Big Fry" (5.5 oz.)	500	5	180	30
Wendy's, Biggie (6 oz.)	449	5	271	32
Arby's Cheddar Fries (5 oz.)	399	5	443	35
Arby's, large (5 oz.)	492	6	228	37

Though they're deep-fried and salted, french fries are not quite as bad as their reputation would have them. But a small serving should be more than enough, and ask the clerk to hold the salt. Kudos to Long John Silver's for making salt-free "Fryes" its standard.

HAMBURGERS AND CHEESEBURGERS

COMPANY/PRODUCT	CALORIES	FAT *(tsp.)*	SODIUM *(mg.)*	GLOOM
McDonald's Hamburger	255	2	490	16
Hardee's Hamburger	270	2	490	16
Burger King Hamburger	272	3	505	18
Jack in the Box Hamburger	267	3	556	18
McDonald's McLean Deluxe	320	2	670	18
McDonald's Cheeseburger	305	3	710	22
Hardee's Real Lean Deluxe	340	3	650	22
Burger King Cheeseburger	318	3	661	24
Burger King Burger Buddies	349	4	717	26
Burger King Hamburger Deluxe	344	4	496	27
Dairy Queen Single Hamburger with Cheese	365	4	800	29
McDonald's Quarter Pounder	410	5	650	31
Carl's Jr. Carl's Original Hamburger	460	5	810	31
Hardee's Big Twin	450	6	580	33
Dairy Queen Double Hamburger	460	6	630	37
Wendy's Double Hamburger	520	6	710	39
Jack in the Box Double Cheeseburger	467	6	842	40
Burger King Double Cheeseburger	483	6	851	40
McDonald's Big Mac	500	7	890	41
Hardee's Big Deluxe Burger	500	7	760	42
Hardee's Quarter-Pound Cheeseburger	500	7	1060	42

COMPANY/PRODUCT	CALORIES	FAT (tsp.)	SODIUM (mg.)	GLOOM
McDonald's Quarter Pounder with Cheese	510	7	1090	**43**
Burger King Bacon Double Cheeseburger	515	7	748	**45**
Wendy's Big Classic	570	7	1085	**48**
Burger King Whopper	614	8	865	**49**
Jack in the Box Jumbo Jack	584	8	733	**50**
Hardee's Bacon Cheeseburger	610	9	1030	**54**
Wendy's Big Classic with Cheese	640	9	1345	**56**
Carl's Jr. Western Bacon Cheeseburger	730	9	1490	**59**
Burger King Whopper with Cheese	706	10	1177	**61**
Dairy Queen DQ Homestyle Ultimate Burger	700	11	1110	**63**
Wendy's Double Big Classic	750	10	1295	**64**
Burger King Double Whopper	844	12	933	**72**
Wendy's Double Big Classic with Cheese	820	12	1555	**72**
Burger King Double Whopper with Cheese	935	14	1245	**83**
Jack in the Box Ultimate Cheeseburger	942	16	1176	**88**
Carl's Jr. Double Western Bacon Cheeseburger	1030	14	1810	**91**

Befriend your arteries by choosing small burgers and skipping the "special sauces." Cheeseburgers provide some calcium, but skim milk, yogurt, and green vegetables are much better, lower-calorie sources.

ROAST BEEF

COMPANY/PRODUCT	CALORIES	FAT (tsp.)	SODIUM (mg.)	GLOOM
Dairy Queen BBQ Beef Sandwich	225	1	700	10
Arby's Junior Roast Beef	218	2	345	16
Subway Roast Beef Sandwich, 6 in.	375	3	839	16
Arby's French Dip (roast beef sandwich)	345	3	678	20
Hardee's Roast Beef Sandwich (RR)	350	3	732	20
Hardee's Roast Beef Sandwich, large (RR)	373	3	840	21
Hardee's Regular Roast Beef	310	3	930	22
Arby's Regular Roast Beef	353	3	588	23
Hardee's Roast Beef Sandwich with Cheese (RR)	403	3	954	27
Hardee's Big Roast Beef	360	3	1150	28
Hardee's Roast Beef Sandwich with Cheese, large (RR)	427	4	1062	28
Arby's French Dip 'N Swiss (roast beef sandwich)	425	4	1078	32
Arby's Beef 'N Cheddar	451	5	955	33
Carl's Jr. Roast Beef Deluxe Sandwich	540	6	1340	38
Arby's Super Roast Beef	529	6	798	38
Arby's Philly Beef 'N Swiss	498	6	1194	40
Arby's Giant Roast Beef	530	6	908	40
Arby's Bac 'N Cheddar Deluxe	532	7	1672	52

Plain roast beef is lower in fat than most hamburger meat, but a sandwich, topped with bacon, cheese, or sauces, may be quite fatty. Hardee's sandwiches marked "(RR)" are available in the New York-Washington region; they are made from the very lean roast beef that has been used by Roy Rogers restaurants.

SHAKES AND MALTS

COMPANY/PRODUCT	CALORIES	FAT (tsp.)	SODIUM (mg.)	SUGAR (tsp).	GLOOM
McDonald's, Low-fat (average)	310	<1	193	9	7
Carl's Jr., regular	350	2	230	10	14
Jack in the Box (average)	323	2	247	10	15
Hardee's (average)	433	2	320	12	19
Arby's, Vanilla	330	3	281	7	20
Arby's, Jamocha	368	2	262	8	20
Arby's, Chocolate	451	3	341	12	24
Dairy Queen Malt, regular	610	3	230	9	25
Burger King Chocolate, large	472	3	286	12	26
Dairy Queen Shake, regular	520	3	230	11	26
Dairy Queen Shake, large	600	4	260	13	30
Wendy's Frosty Dairy Dessert, medium	520	4	286	7	30
Arby's Snickers Polar Swirl	511	4	351	9	33
Wendy's Frosty Dairy Dessert, large	680	5	374	9	40
Arby's Peanut Butter Cup Polar Swirl	517	5	385	14	41
Dairy Queen Heath Blizzard, regular	820	8	410	14	58

Typical shakes contain about as much sugar (figures shown are estimates) as a can of cola, as much fat as a glass of milk, and as many calories as a cola and milk combined. However, they are all good sources of calcium. McDonald's was first to offer a low-fat shake. For some companies, the average of several flavors is shown.

ARBY'S

Over 25 years ago, Arby's put Sunday dinner's roast beef on a kaiser roll and introduced an upscale sandwich—fancier than the hamburger and just as portable. While the chain has introduced salads, chicken sandwiches, and sub sandwiches, roast-beef sandwiches are still its top sellers. Featuring roast beef instead of hamburgers clearly sets Arby's apart from the burger specialists.

Founded in 1964 by the Raffel brothers (it's the initials R.B. that yield Arby's name), Arby's started with one store in Boardman, Ohio, that sold roast-beef sandwiches and potato chips. Arby's is now the tenth-largest fast-food franchiser in the world, according to *Restaurants & Institutions,* with sales exceeding $1.4 billion in 1990. It has 2,400 outlets in the United States and about 100 overseas. In recent years, it has been one of the fastest-growing chains.

Making a uniform, predictable, sliced-meat product is facilitated by judicious processing. In Arby's factories, lean beef is "chunked" and combined with water, salt, and sodium phosphate. This process leads to a relatively high sodium content and a higher fat content than Roy Rogers' roast beef (available at Hardee's restaurants in the mid-Atlantic region). Still, Arby's roast beef is significantly leaner than typical hamburger meat and in general a healthier basis for a sandwich. But roast beef at a fast-food restaurant is eaten as part of a sandwich, not by itself, and the sandwiches are not all that different from hamburger sandwiches. Gloom ratings run from 20 for the French Dip to 52 for the Bac 'N Cheddar Deluxe.

The French Dip gets 32 percent of its 345 calories from just under 3 teaspoons of fat. As for some of the others, fat provides 38 percent of the 353 calories in a Regular Roast Beef, and 40 percent of the 451 calories in the Beef 'N Cheddar. The Super Roast Beef gets just under 50 percent of its 529 calories from fat, while the Bac 'N Cheddar Deluxe gets just over half of its 532 calories from fat. By contrast, a Quarter Pounder hamburger sandwich at McDonald's gets 44 percent of its 410 calories from fat. You can easily cut calories from the Super and Beef 'N Cheddar sandwiches by asking the clerk to leave off the ranch dressing.

Adding a large order of fries and Coke to a Giant Roast Beef sandwich gives you 1,166 calories and 81 Gloom points. If instead you had the Regular Roast Beef with Potato Cakes and a side salad with Weight Watcher's salad dressing, you'd have 611 calories and 48 Gloom points.

Arby's recently added a line of Light Roast Beef, Roast Chicken, and Roast Turkey Deluxe sandwiches, all of which are under 300 calories and 30 percent or less calories from fat. The roast chicken sandwich, for instance, provides 253 calories and less than 1 teaspoonful of fat. The Gloom ratings of the Light Deluxe sandwiches are between 13 (chicken) and 18 (roast beef). These sandwiches can be the centerpiece of an excellent meal.

Arby's also offers a Grilled Chicken Barbecue sandwich with a Gloom rating of 25. The Hot Ham 'N Cheese has a Gloom rating of 27, the Roast Chicken Deluxe 27, and Turkey Deluxe a rating of 28. You can improve the deluxes by skipping the low-calorie mayonnaise, but you can't do much about their salt content of about one-half teaspoon or more.

Some of the other chicken sandwiches aren't nearly as good. For instance, the Roast Chicken Club provides 513 calories, almost 7 teaspoons of fat, and 43 Gloom points. But even that isn't bad compared to the 658-calorie Chicken Cordon Bleu. That sandwich provides more than 8 teaspoons of fat, almost a full teaspoon of salt, and a Gloom rating of 60. Looking on the brighter side, though, a giant hamburger sandwich, such as Burger King's Double Whopper or Wendy's Double Big Classic with Cheese, is far worse, if that's any comfort.

A meal combination of the Grilled Chicken Barbecue with a side salad and 2 percent low-fat milk would give you 524 calories and 28 Gloom points. Contrast that with a diet-buster special consisting of Chicken Cordon Bleu, Curly Fries, Jamocha Shake, and Cherry Turnover: 1,643 calories, 19 teaspoons of fat, well over a teaspoon of salt, and 132 Gloom points.

A big baked potato from Arby's (sold at about one out of three outlets) is one of the best foods you can get anywhere. But the Gloom rating zooms from 2 to 48 for the Deluxe Baked Potato, which is loaded with butter, sour cream, cheese, and other goodies. The healthiest of the stuffed baked potatoes is the Broccoli & Cheddar variety, which has 417 calories and 21 Gloom points. If you order a stuffed baked potato, you can save yourself some Gloom points by removing some of the fatty sauce and keeping the vegetables.

Arby's has largely switched from salad bars to prepackaged salads which are both more hygienic and more profitable. All of the salads—Chef, Garden, Roast Chicken, Side—are excellent. Whichever salad you choose, ask for one of the low-calorie dressing.

A meal consisting of a chef salad with Weight Watcher's dressing, plain baked potato, and orange juice would give you 568 calories and 27 Gloom points.

You may find a choice of soups at the Arby's you patronize. Tomato Florentine, Lumberjack Mixed Vegetable, and Old Fashioned Chicken Noodle are several of the healthier choices. Unfortunately, all the soups rely on about half a teaspoon of salt for much of their flavor.

ARBY'S
COMPLETE NUTRITIONAL VALUES[1]

	WEIGHT[2] (gm.)	CALORIES	PROTEIN (gm.)
Arby's Sauce	28	30	0
Bac 'N Cheddar Deluxe	229	532	29
Bacon Platter	217	869	17
Baked Potato, Broccoli & Cheddar	340	417	10
Baked Potato, Butter/Margarine and Sour Cream	312	463	8
Baked Potato, Deluxe	348	621	17
Baked Potato, Mushroom and Cheese	347	515	15
✔ Baked Potato, Plain	241	240	6
Beef 'N Cheddar	198	451	25
Biscuit, Bacon	98	318	7
Biscuit, Ham	124	323	13
Biscuit, Plain	82	280	6
Biscuit, Sausage	118	460	12

Some Arby's are open for breakfast. Your best bet is a glass of orange juice, blueberry muffin, and 2 percent low-fat milk. They also offer either biscuits (mostly in the South) or croissant sandwiches. The croissants tend to be rich in fat and calories, while the biscuits are loaded with sodium. The plain biscuit and croissant each have about 270 calories and 24 Gloom points, while the Sausage and Egg Croissant has 519 calories and 69 Gloom points. (Can you just feel your arteries closing?)

Arby's has shown increased interest in serving its customers' nutritional needs—it was the first chain to disclose ingredients, and in 1989 it stopped frying foods in beef fat. It is currently researching a cooking process that eliminates deep-fat fryers and frying oils. What a breakthrough that would be!

CARBOHYDRATES (gm.)	ADDED SUGAR³ (gm.)	FAT⁴ (gm.)	FAT % CALORIES	SATURATED FAT (gm.)	CHOLESTEROL (mg.)	SODIUM (mg.)	VITAMIN A (% U.S. RDA)	VITAMIN C (% U.S. RDA)	IRON (% U.S. RDA)	CALCIUM (% U.S. RDA)	GLOOM
6	4	0	9	0	0	227	0	0	10	0	2
35	5	33	55	8	83	1672	0	2	25	15	52
49	0	32	33	10	366	1051	8	6	20	6	73
55	0	18	39	7	22	361	6	75	15	10	21
53	0	25	49	12	40	203	4	55	15	10	31
59	0	36	53	18	58	605	10	55	15	15	48
57	0	27	47	6	47	923	15	55	15	25	35
50	0	2	7	0	0	58	0	55	15	0	2
43	5	20	40	7	52	955	4	0	20	2	33
35	0	18	51	4	8	904	0	0	15	10	30
34	0	17	46	4	21	1169	0	0	15	10	30
34	0	15	48	3	0	730	0	0	15	10	24
35	0	32	62	9	60	1000	0	0	20	10	51

ARBY'S

COMPLETE NUTRITIONAL VALUES[1] CONTINUED	WEIGHT[2] (gm.)	CALORIES	PROTEIN (gm.)
✔ Blueberry Muffin	71	200	3
Butterfinger Polar Swirl	329	457	12
Cheddar Fries	142	399	6
Cheese Cake	85	306	5
Chicken Breast Sandwich	184	489	23
Chicken Cordon Bleu	216	658	31
✔ Chicken Fajita Pita	312	256	15
Chocolate Chip Cookie	27	130	2
Cinnamon Nut Danish	99	340	7
Coca-Cola Classic, 12 fl. oz.	358	144	0
Croissant, Bacon and Egg	113	389	12
Croissant, Ham and Cheese	119	345	16
Croissant, Mushroom and Cheese	148	493	13
Croissant, Sausage and Egg	142	519	17
Croissant (plain)	63	260	6
Croutons	14	59	2
Curly Fries	99	337	4
Egg Platter	201	460	15
Fish Fillet Sandwich	193	537	21
✔ French Dip (roast beef sandwich)	150	345	24
French Dip 'N Swiss (roast beef sandwich)	178	425	30
French Fries, large	142	492	4
French Fries, medium	114	394	3
French Fries, small	71	246	2
Grilled Chicken Barbecue	185	378	21

CARBOHYDRATES (gm.)	ADDED SUGAR[3] (gm.)	FAT[4] (gm.)	FAT % CALORIES	SATURATED FAT (gm.)	CHOLESTEROL (mg.)	SODIUM (mg.)	VITAMIN A (% U.S. RDA)	VITAMIN C (% U.S. RDA)	IRON (% U.S. RDA)	CALCIUM (% U.S. RDA)	GLOOM
34	15	6	25	2	22	269	0	0	6	4	13
62	40	18	36	8	28	318	4	0	2	25	31
46	0	22	49	9	9	443	0	0	8	8	35
21	18	23	67	7	95	220	20	0	4	6	35
48	5	26	47	4	45	1019	0	8	20	8	36
50	5	37	50	9	65	1824	0	0	20	15	60
32	—	9	32	—	33	787	—	—	—	—	—
17	10	4	28	2	0	95	0	0	2	0	8
59	16	9	25	2	0	230	0	0	15	4	16
38	27	0	0	0	0	15	0	0	0	0	4
30	0	26	61	14	221	582	4	0	20	4	49
29	0	21	54	12	90	939	0	0	15	15	36
34	0	38	69	15	116	935	0	0	15	20	60
29	0	39	68	19	271	632	4	0	20	6	69
28	0	16	54	10	49	300	0	0	15	4	25
8	0	2	34	0	1	155	0	0	2	0	4
43	0	18	47	7	0	167	0	0	8	2	27
45	0	24	47	7	346	591	8	6	20	6	51
47	5	29	49	6	79	994	0	0	20	8	46
34	5	12	32	6	5	678	0	0	15	2	20
36	5	18	39	8	87	1078	4	0	15	25	32
60	0	26	48	6	0	228	0	12	12	0	37
48	0	21	48	5	0	182	0	10	10	0	29
30	0	13	48	3	0	114	0	6	6	0	18
44	0	14	34	4	44	1059	0	0	20	10	25

ARBY'S

COMPLETE NUTRITIONAL VALUES[1] CONTINUED

	WEIGHT[2] (gm.)	CALORIES	PROTEIN (gm.)
Grilled Chicken Deluxe	228	426	21
Ham Platter	258	518	24
Heath Polar Swirl	329	543	11
Horsey Sauce	14	55	0
Hot Chocolate, 8 fl. oz.	244	110	2
Hot Ham 'N Cheese	162	330	23
✔ Light Roast Beef Deluxe	182	296	18
✔ Light Roast Chicken Deluxe	189	253	17
Light Roast Turkey Deluxe	189	249	19
Maple Syrup	43	120	0
✔ Milk, 2% Low-fat, 8 fl. oz.	244	121	8
✔ Orange Juice, 6 fl. oz.	180	82	1
Oreo Polar Swirl	329	482	10
Peanut Butter Cup Polar Swirl	329	517	14
Philly Beef 'N Swiss	196	498	26
Potato Cakes	85	204	2
Roast Beef, Giant	227	530	36
✔ Roast Beef, Junior	85	218	13
Roast Beef, Regular	147	353	22
Roast Beef, Super	246	529	33
Roast Chicken Club	234	513	31
Roast Chicken Deluxe	208	373	17
Salad Dressing, Blue Cheese, 2 fl. oz.	57	295	2
Salad Dressing, Buttermilk Ranch, 2 fl. oz.	57	349	0
Salad Dressing, Honey French, 2 fl. oz.	65	322	0

CARBOHYDRATES (gm.)	ADDED SUGAR[3] (gm.)	FAT[4] (gm.)	FAT % CALORIES	SATURATED FAT (gm.)	CHOLESTEROL (mg.)	SODIUM (mg.)	VITAMIN A (% U.S. RDA)	VITAMIN C (% U.S. RDA)	IRON (% U.S. RDA)	CALCIUM (% U.S. RDA)	GLOOM
39	0	21	45	5	44	877	0	0	20	10	32
45	0	26	46	8	374	1177	8	6	20	6	58
76	38	22	36	5	39	346	4	0	0	25	38
3	0	5	82	1	1	105	0	0	0	2	8
23	13	1	10	1	0	120	0	0	2	6	4
33	5	14	38	4	45	1350	4	0	15	10	27
33	—	10	30	—	42	826	6	2	20	0	18
33	—	5	17	—	39	874	6	2	20	0	13
33	—	4	15	—	30	1172	6	2	20	0	14
29	29	0	1	0	0	52	0	0	0	0	5
12	0	4	33	3	18	122	10	4	0	30	5
20	0	0	0	0	0	2	0	119	0	0	0
66	38	20	37	10	35	521	4	0	4	25	36
61	54	24	42	8	34	385	4	0	0	25	41
37	5	26	47	6	91	1194	6	4	20	30	40
20	0	12	53	2	0	397	0	15	8	0	17
41	5	27	46	10	78	908	0	0	35	6	40
21	5	11	44	3	23	345	0	0	10	2	16
32	5	15	38	7	39	588	0	0	20	4	23
46	5	28	48	8	47	798	6	2	25	6	38
40	5	29	51	5	75	1423	0	0	20	15	43
37	5	19	47	3	2	913	0	0	20	4	27
2	8	31	95	6	50	489	2	0	2	6	48
2	8	38	99	6	6	471	0	0	0	0	59
22	20	27	75	4	0	486	8	2	10	0	38

ARBY'S

COMPLETE NUTRITIONAL VALUES[1] CONTINUED

	WEIGHT[2] (gm.)	CALORIES	PROTEIN (gm.)
Salad Dressing, Thousand Island, 2 fl. oz.	62	298	0
Salad Dressing, Weight Watchers Creamy French, 1 fl. oz.	30	29	—
Salad Dressing, Weight Watchers Creamy Italian, 1 fl. oz.	30	29	—
Salad, Chef	383	217	20
Salad, Garden	295	109	7
Salad, Roast Chicken	383	172	15
✔ Salad, Side	150	25	2
Sausage Platter	238	640	21
Shake, Chocolate	340	451	10
Shake, Jamocha	326	368	9
Shake, Vanilla	312	330	10
Snickers Polar Swirl	329	511	12
Soup, Beef with Vegetables and Barley, 6 fl. oz.	244	96	5
Soup, Boston Clam Chowder, 6 fl. oz.	244	207	10
Soup, Cream of Broccoli, 6 fl. oz.	244	180	8
Soup, French Onion, 6 fl. oz.	244	67	2
Soup, Lumberjack Mixed Vegetable, 6 fl. oz.	244	89	2
Soup, Old Fashioned Chicken Noodle, 6 fl. oz.	244	99	6
Soup, Pilgrim's Corn Chowder, 6 fl. oz.	244	193	10
Soup, Split Pea with Ham, 6 fl. oz.	244	200	8
Soup, Tomato Florentine, 6 fl. oz.	244	84	3
Soup, Wisconsin Cheese, 6 fl. oz.	244	287	9
Sub Deluxe	226	482	24

CARBOHYDRATES (gm.)	ADDED SUGAR[3] (gm.)	FAT[4] (gm.)	FAT % CALORIES	SATURATED FAT (gm.)	CHOLESTEROL (mg.)	SODIUM (mg.)	VITAMIN A (% U.S. RDA)	VITAMIN C (% U.S. RDA)	IRON (% U.S. RDA)	CALCIUM (% U.S. RDA)	GLOOM
10	8	29	88	4	24	493	2	2	4	0	44
–	–	3	93	–	–	–	–	–	–	–	4
–	–	3	93	–	–	–	–	–	–	–	4
11	0	11	44	–	172	706	40	40	15	10	19
10	0	5	43	–	12	134	35	35	10	10	5
12	0	7	35	–	45	562	–	–	–	–	–
4	0	0	0	0	0	30	20	6	4	4	0
46	0	41	58	13	406	861	8	6	20	8	79
76	48	12	23	3	36	341	6	0	4	25	24
59	33	10	26	2	35	262	6	0	0	25	20
46	28	11	31	4	32	281	6	0	0	30	20
73	35	19	33	7	33	351	6	0	0	25	33
14	0	3	26	1	10	996	30	8	8	2	8
18	0	11	46	4	28	1157	10	6	10	20	19
19	0	8	40	5	3	1113	10	15	4	30	15
7	0	3	42	0	0	1248	2	4	4	2	12
13	0	4	36	2	4	1075	50	15	6	2	9
15	1	2	16	0	25	929	20	0	8	0	9
18	1	11	49	4	28	1157	35	6	4	15	18
21	1	10	43	5	30	1029	30	2	12	2	18
15	0	1	16	1	2	910	20	20	8	4	6
19	0	19	59	8	31	1129	6	4	4	35	30
38	5	26	49	5	50	1530	0	4	20	15	42

ARBY'S

COMPLETE NUTRITIONAL VALUES[1] CONTINUED	WEIGHT[2] (gm.)	CALORIES	PROTEIN (gm.)
Toastix	99	420	8
Turkey Deluxe	221	399	27
Turnover, Apple	85	303	4
Turnover, Blueberry	85	320	3
Turnover, Cherry	85	280	5

Arby's Meals

	WEIGHT[2] (gm.)	CALORIES	PROTEIN (gm.)
✔ Blueberry Muffin, Orange Juice	251	282	4
Toastix with Maple Syrup, Orange Juice, Coffee	566	625	9
Sausage Biscuit, Potato Cakes, Hot Chocolate	447	774	16
✔ Light Roast Chicken Deluxe, Garden Salad, Orange Juice	664	444	25
✔ French Dip, Baked Potato (plain), Orange Juice	571	667	31
Plain Baked Potato, Chef Salad with Light Italian Dressing, Orange Juice	795	555	28
Grilled Chicken Barbecue, Side Salad, 2% Low-fat Milk	579	524	31
Turkey Deluxe, Tomato Florentine Soup, Plain Baked Potato	706	723	36
Regular Roast Beef, Potato Cakes, Side Salad with Light Italian Dressing	444	605	26
Giant Roast Beef, Large Fries, Coca-Cola Classic	727	1166	41
Chicken Cordon Bleu, Curly Fries, Jamocha Shake, Cherry Turnover	726	1643	50

1. *A dash means that data not available.*
2. *To convert grams to ounces (weight), divide by 28.35; to convert grams to fluid ounces (volume), divide by 29.6.*

CARBOHYDRATES (gm.)	ADDED SUGAR[3] (gm.)	FAT[4] (gm.)	FAT % CALORIES	SATURATED FAT (gm.)	CHOLESTEROL (mg.)	SODIUM (mg.)	VITAMIN A (% U.S. RDA)	VITAMIN C (% U.S. RDA)	IRON (% U.S. RDA)	CALCIUM (% U.S. RDA)	GLOOM
43	0	25	54	5	20	440	0	0	15	0	38
35	5	20	46	4	39	1047	6	8	15	8	28
27	13	18	54	7	0	178	2	2	4	0	29
32	8	19	53	6	0	240	0	10	4	0	29
25	11	18	57	5	0	200	0	8	4	0	26
54	15	6	18	2	22	271	0	119	6	4	9
92	29	25	36	5	20	497	0	119	15	0	33
78	13	45	52	12	60	1517	0	15	30	16	73
63	0	10	20	—	51	1010	41	156	30	10	15
104	5	14	19	6	5	738	0	174	30	2	18
77	3	14	23	5	115	1890	30	184	23	15	27
61	0	19	32	7	62	1211	30	10	24	44	28
100	5	24	29	5	41	2015	26	83	38	12	36
59	8	28	42	10	39	2125	20	21	32	8	48
139	32	54	41	17	78	1151	0	12	47	6	81
178	49	83	45	24	100	2453	6	8	32	42	132

3. To convert grams of sugar to teaspoons of sugar, divide by 4.0.
4. To convert grams of fat to teaspoons of fat, divide by 4.4.

BASKIN-ROBBINS

In the past, Baskin-Robbins, which boasts 2,400 units in the United States and 1,100 more in 43 countries overseas (including three in the Soviet Union), had about as much to do with health as Attila the Hun had to do with peace and justice. But these are the 1990s, and even Baskin-Robbins is catering to the health- or weight-conscious.

A regular scoop of Baskin-Robbins' Red Raspberry Sorbet provides 140 calories and zero fat. Five ounces of a tasty nonfat frozen yogurt that is available at many outlets has only 100 to 125 calories. Some outlets feature a low-fat frozen yogurt that has only 1 teaspoon of fat and 150 calories per 5-ounce serving. A standard serving of Sorbet Fruit Whip (a soft-serve purée) is also fat-free, with just 80 calories. Most of the calories in those products come from sugar.

Baskin-Robbins also has a line of sugar-free, low-fat frozen dairy products called Sugar Free Dairy Dessert, which are artificially sweetened with NutraSweet. A half-cup of the Chunky Banana flavor has just 100 calories and 1 gram of fat. What's more, many Baskin-Robbins have begun offering Fat Free Just Peachy and Fat Free Just Chocolate Vanilla Twist ice-cream-like products, also with 100 calories per half-cup serving. No longer does an evening visit to Baskin-Robbins mean a morning visit to Weight Watchers.

Compare the above numbers to the 240 calories and 3 teaspoons of fat in a scoop of their vanilla ice cream or the 310 calories and 4 teaspoons of fat in a scoop of Chocolate Raspberry Truffle ice cream.

BASKIN-ROBBINS

COMPLETE NUTRITIONAL VALUES[1]	WEIGHT[2] (gm.)	CALORIES	PROTEIN (gm.)	
Cone (only), Sugar	–	60	1	
Cone (only), Waffle	–	140	3	
✔ Fat Free Just Peachy (frozen dairy dessert), ½ cup	–	100	–	

To all of those calorie counts, add another 60 calories for a sugar cone and 140 calories for the waffle cone. Well, at least ice cream is low in sodium and a decent source of calcium.

Baskin-Robbins was founded in 1945 by Irv Robbins and his brother-in-law Burton Baskin in Glendale, California. It is now owned by a British conglomerate, Allied Lyons. A Baskin-Robbins official told us that the company has developed over 600 different flavors, 31 of which are in shops at any given time. But the company cannot or will not provide customers with lists of ingredients, so if you have allergies, avoid the stranger-sounding flavors, ask the clerk to read you the ingredients that are listed (often illegibly) on container lids, or write to the company's headquarters (see Appendix for address). A company brochure says that sulfite-sensitive people should avoid products containing cherries, coconut, and other processed fruit-cocktail-type ingredients and toppings. Flavors in which eggs are used include lemon custard, orange custard, eggnog, French vanilla, vanilla, and those with cookie or cake pieces or marshmallow ribbons.

Even though Baskin-Robbins sells over $600 million worth of desserts a year, it's never gotten around to measuring the vitamin or mineral content of its products, and won't disclose their sugar content. The sugar values shown in the chart are our estimates. We have not included Gloom ratings, but estimate that scores would range from 2 for a small, nonfat yogurt to 13 for a large, low-fat yogurt. A scoop of the light ice creams, sorbets, and sherbet would have Gloom ratings between about 5 and 10. A scoop of the regular ice creams would likely be about 25.

CARBOHYDRATES (gm.)	ADDED SUGAR[3, 5] (gm.)	FAT[4] (gm.)	FAT % CALORIES	SATURATED FAT (gm.)	CHOLESTEROL (mg.)	SODIUM (mg.)	VITAMIN A (% U.S. RDA)	VITAMIN C (% U.S. RDA)	IRON (% U.S. RDA)	CALCIUM (% U.S. RDA)	GLOOM[5]
11	3	1	15	—	0	45	—	—	—	0	—
28	3	2	13	—	0	5	—	—	—	0	—
—	—	0	0	0	0	—	—	—	—	—	—

BASKIN-ROBBINS

COMPLETE NUTRITIONAL VALUES[1] CONTINUED	WEIGHT[2] (gm.)	CALORIES	PROTEIN (gm.)
✔ Fat Free Just Chocolate Vanilla Twist (frozen dairy dessert), ½ cup	—	100	—
Frozen Yogurt, Low-fat			
Chocolate, large (9 oz.)	—	315	9
Chocolate, medium (7 oz.)	—	246	7
Chocolate, small (5 oz.)	—	175	5
Strawberry, large (9 oz.)	—	270	9
Strawberry, medium (7 oz.)	—	211	7
Strawberry, small (5 oz.)	—	150	5
Vanilla, large (9 oz.)	—	270	9
Vanilla, medium (7 oz.)	—	211	7
Vanilla, small (5 oz.)	—	150	5
Frozen Yogurt, Nonfat			
Coconut, large (9 oz.)	—	180	9
Coconut, medium (7 oz.)	—	140	7
✔ Coconut, small (5 oz.)	—	100	5
Raspberry, large (9 oz.)	—	225	9
Raspberry, medium (7 oz.)	—	164	7
✔ Raspberry, small (5 oz.)	—	125	5
Strawberry, large (9 oz.)	—	225	9
Strawberry, medium (7 oz.)	—	176	7
✔ Strawberry, small (5 oz.)	—	125	5
Ice Cream			
Chocolate Chip, 1 scoop	114	260	4
Chocolate Raspberry Truffle, 1 scoop	114	310	4

CARBOHYDRATES (gm.)	ADDED SUGAR[3] (gm.)	FAT[4] (gm.)	FAT % CALORIES	SATURATED FAT (gm.)	CHOLESTEROL (mg.)	SODIUM (mg.)	VITAMIN A (% U.S. RDA)	VITAMIN C (% U.S. RDA)	IRON (% U.S. RDA)	CALCIUM (% U.S. RDA)	GLOOM[5]
–	–	0	0	0	0	–	–	–	–	–	–
54	39	9	26	–	9	90	4	–	–	34	–
42	30	7	26	–	7	70	3	–	–	26	–
30	23	5	26	–	5	50	2	–	–	19	–
54	34	9	30	–	9	90	4	–	–	34	–
42	27	7	30	–	7	70	3	–	–	26	–
30	19	5	30	–	5	50	2	–	–	19	–
54	34	9	30	–	9	90	4	–	–	34	–
42	27	7	30	–	7	70	3	–	–	26	–
30	19	5	30	–	5	50	2	–	–	19	–
45	23	0	0	–	0	90	–	–	–	22	–
35	18	0	0	–	0	70	–	–	–	17	–
25	13	0	0	–	0	50	0	–	–	12	–
45	28	0	0	–	0	90	–	–	–	22	–
35	22	0	0	–	0	70	–	–	–	17	–
25	16	0	0	–	0	50	0	–	–	12	–
45	28	0	0	–	0	90	–	–	–	22	–
35	22	0	0	–	0	70	–	–	–	17	–
25	16	0	0	–	0	50	0	–	–	12	–
27	16	15	52	–	40	110	–	–	–	15	–
35	20	17	49	–	45	115	–	–	–	10	–

BASKIN-ROBBINS

COMPLETE NUTRITIONAL VALUES[1] CONTINUED	WEIGHT[2] (gm.)	CALORIES	PROTEIN (gm.)
Chocolate, 1 scoop	114	270	5
French Vanilla, 1 scoop	114	280	4
Jamoca Almond Fudge, 1 scoop	114	270	5
Pralines 'n Cream, 1 scoop	114	280	4
Rocky Road, 1 scoop	114	300	5
Vanilla, 1 scoop	114	240	4
Very Berry Strawberry, 1 scoop	114	220	3
World Class Chocolate, 1 scoop	—	280	5
Ice, Daiquiri, 1 scoop	114	140	0
Light, Chocolate Caramel Nut, ½ cup	—	130	3
Light, Espresso and Cream, ½ cup	—	120	3
Light, Praline Dream, ½ cup	—	130	3
✔ Light, Strawberry Royal, ½ cup	—	110	2
Sherbet, Rainbow, 1 scoop	114	160	1
Sorbet Fruit Whip	—	80	0
Sorbet, Red Raspberry, 1 scoop	114	140	0
Sugar Free Dairy Dessert, Chunky Banana, ½ cup	113	100	3

1. A dash means that data not available.
2. To convert grams to ounces (weight), divide by 28.35; to convert grams to fluid ounces (volume), divide by 29.6.
3. To convert grams of sugar to teaspoons of sugar, divide by 4.0.

CARBOHYDRATES (gm.)	ADDED SUGAR[3] (gm.)	FAT[4] (gm.)	FAT % CALORIES	SATURATED FAT (gm.)	CHOLESTEROL (mg.)	SODIUM (mg.)	VITAMIN A (% U.S. RDA)	VITAMIN C (% U.S. RDA)	IRON (% U.S. RDA)	CALCIUM (% U.S. RDA)	GLOOM[5]
32	19	14	47	—	37	160	—	—	—	15	—
25	16	18	58	—	90	90	—	—	—	15	—
30	18	14	47	—	32	115	—	—	—	10	—
35	20	14	45	—	36	180	—	—	—	15	—
39	22	14	42	—	32	135	—	—	—	10	—
24	16	14	53	—	52	115	—	—	—	15	—
30	14	10	41	—	30	95	—	—	—	10	—
35	20	14	45	—	36	145	—	—	—	15	—
35	35	0	0	0	0	15	—	—	—	0	—
19	12	5	35	3	8	0	—	—	—	8	—
15	9	5	38	3	12	0	—	—	—	10	—
17	10	6	42	4	7	0	—	—	—	10	—
19	12	3	25	2	9	0	—	—	—	8	—
34	32	2	11	—	6	85	—	—	—	4	—
24	20	0	0	0	0	20	—	—	—	0	—
34	30	0	0	0	0	25	—	—	—	0	—
20	0	1	9	—	3	50	—	—	—	8	—

4. *To convert grams of fat to teaspoons of fat, divide by 4.4.*
5. *Baskin-Robbins provides too little nutrition information to calculate Gloom ratings. The sugar values are the author's estimates.*

BURGER KING

Now with over 6,000 restaurants worldwide, Burger King has been the home of flame-broiled burgers since its founding in 1954 in Miami. Burger King, which is currently owned by Grand Metropolitan, a British liquor conglomerate, has been playing a valiant, but futile, game of catch-up with McDonald's.

Burger King has built its reputation on flame-broiled and custom-made ("Have it your way") hamburgers. What does flame-broiling do for a burger? It adds a unique flavor not found in those from other restaurants. However, our tests found that the meat in Burger King's hamburgers contained slightly more fat than McDonald's burgers—Burger King burgers are just over 21 percent fat, while McDonald's are 19 percent fat.

If you are thinking of eating a fast-food breakfast, Burger King can present quite a challenge especially if you are a nutrionally-minded customer. People who want to avoid calories, fat, and/or cholesterol should avoid the Croissan'wiches—the soggy bulwark of the Burger King breakfast menu.

All the croissant sandwiches come with egg and cheese; the meat extras are all traditional, salty breakfast meats—ham, bacon, and sausage. Unfortunately, there is no way to turn croissants into a prudent breakfast choice; adding salty meats only increases their Gloom rating. As far as taste and texture are concerned, a Croissan'wich is to a real croissant as a piece of Wonder Bread is to a chewy, fresh-baked bagel. If you must have a Croissan'wich, the "best" is a plain one (Gloom = 16) or the egg-and-cheese variety without meat (Gloom = 39), the worst is the one with sausage, egg, and cheese (9 teaspoons of fat; Gloom = 69).

A few Burger King outlets offer bagel sandwiches, which are a tad lower in fat than the Croissan'wiches. Some other outlets offer biscuits, which come either plain or with bacon or sausage, with or without egg. They have about the same Gloom ratings as Croissan'wiches—or worse.

Your best overall breakfast combination is a plain croissant with a bit of cream cheese, orange juice, and low-fat milk (total Gloom of 17)—or a walk down the street to McDonald's.

For lunch, probably your best bet is Burger King's BK Broiler

Chicken Sandwich, flame-broiled (chunked) chicken white meat served on a bun with lettuce and a slice of tomato (Gloom = 15). The sandwich weighs in at 267 calories and less than 2 teaspoons of fat. (The bun is called an "oat-bran bun," but Burger King won't say how much oat bran the bun contains; the ingredient list indicates that there is more sugar than oat bran.) Adding a side salad and orange juice to the BK Broiler Chicken Sandwich gives you a 374-calorie meal with 12 Gloom points.

On the other hand, the Chicken Sandwich is the fattiest chicken product in fast-food land, with 9 teaspoons of fat, 685 calories, and 61 Gloom points. Again, you can slash the fat by asking the clerk to hold the mayonnaise on this breaded and fried sandwich.

Burger King's other chicken entry is Chicken Tenders, which are tastier than McDonald's Chicken McNuggets, being made from strips of chicken breast rather than ground-up chicken. The Barbecue and Sweet & Sour Dipping Sauces are the two lowest-fat sauces. The plain Tenders have a Gloom of 21.

Burger King's Ocean Catch Fish Filet sandwich (Gloom = 37) is alas, breaded and fried, so forget about the fish choice being low in fat. It gets about half of its calories from fat and has almost as much fat as the Double Cheeseburger. But if you skip the tartar sauce on the Ocean Catch Fish Filet, you bring the fat percentage down to 27 and have quite a decent sandwich.

A few Burger Kings have salad bars, which give you an all-you-can-eat opportunity to load up on vitamin-rich broccoli, carrots, and other foods. But most outlets have switched to more profitable packaged salads. The best ones are the Chunky Chicken and Garden salads, and the 178-calorie Chef Salad (Gloom = 13). The dressings, made by Newman's Own, include a reduced-calorie Italian dressing (but still 170 calories per serving).

At our local Burger King, the Chunky Chicken Salad, priced at $3.49, is the most expensive item on the menu — 10 cents more than the Double Whopper with Cheese. That's an awful lot to pay for an item that contains 8 ounces of iceberg lettuce but only 3 ounces of chicken.

So far, we've avoided the hamburger department, still the foundation of Burger King's business. As of mid-1991, Burger King did not yet have a low-fat burger to compete with McDonald's McLean

Deluxe and Hardee's Real Lean Deluxe. The "small" choice is the plain hamburger, with its 272 calories and Gloom rating of 18. As the burgers get bigger, so do their Gloom ratings. Sandwiches include the cheeseburger (Gloom = 24), bacon double cheeseburger (Gloom = 45), and the regular Whopper, which sells at the rate of two million a day, and is packed with 614 calories and a Gloom rating of 49. A pair of Burger Buddies, which are little cheeseburgers, is nutritionally similar to one regular cheeseburger.

The "extreme burger" is the Double Whopper with Cheese. Its heart-stopping Gloom rating of 83 tips us off to its 935 calories, high sodium content (1,245 milligrams), and 14 teaspoons of fat. This is the fast-food industry's third-fattiest hamburger (Jack in the Box and Carl's Jr. have fattier ones). Just for fun we calculated what a meal consisting of the Double Whopper with Cheese, medium fries, a large strawberry shake, and apple pie would add up to. The totals: 2,187 calories, 25 teaspoons of fat, 22 teaspoons of sugar, and over a teaspoon of salt, leading to a horrific Gloom score of 161.

Believe it or not, Burger King actually has a key on many of its

BURGER KING

COMPLETE NUTRITIONAL VALUES[1]

	WEIGHT[2] (gm.)	CALORIES	PROTEIN (gm.)
Apple Pie	125	311	3
✔ BK Broiler Chicken Sandwich	154	267	22
BK Broiler Sauce	11	37	0
Bacon Bits, 1 packet	3	16	1
Bacon Double Cheeseburger	160	515	32
Bacon Double Cheeseburger Deluxe	195	592	33
Bull's Eye Barbecue Sauce	14	22	0
Burger Buddies	129	349	18
Burger King A.M. Express Dip	28	84	0
Cheese, Processed American	25	92	5

144

cash registers for a "Veggie Whopper." Don't expect to have a sandwich made with a large patty of oats and mushrooms. Rather, the Veggie Whopper is simply a toasted bun with lettuce, tomato, and mayonnaise, for about $1.09. That's better than McDonald's, which charges non-meat-eaters the same for a Big Mac with or without meat. You can also get a Veggie Whopper with Cheese for a few cents more.

Shakes at Burger King are higher in calories and far fattier than at McDonald's, though they're loaded with an estimated 9 or more teaspoons of sugar at both restaurants. As at any restaurant, water remains the lowest-calorie and cheapest beverage available.

Burger King pies are (packaged) slices of regular baked pies, rather than the fried pies that McDonald's, Church's, and several other chains offer. Nutritionally, though, they end up about the same.

Burger King provides a handy brochure with both nutrition and ingredient information, and in 1990 the company posted nutrition charts in all its units. Persons sensitive to MSG should avoid all the chicken products, Fish Tenders, and breakfast sausage.

CARBOHYDRATES (gm.)	ADDED SUGAR[3] (gm.)	FAT[4] (gm.)	FAT % CALORIES	SATURATED FAT (gm.)	CHOLESTEROL (mg.)	SODIUM (mg.)	VITAMIN A (% U.S. RDA)	VITAMIN C (% U.S. RDA)	IRON (% U.S. RDA)	CALCIUM (% U.S. RDA)	GLOOM
44	16	14	41	4	4	412	0	8	7	0	26
28	5	8	27	2	45	728	4	6	14	5	15
1	0	4	97	1	5	74	0	0	0	0	7
0	0	1	56	0	5	0	0	0	0	0	1
26	5	31	54	14	105	748	8	0	21	18	45
28	5	39	59	16	111	804	12	5	21	18	54
5	0	0	0	0	0	47	0	0	0	0	1
31	0	17	44	7	52	717	9	8	19	11	26
21	0	0	0	0	0	18	0	0	0	0	0
1	0	7	68	5	25	312	8	0	0	14	10

BURGER KING

COMPLETE NUTRITIONAL VALUES[1] CONTINUED

	WEIGHT[2] (gm.)	CALORIES	PROTEIN (gm.)
Cheeseburger	121	318	17
Cheeseburger Deluxe	151	390	18
Chicken Sandwich (fried)	229	685	26
Chicken Tenders, 6 pieces	90	236	16
Coke, Kids Club, 10 fl. oz.	300	120	0
Coke, large (27 fl. oz.)	810	324	0
Coke, medium (18 fl. oz.)	540	216	0
Coke, small (13 fl. oz.)	390	156	0
Cream Cheese	28	98	2
Croissan'wich with Bacon, Egg and Cheese	118	361	15
Croissan'wich with Egg and Cheese	110	315	13
Croissan'wich with Ham, Egg and Cheese	144	346	19
Croissan'wich with Sausage, Egg and Cheese	159	534	21
Croissant (plain)	41	180	4
Croutons, 1 packet	7	31	1
Diet Coke, medium (18 fl. oz.)	540	1	0
Dipping Sauce, Barbecue	28	36	0
Dipping Sauce, Honey	28	91	0
Dipping Sauce, Ranch	28	171	0
Dipping Sauce, Sweet & Sour	28	45	0
Double Cheeseburger	172	483	30
French Fries, medium	116	372	5
French Toast Sticks	141	538	10
✔ Frozen Yogurt, Breyers Vanilla	65	120	2

CARBOHYDRATES (gm.)	ADDED SUGAR[3] (gm.)	FAT[4] (gm.)	FAT % CALORIES	SATURATED FAT (gm.)	CHOLESTEROL (mg.)	SODIUM (mg.)	VITAMIN A (% U.S. RDA)	VITAMIN C (% U.S. RDA)	IRON (% U.S. RDA)	CALCIUM (% U.S. RDA)	GLOOM
28	5	15	42	7	50	661	7	5	15	11	24
29	5	23	53	8	56	652	10	9	15	11	33
56	5	40	53	8	82	1417	3	0	19	8	61
14	0	13	50	3	46	541	0	0	4	0	21
32	32	0	0	0	0	—	0	0	0	0	5
86	86	0	0	0	0	—	0	0	0	0	13
57	57	0	0	0	0	—	0	0	0	0	9
41	41	0	0	0	0	—	0	0	0	0	6
1	0	10	92	5	28	86	7	0	0	2	14
19	5	24	60	8	227	719	10	0	10	14	45
19	5	20	57	7	222	607	10	0	10	14	39
19	5	21	55	7	241	962	10	0	11	15	43
22	6	40	67	13	268	985	10	0	16	15	69
18	5	10	50	2	4	285	0	0	6	3	16
5	0	1	29	—	0	90	0	0	0	0	2
0	0	0	0	0	0	—	0	0	0	0	0
9	8	0	0	0	0	397	3	4	0	0	4
23	23	0	0	0	0	12	0	0	0	0	4
2	1	18	95	3	0	208	0	0	0	0	27
11	9	0	0	0	0	52	0	0	0	0	2
29	5	27	50	13	100	851	11	5	21	18	40
43	0	20	48	5	0	238	0	5	7	0	30
53	2	32	54	5	80	537	0	0	16	8	48
20	9	3	23	2	10	40	2	0	0	8	6

BURGER KING

COMPLETE NUTRITIONAL VALUES[1] CONTINUED	WEIGHT[2] (gm.)	CALORIES	PROTEIN (gm.)
✔ Frozen Yogurt, Breyers Chocolate	65	130	3
✔ Hamburger	108	272	15
Hamburger Deluxe	138	344	15
Mayonnaise, 2 tbsp.	28	194	0
✔ Milk, 2% Low-fat, 8 fl. oz.	244	121	8
Milk, Whole, 8 fl. oz.	244	157	8
Mini Muffins, Blueberry	95	292	4
Mustard	3	2	0
Ocean Catch Fish Filet	194	495	20
Onion Rings	97	339	5
✔ Orange Juice, 6 fl. oz.	183	82	1
Salad Dressing, Bleu Cheese, 2 fl. oz.	59	300	3
Salad Dressing, French, 2 fl. oz.	64	290	0
Salad Dressing, Olive Oil and Vinegar, 2 fl. oz.	56	310	0
Salad Dressing, Ranch, 2 fl. oz.	57	350	1
✔ Salad Dressing, Reduced Calorie Light Italian, 2 fl. oz.	59	170	0
Salad Dressing, Thousand Island, 2 fl. oz.	63	290	1
Salad, Chef	273	178	17
✔ Salad, Chunky Chicken	258	142	20
Salad, Garden	223	95	6
✔ Salad, Side	135	25	1
Sausage Breakfast Buddy (test product)	84	255	11
Shake, Chocolate, large	410	472	13
Shake, Chocolate (with syrup), large	438	598	14

CARBOHYDRATES (gm.)	ADDED SUGAR[3] (gm.)	FAT[4] (gm.)	FAT % CALORIES	SATURATED FAT (gm.)	CHOLESTEROL (mg.)	SODIUM (mg.)	VITAMIN A (% U.S. RDA)	VITAMIN C (% U.S. RDA)	IRON (% U.S. RDA)	CALCIUM (% U.S. RDA)	GLOOM
21	9	3	21	2	10	40	2	0	0	8	6
28	5	11	36	4	37	505	3	5	15	4	18
28	5	19	50	6	43	496	5	9	15	4	27
2	0	21	97	4	16	142	0	0	0	0	32
12	0	5	37	3	18	122	10	4	0	30	6
11	0	9	52	6	35	119	7	6	0	29	11
37	18	14	43	3	72	244	0	0	7	4	26
0	0	0	0	0	0	34	0	0	0	0	0
49	5	25	45	4	57	879	0	4	14	6	37
38	0	19	50	5	0	628	15	0	3	11	28
20	0	0	0	0	0	2	3	119	0	0	0
2	0	32	96	7	58	512	0	0	0	0	53
23	22	22	68	3	0	400	31	0	0	0	31
2	0	33	96	5	0	214	0	0	0	0	53
4	0	37	95	7	20	316	0	0	0	0	56
3	3	18	95	3	0	762	0	0	0	0	34
15	0	26	81	5	36	403	64	0	0	0	29
7	0	9	46	4	103	568	95	25	9	16	13
8	0	4	25	1	49	443	92	34	7	4	6
8	0	5	47	3	15	125	100	58	6	15	4
5	0	0	0	0	0	27	88	20	3	3	0
15	0	16	56	6	127	492	5	0	10	8	28
71	49	15	29	9	45	286	10	6	6	45	26
103	80	14	21	9	44	357	0	0	2	42	32

BURGER KING

COMPLETE NUTRITIONAL VALUES[1] CONTINUED

	WEIGHT[2] (gm.)	CALORIES	PROTEIN (gm.)
Shake, Strawberry (with syrup), large	438	569	12
Shake, Vanilla, regular	293	345	9
Shake, Vanilla, large	410	483	13
Snickers Ice Cream Bar	57	220	5
Sprite, medium (18 fl. oz.)	540	216	0
Tartar Sauce	28	134	0
Tater Tenders	71	213	2
Whopper	270	614	27
Whopper with Cheese	294	706	32
Whopper, Double	351	844	46
Whopper, Double, with Cheese	375	935	51
Burger King Meals			
Croissant (plain), Orange Juice, 2% Low-fat Milk	468	383	13
French Toast Sticks, Burger King A.M. Express Dip, 2% Low-fat Milk, Orange Juice	596	825	19
Croissan'wich with Sausage, Egg and Cheese, 2% Low-fat Milk	403	655	29
✔ BK Broiler Chicken Sandwich, Side Salad (no dressing), Orange Juice	472	374	24
Chicken Tenders, 6 Pieces, BBQ Dipping Sauce, Side Salad with Reduced Calorie Light Italian Dressing	312	467	17
Chunky Chicken Salad with Thousand Island Dressing, Onion Rings, Diet Coke, medium	1078	772	26

CARBOHYDRATES (gm.)	ADDED SUGAR[3] (gm.)	FAT[4] (gm.)	FAT % CALORIES	SATURATED FAT (gm.)	CHOLESTEROL (mg.)	SODIUM (mg.)	VITAMIN A (% U.S. RDA)	VITAMIN C (% U.S. RDA)	IRON (% U.S. RDA)	CALCIUM (% U.S. RDA)	GLOOM
99	66	14	22	8	44	319	0	2	0	43	30
53	35	11	29	6	34	220	0	0	0	32	20
74	49	15	28	9	47	308	0	0	0	44	29
20	—	14	57	7	15	65	2	0	2	6	21
54	54	0	0	0	0	—	0	0	0	0	8
2	0	14	94	2	20	202	0	0	0	0	23
25	0	12	51	3	0	318	12	9	2	0	17
45	5	36	53	12	90	865	11	20	27	8	49
47	5	44	56	16	115	1177	19	20	27	22	61
45	5	53	57	19	169	933	11	20	40	9	72
47	5	61	59	24	194	1245	19	20	40	24	83
50	5	15	35	4	22	409	13	123	6	33	17
106	2	37	40	8	98	679	13	123	16	38	45
34	6	45	62	16	286	1107	20	4	16	45	71
53	5	8	19	2	45	757	95	145	17	8	12
31	11	31	60	6	46	1727	91	24	7	3	42
61	0	49	57	11	85	1474	171	34	10	15	57

BURGER KING

COMPLETE NUTRITIONAL VALUES[1] CONTINUED	WEIGHT[2] (gm.)	CALORIES	PROTEIN (gm.)
Whopper, Side Salad with French Dressing, Orange Juice	652	1011	29
Garden Salad with Reduced Calorie Light Italian Dressing, Chicken Tenders, 6 pieces, BBQ Dipping Sauce, Vanilla Shake, large	810	1020	35
Ocean Catch Fish Fillet, Side Salad with Bleu Cheese Dressing, Fries, medium, Coke, medium	1044	1408	29
Double Whopper with Cheese, French Fries, medium, Strawberry Shake, large, Apple Pie	1054	2187	71

1. A dash means that data not available.
2. To convert grams to ounces (weight), divide by 28.35; to convert grams to fluid ounces (volume), divide by 29.6.

CARL'S JR.

Based in Anaheim, California, Carl's Jr. claims to have introduced salad bars to fast-food restaurants back in 1977. This chain's 572 outlets in California, Arizona, Nevada, Oregon, and Japan serve several foods with low Gloom ratings, in addition to typical unhealthful fast food. The restaurant offers modified table service—servers bring food to the table after patrons place their order at the counter. And in 1991, 426 of the company-owned outlets adopted a no-smoking policy for a fresher-smelling environment.

For breakfast, Carl's Jr. offers a range of choices. At the low-calorie end, an English muffin with margarine provides 190 calories and 7 Gloom points. A bran muffin will cost you 310 calories, only one-fifth of them coming from fat. The hotcakes, too, are a reasonable choice, especially when you push most of the margarine to the side, but too bad they contain 950 milligrams of sodium. Wash any of these breakfast items down with low-fat (1 percent) milk or orange juice.

CARBOHYDRATES (gm.)	ADDED SUGAR³ (gm.)	FAT⁴ (gm.)	FAT % CALORIES	SATURATED FAT (gm.)	CHOLESTEROL (mg.)	SODIUM (mg.)	VITAMIN A (% U.S. RDA)	VITAMIN C (% U.S. RDA)	IRON (% U.S. RDA)	CALCIUM (% U.S. RDA)	GLOOM
93	27	58	52	15	90	1294	133	159	30	11	66
108	60	51	45	18	108	2133	103	62	10	59	74
156	62	77	49	16	115	1656	88	29	24	9	113
233	87	109	45	41	242	2214	19	35	54	67	169

3. To convert grams of sugar to teaspoons of sugar, divide by 4.0.
4. To convert grams of fat to teaspoons of fat, divide by 4.4.

Steer clear of the French Toast Dips; even without syrup, they'll give you 490 calories, almost half from fat, and 41 Gloom points. The Hash Brown Nuggets offer 270 calories and a salty, greasy taste you probably won't want twice.

If you want eggs, the plain Sunrise Sandwich is the best choice (300 calories; Gloom = 25). The Sunrise Sandwiches with meat are high in sodium (700 milligrams for the bacon version, 1,070 milligrams for the sausage version). The sausage version's 7 teaspoons of fat contribute to a gory Gloom rating of 51. The Breakfast Burrito, which contains scrambled eggs, bacon, and Cheddar cheese, not surprisingly, provides an entire day's worth of cholesterol.

The best breakfasts might be either a bran muffin and orange juice (400 calories, Gloom = 12) or, if you're hungrier, hotcakes with some margarine, orange juice, and low-fat milk (738 calories, Gloom = 35). On the other hand, your diet may go down the tubes if you have a Sunrise Sandwich with Sausage, Hash Brown Nuggets, and low-fat milk, which gives you 898 calories and 82 Gloom points.

In case you've decided to pass up Carl's Jr. for breakfast, how about lunch? "Why buy chicken for the health aspects and dip it in fat?" said one spokesperson for Carl's Jr. So the chain "charbroils" the chicken for its tasty Charbroiler BBQ Chicken Sandwich and serves it with lots of lettuce, tomato, and barbecue sauce on a honey wheat bun (not to be confused with whole wheat). It has a hefty amount of sodium—680 milligrams—but gets only 17 percent of its 310 calories from fat, according to the company. The Charbroiler Chicken Club Sandwich has 570 calories, four times as much fat, and one-third more sodium than the BBQ version. On the Gloom scale, the Charbroiler Chicken Club Sandwich rings up 42 points, compared to just 12 points for the Charbroiler BBQ. The Santa Fe Chicken Sandwich provides about twice the fat, sodium, and Gloom of the Charbroiler BBQ Chicken Sandwich.

For lunch or dinner, healthier offerings include prepackaged salads with reduced-calorie dressing. The Charbroiler Chicken Salad and the garden salad have fewer calories and much less fat than any hamburger offering. Carl's salad bars offer strawberries, melon, or other fresh fruit seasonally, a real plus.

Big baked potatoes are another Carl's Jr. offering. Depending upon whether you get one plain or loaded, you'll have a healthy or fatty meal. The Lite (plain) Potato has a terrific Gloom rating of 1. The Bacon & Cheese Potato, which weighs almost a pound, has 730 calories, 10 teaspoons of fat, and a Gloom rating of 57, that's worse than many double cheeseburgers.

The health-club special at Carl's Jr. would be a garden salad with reduced-calorie dressing and a Lite Potato. That meal provides just 420 calories, 2 teaspoons of fat, and 11 Gloom points.

The roast beef itself at Carl's Jr. starts out 95 percent lean and natural. But the roast-beef sandwiches, each made with 2½ ounces of beef, are oozing with calories and fat from both the beef and the toppings. The Roast Beef Deluxe Sandwich offers 540 calories and 6 teaspoons of fat (Gloom = 38). Adding onion rings and a small shake would bring you up to over 1,300 calories, more than a teaspoon of salt, and 89 Gloom points.

A burger's a burger the world around, and there's nothing wildly different about the burger situation at Carl's Jr. The best burger deal nutritionally is the Happy Star Hamburger, which is also the center of the children's package. This medium-size burger, served with mustard, ketchup, and pickles, provides 320 calories and 590 milligrams of sodium, and gets about 40 percent of its calories from fat.

The bigger the burger, of course, the gloomier the picture. The Super Star Hamburger contains 820 calories, nearly 60 percent of them from the 53 grams of fat (12 teaspoons); its Gloom rating is 60. Even worse is the double western bacon cheeseburger. Its 1,030 calories — the highest of all fast-food sandwiches — and 63 grams (14 teaspoons) of fat give it a shockingly high Gloom rating of 91. Obviously, you can cut the calories, fat, and sodium by getting all these burgers without the bacon or cheese or mayonnaise.

As an example of how awful fast-food meals can be, consider that double western bacon cheeseburger eaten with a large fries, large shake, and a fudge brownie: The 2,465 calories come with 27 teaspoons of fat, 2,580 milligrams of sodium . . . and 176 Gloom points. That's as much, or more, calories, fat, and sodium as one should consume in an entire day.

A better, but still disastrous, meal would be a Famous Star Hamburger, zucchini, cookie, and a carbonated beverage, which delivers 1,570 calories and 124 Gloom points. A more normal burger meal would consist of a Happy Star Hamburger, regular order of fries, and soda that provides under a thousand calories (as if that's not a huge amount!), 8 teaspoons of fat, and 61 Gloom points.

Carl's Jr. offers the standard breaded and fried Carl's Catch Fish Sandwich. The fish patty contains about as much breading as fish, and may slide off the bun because of all the mayonnaise loaded on top and bottom. The sandwich, which contains cheese, lettuce, and tomato, provides 7 teaspoons of fat and a Gloom rating of 47. Hold the mayo!

As fast-food joints are wont to do, Carl's Jr. fries more than it needs to. For instance, burritos are customarily not fried, but Jr. Crisp Burritos, which are filled with a little beans and cheese, are deep-fat fried. A friendly clerk microwaved the burritos for us for thirty seconds, and everyone agreed that they were much less greasy and tasted fine.

A regular shake provides 350 calories. If you choose to have a large shake, make sure you have someone to share it with. The 28-ounce version provides 459 calories and is loaded with sugar — we estimate about 13 teaspoons.

Carl's Jr. advertises the calorie content of its more healthful items right on the menu board. It has also replaced beef fat with vegetable shortening for frying all its food. The company even ran full-page, color magazine ads touting that fact when many of its competitors were frying in beef fat. It is unfortunate, however, that Carl's Jr. feels compelled to bread and fry zucchini, turning a fat-free green vegetable into yet another greasy, 390-calorie dish with a Gloom rating of 38. At least the breading is easy to peel off.

Carl's Jr. outlets offer nutrition brochures, which you should ask

CARL'S JR.

COMPLETE NUTRITIONAL VALUES

	WEIGHT[1] (gm.)	CALORIES	PROTEIN (gm.)
Bacon, 2 strips	8	45	3
Baked Potato, Bacon & Cheese	424	730	26
Baked Potato, Broccoli & Cheese	427	590	18
Baked Potato, Cheese	415	690	23
Baked Potato, Fiesta	456	720	31
Baked Potato, Lite	286	290	9
Baked Potato, Sour Cream & Chive	353	470	11
Breakfast Burrito	151	430	22
Carbonated Beverage, regular (20 fl. oz.)	600	240	0
Carbonated Beverage, diet, regular (20 fl. oz.)	600	2	0
Carl's Catch Fish Sandwich	212	560	17
✔ Charbroiler BBQ Chicken Sandwich	192	310	25
Charbroiler Chicken Club Sandwich	250	570	35
Chocolate Chip Cookie, 2½ oz.	71	330	4
Cinnamon Rolls	114	460	7
Country Fried Steak Sandwich	228	720	20

for. Also, you can glance through a booklet of ingredient listings, but you'll have to write headquarters for your own copy. Yellow No. 5 dye, to which some people are allergic, is present in the cheese sauce, chili pepper, chocolate shake, cinnamon roll, dill pickle, fruit danish, and sweet relish. And you'll be interested to know that as much as 10 percent of the roast beef comes from an injection of water, salt, sodium tripolyphosphate, hydrolyzed vegetable protein (contains MSG), and spices. The au jus flavoring contains MSG and a dozen other flavorings and flavor enhancers. Up to 16 percent of the Charbroiler BBQ Chicken breast consists of a flavoring solution.

CARBOHYDRATES (gm.)	ADDED SUGAR[2] (gm.)	FAT[3] (gm.)	FAT % CALORIES	SATURATED FAT (gm.)	CHOLESTEROL (mg.)	SODIUM (mg.)	VITAMIN A (% U.S. RDA)	VITAMIN C (% U.S. RDA)	IRON (% U.S. RDA)	CALCIUM (% U.S. RDA)	GLOOM
0	0	4	80	1	5	150	0	6	0	0	5
60	0	43	53	15	45	1670	15	85	20	20	57
60	0	31	47	11	25	830	20	90	30	30	35
70	0	36	47	15	40	1160	30	75	10	45	45
64	0	38	48	14	15	1470	45	100	30	30	45
60	0	1	3	0	0	60	0	50	15	4	1
64	0	19	36	7	20	180	30	2	15	10	24
29	0	26	54	12	285	740	30	0	25	35	44
62	62	0	0	0	0	35	0	0	0	0	10
0	0	0	0	0	0	15	0	0	0	0	0
54	5	30	48	4	5	1220	0	0	20	20	47
34	5	6	17	2	30	680	30	0	10	10	12
42	5	29	46	8	60	1160	4	2	20	35	42
41	18	17	46	7	5	170	4	0	10	4	27
70	18	18	35	1	0	230	0	0	10	0	26
61	5	43	54	11	50	1420	5	2	20	20	66

CARL'S JR.

COMPLETE NUTRITIONAL VALUES CONTINUED	WEIGHT (gm.)	CALORIES	PROTEIN (gm.)
Danish (varieties)	114	520	7
Dog, All Star Chili	255	720	24
Dog, All Star Hot	198	540	16
Double Western Bacon Cheeseburger	329	1030	56
✔ English Muffin with margarine	62	190	6
French Fries, Criss-Cut	92	330	4
French Fries, large	168	546	5
French Fries, regular	126	420	4
French Fries, small	84	281	3
French Toast Dips, no syrup	124	490	8
Fudge Brownie	92	430	6
Fudge Brownie Mousse Cake	113	400	5
Guacamole	28	50	1
Hamburger, Carl's Original	194	460	25
Hamburger, Famous Star	244	610	26
Hamburger, Happy Star	122	320	17
Hamburger, Super Star	319	820	43
Hash Brown Nuggets	95	270	3
Hot Cakes with margarine (no syrup)	188	510	11
Jr. Crisp Burritos, 3	129	420	15
✔ Milk, 1% Low-fat, 10 fl. oz.	311	138	11
Muffin, Blueberry	120	340	5
✔ Muffin, Bran	135	310	6
Onion Rings	151	520	9
✔ Orange Juice, small (8 fl. oz.)	249	90	2

CARBOHYDRATES (gm.)	ADDED SUGAR[2] (gm.)	FAT[3] (gm.)	FAT % CALORIES	SATURATED FAT (gm.)	CHOLESTEROL (mg.)	SODIUM (mg.)	VITAMIN A (% U.S. RDA)	VITAMIN C (% U.S. RDA)	IRON (% U.S. RDA)	CALCIUM (% U.S. RDA)	GLOOM
73	20	16	28	1	0	230	0	0	10	0	24
51	4	47	59	19	15	1530	0	2	20	12	74
41	4	35	58	13	5	1130	0	0	15	10	56
58	5	63	55	32	145	1810	10	10	35	50	91
30	5	5	24	1	0	280	20	0	8	10	7
27	0	22	60	3	0	890	0	0	10	0	39
70	0	26	43	7	0	260	0	13	8	0	40
54	0	20	43	5	0	200	0	10	6	0	29
36	0	13	42	3	0	134	0	7	4	0	19
55	0	26	48	6	40	620	10	0	15	6	41
64	35	19	40	5	0	210	0	0	10	0	30
42	23	23	52	11	110	85	0	0	8	8	42
5	0	4	72	1	0	100	2	0	0	4	5
46	5	20	39	9	50	810	6	6	10	20	31
42	5	38	56	13	50	890	10	6	20	30	51
33	5	14	39	5	35	590	2	2	15	15	21
41	5	53	58	24	105	1210	15	10	35	35	60
27	0	17	57	4	5	410	2	0	4	2	28
61	0	24	42	5	10	950	30	0	10	25	33
39	0	21	45	9	195	600	18	0	18	12	38
15	0	2	30	2	12	160	12	5	0	35	3
61	20	9	24	1	45	300	4	0	20	4	17
52	11	7	20	0	60	370	0	0	10	0	17
63	0	26	45	6	0	960	0	4	20	4	43
21	0	1	10	0	0	2	4	120	2	2	1

CARL'S JR.

COMPLETE NUTRITIONAL VALUES CONTINUED	WEIGHT[1] (gm.)	CALORIES	PROTEIN (gm.)
Raspberry Cheesecake	99	310	7
Roast Beef Club Sandwich	271	620	30
Roast Beef Deluxe Sandwich	265	540	28
Salad Dressing, Blue Cheese, 2 oz. packet	56	300	2
Salad Dressing, House, 2 oz. packet	56	220	2
Salad Dressing, Italian, 2 oz. packet	56	240	0
✔ Salad Dressing, Reduced Calorie French, 2 oz. packet	56	80	0
Salad Dressing, Thousand Island, 2 oz. packet	56	220	0
✔ Salad, Charbroiler Chicken	342	200	24
✔ Salad, Garden	137	50	3
Salsa	28	8	0
✔ Santa Fe Chicken Sandwich	220	540	30
Sausage, 1 patty	45	190	8
Shake, large	429	459	14
Shake, regular	330	350	11
Shake, small	251	268	8
Sunrise Sandwich	116	300	15
Sunrise Sandwich with Bacon	124	345	18
Sunrise Sandwich with Sausage	161	490	23
Taco Sauce	28	8	0
Western Bacon Cheeseburger	232	730	34
Zucchini (breaded, fried)	168	390	7
Carl's Jr. Meals			
✔ Bran Muffin, Orange Juice, small	384	400	8

CARBOHYDRATES (gm.)	ADDED SUGAR[2] (gm.)	FAT[3] (gm.)	FAT % CALORIES	SATURATED FAT (gm.)	CHOLESTEROL (mg.)	SODIUM (mg.)	VITAMIN A (% U.S. RDA)	VITAMIN C (% U.S. RDA)	IRON (% U.S. RDA)	CALCIUM (% U.S. RDA)	GLOOM
32	27	17	49	8	60	200	10	0	4	15	29
48	5	34	49	11	45	1950	15	4	25	35	51
46	5	26	43	10	40	1340	15	10	20	40	38
0	0	30	90	6	40	500	0	0	0	0	52
4	0	22	90	6	20	340	0	0	0	4	36
2	4	26	98	4	0	420	0	0	0	0	42
10	0	4	45	0	0	580	0	0	0	0	12
8	0	22	90	6	10	400	4	0	4	0	35
8	0	8	36	4	70	300	150	8	10	15	15
4	0	3	54	2	5	75	80	2	2	8	2
2	0	0	0	0	0	210	50	0	0	0	1
75	5	13	22	3	40	1180	6	10	20	25	24
0	1	18	85	5	30	520	0	0	6	0	28
79	51	9	18	5	19	300	0	0	13	59	18
61	39	7	18	4	15	230	0	0	10	45	14
46	30	5	17	3	11	175	0	0	8	35	10
31	5	13	39	6	160	550	20	0	10	15	25
31	5	17	44	7	165	700	20	6	10	15	30
31	6	31	57	11	190	1070	20	0	16	15	51
2	0	0	0	0	0	160	0	0	0	0	2
59	5	39	48	20	90	1490	8	6	30	40	59
38	3	23	53	6	0	1040	20	0	10	4	38
73	11	8	18	0	60	372	4	120	12	2	12

CARL'S JR.

COMPLETE NUTRITIONAL VALUES CONTINUED	WEIGHT[1] (gm.)	CALORIES	PROTEIN (gm.)
✔ Hot Cakes with Margarine (no syrup), 1% Low-fat Milk, Orange Juice, small	748	738	24
French Toast Dips (no syrup), Bacon, 2 strips, Orange Juice, small	381	625	13
Sunrise Sandwich with Sausage, Hash Brown Nuggets, 1% Low-fat Milk	567	898	37
✔ Lite Potato, Garden Salad with Reduced Calorie French Dressing, 2 oz. packet	479	420	12
Charbroiler BBQ Chicken Sandwich, Garden Salad, Vanilla Shake, small	580	628	36
Happy Star Hamburger, French Fries, regular, Soda, regular	848	980	21
Santa Fe Chicken Sandwich, Jr. Crisp Burritos, 3, with Salsa, Soda, regular	977	1208	45
Roast Beef Deluxe Sandwich, Onion Rings, Shake, small	667	1328	45
Famous Star Hamburger, Zucchini, Chocolate Chip Cookie, Soda, regular	1083	1570	37
All-Star Chili Dog, Criss-Cut Fries, Soda, regular, Raspberry Cheesecake	1046	1600	35
Double Western Bacon Cheeseburger, French Fries, large, Shake, large, Fudge Brownie	1018	2465	81

1. To convert grams to ounces (weight), divide by 28.35; to convert grams to fluid ounces (volume), divide by 29.6.

CARBOHYDRATES (gm.)	ADDED SUGAR[2] (gm.)	FAT[3] (gm.)	FAT % CALORIES	SATURATED FAT (gm.)	CHOLESTEROL (mg.)	SODIUM (mg.)	VITAMIN A (% U.S. RDA)	VITAMIN C (% U.S. RDA)	IRON (% U.S. RDA)	CALCIUM (% U.S. RDA)	GLOOM
97	0	27	33	7	22	1112	46	125	12	62	32
76	0	31	45	7	45	772	14	126	17	8	39
73	6	50	50	17	207	1640	34	5	20	52	78
74	0	8	17	2	5	715	80	52	17	12	11
84	35	14	20	7	46	930	110	2	20	53	24
149	67	34	31	10	35	825	2	12	21	15	61
178	67	34	25	12	235	2025	74	10	38	37	70
155	35	57	39	19	51	2475	15	14	48	79	89
183	88	78	45	26	55	2135	34	6	40	38	124
172	93	86	48	30	75	2655	10	2	34	27	154
271	91	117	43	49	164	2580	10	23	66	109	176

2. *To convert grams of sugar to teaspoons of sugar, divide by 4.0.*
3. *To convert grams of fat to teaspoons of fat, divide by 4.4.*

CHURCH'S FRIED CHICKEN

Church's has been number three in fast-food-chicken sales, offering fried chicken and 15-cent jalapeño peppers—the hottest morsel in any fast-food restaurant. Church's was popular in the early 1980s, but shrunk by one-third in the late eighties, enabling fast-growing Popeyes to buy it out.

Church's has not updated its nutrition information for over five years. We've presented the old data on the assumption that any

CHURCH'S FRIED CHICKEN* COMPLETE NUTRITIONAL VALUES[1]	WEIGHT[2] (gm.)	CALORIES	PROTEIN (gm.)
Apple Pie, 3 oz.	85	300	2
Coleslaw	85	83	1
✔ Corn on the Cob, buttered, 9 oz.	256	165	5
Foot-Long Chicken Sandwich	- no data -		
French Fries, regular, 3 oz.	85	256	4
Fried Chicken Breast**	93	278	21
Fried Chicken Drumstick	56	147	13
Fried Chicken Thigh	93	305	19
Fried Chicken Wing	97	303	22
Jalapeño Pepper	17	4	0
Okra, fried	- no data -		

* Church's has not updated this nutrition information in over five years. (Therefore, they didn't provide information on their newer products such as the Foot-Long Chicken Sandwich.)
** Chicken breast is an excellent choice, if the breading and batter is removed.
1. A dash means that data not available.

changes have been minor. Church's provides too little information to calculate Glooms. The company never did disclose ingredient information.

"Best choices" are few and far between. The chicken, fish, pies, okra—practically everything—is deep-fat fried in beef-fat shortening and gets about 60 percent of its calories from fat.

Based on limited sampling, we'd say that the chicken is salty and greasy, the mashed potatoes largely tasteless, the biscuits weak, the fries ordinary, the coleslaw bland. The servers seemed basically

CARBOHYDRATES (gm.)	ADDED SUGAR[3] (mg.)	FAT[4] (gm.)	FAT % CALORIES	SATURATED FAT (gm.)	CHOLESTEROL (mg.)	SODIUM (mg.)	VITAMIN A (% U.S. RDA)	VITAMIN C (% U.S. RDA)	IRON (% U.S. RDA)	CALCIUM (% U.S. RDA)	GLOOM[5]
31	14	19	57	—	—	—	—	—	—	—	—
6	0	7	76	—	—	—	—	—	—	—	—
29	0	3	16	—	—	—	—	—	—	—	—
					- no data -						
31	0	13	46	—	—	—	—	—	—	—	—
9	0	17	55	—	—	—	—	—	—	—	—
5	0	9	55	—	—	—	—	—	—	—	—
9	0	22	65	—	—	—	—	—	—	—	—
9	0	20	59	—	—	—	—	—	—	—	—
1	0	0	0	0	0	—	—	—	—	—	—
					- no data -						

2. To convert grams to ounces (weight), divide by 28.35; to convert grams to fluid ounces (volume), divide by 29.6.
3. To convert grams of sugar to teaspoons of sugar, divide by 4.0.
4. To convert grams of fat to teaspoons of fat, divide by 4.4.
5. Too little information is available to calculate Gloom ratings.

disinterested in serving, and the molded plastic benches were particularly uncomfortable. If we were stuck eating at Church's, we would have chicken breast and drumstick (removing the skin and breading), corn on the cob (without the imitation butter), coleslaw, and a large glass of water to help deal with a hot jalapeño pepper. That wouldn't be too bad a meal. It would provide about 500 calories.

Church's new Foot-Long Chicken Sandwich includes two salty, breaded, fried patties, a double slather of mayonnaise, and some shredded lettuce and pickle slices on a floppy white-bread bun. We estimate that it provides 800 calories. Church's, like other chains, is developing a nonfried chicken product, which may be available at some outlets.

DAIRY QUEEN/BRAZIER

In the old days, going all the way back to 1940, Dairy Queen was famous for its soft-serve ice milk and not as a fast-food restaurant. Now, with quarter-pound hot dogs and double burgers contributing to its $2 billion in annual sales, Dairy Queen is obviously more than an ice-milk store — or 5,200 ice-milk stores.

Many Dairy Queen outlets, called Dairy Queen/Brazier, now sell typical fast foods, including fried fish, hamburgers, and french fries. All of these, combined with the famous soft-serve, keep Dairy Queen in the top ten largest chains.

In 1990 Dairy Queen proved that it had heard of the word "health" when some outlets began offering nonfat, cholesterol-free frozen yogurt products. A regular size cup of yogurt has 170 calories, though optional candy-like toppings can really boost the calories.

Frozen yogurt aside, Dairy Queen offers a wide array of not-so-healthy desserts, such as the slush-like Mr. Misty and the 710-calorie Peanut Buster Parfait. For example, a small Mr. Misty has 12 teaspoons of sugar. We could go on and on about Dairy Queen sugar-filled cones and cakes, Blizzards and bars, and other sweet, gooey products, but you never thought they were healthy.

Compared to those desserts, Dairy Queen's basic soft vanilla ice milk is real health food. Fat provides only one-quarter of the calories. The small cone provides a reasonable 140 calories — the same as the

DQ (ice milk) Sandwich—while the large one is worth 340 calories. Sugar is still present in ample quantity—an estimated 2 teaspoons in the small and 6 in the large cone. That leads to a Gloom rating of 7 for the small cone and 17 for the large. Dipping the cone in chocolate coating may be heavenly to chocoholics, but it adds 100 calories and 2 teaspoons of fat to a regular-size chocolate cone.

Eating healthfully at Dairy Queen restaurants got a lot easier when the Grilled Chicken Fillet Sandwich, made with breast fillet, was added to the menu. This 300-calorie sandwich comes with lettuce, tomato, and reduced-calorie mayonnaise. Just 24 percent of its calories come from fat; it has a Gloom rating of 16. The breaded Chicken Fillet Sandwich, is fried and has the expected high fat content and Gloom rating (32). The fried Fish Fillet Sandwich, with a teaspoon less fat, is a bit better. The most nutritious sandwich, but not at all restaurants, is the BBQ Beef Sandwich. This little sandwich weighs less than a regular McDonald's hamburger. Its vital statistics are 225 calories, 1 teaspoon of fat, and 10 Gloom points.

If you crave a burger, Dairy Queen will give you a 310-calorie single (Gloom = 22) or 460-calorie double (Gloom = 37). Adding cheese adds calories, so the double hamburger with cheese tips the scales at 570 calories and 50 Gloom points. We were ready to praise Dairy Queen for dropping its triple hamburger and triple cheeseburger, but then we tripped over its DQ Homestyle Ultimate Burger, whose 700 calories, 11 teaspoons of fat, and 63 Gloom points make it one of the unhealthiest items in any restaurant.

Dairy Queen has the distinction of doing more with hot dogs than any other major chain. The regular hot dog sandwich provides 4 teaspoons of fat, a moderate 280 calories, and 28 Gloom points. Adding cheese raises the Gloom to 35 points. The going gets rougher, though, with the Quarter Pound Super Dog. The quarter pounder is loaded with sodium and is helped down your throat by 9 teaspoons of fat. Those numbers add up to an artery-quivering Gloom rating of 64.

The beverages and desserts can really boost a meal's calorie content. The "regular" chocolate shake has 540 calories—equivalent to more than three cans of Coca-Cola. A large vanilla shake offers your waistline 600 calories.

Malts are the same as shakes, but with added malt flavoring and the extra calories. The regular vanilla malt provides 610 calories. The only silver lining in these dietbusters is that they provide 40 percent of one's daily quota for calcium.

With individual foods like the above, it's not hard to construct ridiculously fattening meals at Dairy Queen. A DQ Homestyle Ultimate Burger, large shake, and large order of french fries add up to 1,690 calories, 18 teaspoons of fat, and a Gloom rating of 119. By contrast, a single hamburger with lettuce and tomato, regular fries, and regular cone for dessert provide a more manageable 845 calories, 8 teaspoons of fat, and 53 Gloom points.

For a healthy meal at Dairy Queen, we suggest you have the Grilled Chicken Sandwich, make yourself a salad of lettuce, tomato, and onions from the Fixins bar, buy a banana that they would ordinarily

DAIRY QUEEN/BRAZIER

COMPLETE NUTRITIONAL VALUES[1]

	WEIGHT[2] (gm.)	CALORIES	PROTEIN (gm.)
✔ BBQ Beef Sandwich	128	225	12
Banana Split	369	510	9
Buster Bar	149	450	11
Chicken Fillet Sandwich (breaded)	191	430	24
Chicken Fillet Sandwich with Cheese (breaded)	205	480	27
Cone, Chocolate, large	213	350	8
Cone, Chocolate, regular	142	230	6
Cone, Vanilla, large	213	340	9
Cone, Vanilla, regular	142	230	6
✔ Cone, Vanilla, small	85	140	4
DQ Homestyle Ultimate Burger	276	700	43
DQ Sandwich	61	140	3
Dilly Bar	85	210	3
Dipped Cone, Chocolate, large	234	525	9

use in a banana split (we paid 35 cents), and drink a glass of water. A Dairy Queen in Virginia was the only fast-food restaurant we've seen that had a water fountain and cup dispenser. (Miracles occur in the unlikeliest places.) Then, as long as you're at Dairy Queen, treat yourself to a small frozen yogurt.

You can get nutrition information about certain Dairy Queen products by writing to the company; it does not provide ingredient listings for any products other than the ice milk. According to wrappers, Yellow No. 5 dye is present in the banana and lemon-lime flavors of Starkiss, and possibly other items. Dairy Queen fries in vegetable shortening.

CARBOHYDRATES (gm.)	ADDED SUGAR[3] (gm.)	FAT[4] (gm.)	FAT % CALORIES	SATURATED FAT (gm.)	CHOLESTEROL (mg.)	SODIUM (mg.)	VITAMIN A (% U.S. RDA)	VITAMIN C (% U.S. RDA)	IRON (% U.S. RDA)	CALCIUM (% U.S. RDA)	GLOOM
34	5	4	16	1	20	700	8	0	20	10	10
93	—	11	19	8	30	250	0	0	20	30	21
40	—	29	58	9	15	220	2	0	8	30	38
37	5	20	42	4	55	760	0	0	10	4	32
38	5	25	47	7	70	980	8	0	10	10	39
54	24	11	28	8	30	170	10	0	6	25	18
36	16	7	27	5	20	115	6	0	4	15	12
53	23	10	26	7	30	140	6	0	6	20	17
36	16	7	27	5	20	95	4	0	4	15	12
22	10	4	26	3	15	60	2	0	2	10	7
30	5	47	60	21	140	1110	20	15	40	20	63
24	9	4	26	2	5	135	0	0	4	6	7
21	7	13	56	6	10	50	0	0	4	25	17
61	33	24	41	12	30	145	6	0	8	45	35

DAIRY QUEEN/BRAZIER

COMPLETE NUTRITIONAL VALUES[1] CONTINUED

	WEIGHT[2] (gm.)	CALORIES	PROTEIN (gm.)
Dipped Cone, Chocolate, regular	156	330	6
Dipped Cone, Chocolate, small	92	190	4
Fish Fillet Sandwich	170	370	16
Fish Fillet Sandwich with Cheese	184	420	19
French Fries, large	128	390	5
French Fries, regular	99	300	4
French Fries, small	71	210	3
✔ Grilled Chicken Fillet Sandwich	184	300	25
Hamburger, Double	198	460	31
Hamburger, Double with Cheese	226	570	37
Hamburger, Single	142	310	17
Hamburger, Single with Cheese	156	365	20
Heath Blizzard, regular	404	820	16
Heath Blizzard, small	291	560	11
Heath Breeze (yogurt), regular	379	680	15
Heath Breeze (yogurt), small	273	450	11
Hot Dog	99	280	9
Hot Dog with Cheese	113	330	12
Hot Dog with Chili	127	320	11
Hot Fudge Brownie Delight	305	710	11
Lettuce	14	2	0
Malt, Vanilla, regular	418	610	13
Mr. Misty Float	411	390	5
Mr. Misty Freeze	411	500	9

CARBOHYDRATES (gm.)	ADDED SUGAR[3] (gm.)	FAT[4] (gm.)	FAT % CALORIES	SATURATED FAT (gm.)	CHOLESTEROL (mg.)	SODIUM (mg.)	VITAMIN A (% U.S. RDA)	VITAMIN C (% U.S. RDA)	IRON (% U.S. RDA)	CALCIUM (% U.S. RDA)	GLOOM
40	22	16	44	8	20	100	4	0	6	30	23
25	13	10	47	5	10	60	2	0	4	20	14
39	5	16	39	3	45	630	0	0	10	4	26
43	5	21	45	6	60	850	8	0	10	10	34
52	—	18	42	4	0	200	0	15	8	0	24
40	—	14	42	3	0	160	0	10	6	0	19
29	—	10	43	2	0	115	0	8	4	0	13
33	5	8	24	2	50	800	2	4	20	6	16
29	5	25	49	12	95	630	0	0	30	4	37
31	5	34	54	18	120	1070	15	0	30	20	50
29	5	13	38	6	45	580	0	0	20	4	22
30	5	18	44	9	60	800	8	0	20	15	29
114	55	36	40	17	60	410	8	0	10	40	58
79	38	23	37	11	40	280	6	0	8	30	37
113	—	21	28	6	15	360	0	0	10	50	35
78	—	12	24	3	10	230	0	0	8	40	20
23	4	16	51	6	25	700	0	0	8	4	28
24	4	21	57	9	35	920	8	0	8	10	35
26	4	19	53	7	30	720	0	0	8	4	33
102	—	29	37	14	35	340	8	0	30	30	47
0	0	0	0	0	0	1	0	0	0	0	0
106	34	14	21	8	45	230	8	0	8	40	25
74	46	7	16	5	20	95	4	0	4	20	17
91	59	12	22	9	30	140	8	0	8	30	24

DAIRY QUEEN/BRAZIER

COMPLETE NUTRITIONAL VALUES[1] CONTINUED	WEIGHT[2] (gm.)	CALORIES	PROTEIN (gm.)
Mr. Misty, large	439	340	0
Mr. Misty, regular	330	250	0
Mr. Misty, small	248	190	0
Nutty Double Fudge	276	580	10
Onion Rings, regular	85	240	4
Peanut Buster Parfait	305	710	16
QC Chocolate Big Scoop	127	310	5
QC Vanilla Big Scoop	127	300	5
✔ Salad Dressing, Reduced Calorie French, 2 oz.	57	90	0
Salad Dressing, Thousand Island, 2 oz.	57	225	0
Salad, Garden	284	200	13
✔ Salad, Side	135	25	1
Shake, Chocolate, regular	397	540	12
Shake, Vanilla, large	461	600	13
Shake, Vanilla, regular	397	520	12
Strawberry Blizzard, regular	383	740	13
Strawberry Blizzard, small	266	500	9
Strawberry Breeze (yogurt), regular	354	590	12
Strawberry Breeze (yogurt), small	248	400	9
Strawberry Waffle Cone Sundae	173	350	8
Sundae, Chocolate, regular	177	300	6
Super Dog, Quarter Pound	198	590	20
Tomato	14	3	0
Yogurt Cone, large	213	260	9

CARBOHYDRATES (gm.)	ADDED SUGAR[3] (gm.)	FAT[4] (gm.)	FAT % CALORIES	SATURATED FAT (gm.)	CHOLESTEROL (mg.)	SODIUM (mg.)	VITAMIN A (% U.S. RDA)	VITAMIN C (% U.S. RDA)	IRON (% U.S. RDA)	CALCIUM (% U.S. RDA)	GLOOM
84	84	0	0	0	0	0	0	0	0	0	13
63	63	0	0	0	0	0	0	0	0	0	9
48	48	0	0	0	0	0	0	0	0	0	7
85	—	22	34	10	35	170	6	0	20	30	34
29	0	12	45	3	0	135	0	0	6	0	17
94	—	32	41	10	30	410	6	0	20	35	43
40	—	14	41	10	35	100	15	0	8	15	22
39	—	14	42	9	35	100	15	0	0	15	22
11	—	5	50	1	0	450	15	0	0	0	9
10	—	21	84	3	25	570	4	4	0	0	33
7	0	59	0	7	185	240	60	35	10	25	18
4	0	0	0	0	0	15	50	25	4	2	0
94	47	14	23	8	45	290	8	0	8	40	27
101	51	16	24	10	50	260	10	0	8	45	30
88	44	14	24	8	45	230	8	0	8	40	26
92	—	16	19	11	50	230	8	40	10	35	28
64	—	12	22	8	35	160	6	25	6	25	21
90	—	1	2	0	5	170	0	40	10	50	7
63	—	0	0	0	5	115	0	25	6	35	4
56	—	12	31	5	20	220	4	10	8	15	20
54	34	7	21	5	20	140	4	0	6	15	15
41	4	38	58	16	60	1360	0	0	15	10	64
1	0	0	0	0	0	0	2	4	0	0	0
56	35	0	0	0	0	115	0	0	6	35	4

DAIRY QUEEN/BRAZIER

COMPLETE NUTRITIONAL VALUES[1] CONTINUED	WEIGHT[2] (gm.)	CALORIES	PROTEIN (gm.)	
✔ Yogurt Cone, regular	142	180	6	
Yogurt Strawberry Sundae, regular	170	200	6	
Yogurt, Cup, large	198	230	8	
✔ Yogurt, Cup, regular	142	170	6	
Dairy Queen Meals				
Grilled Chicken Fillet Sandwich, Side Salad, Reduced Calorie French Dressing	376	415	26	
Single Hamburger with Lettuce and Tomato, French Fries, regular, Vanilla Cone, regular	411	845	27	
Fish Fillet Sandwich, Yogurt Cone, regular, Vanilla Shake, regular	709	1070	34	
Quarter Pound Super Dog, Yogurt Cone, regular, Vanilla Malt, regular	758	1380	39	
Double Hamburger with Cheese, French Fries, large, Vanilla Shake, large	815	1560	55	
DQ Homestyle Ultimate Burger, French Fries, large, Vanilla Shake, large	865	1690	61	

1. *A dash means that data not available.*
2. *To convert grams to ounces (weight), divide by 28.35; to convert grams to fluid ounces (volume), divide by 29.6.*

DOMINO'S PIZZA

Ann Arbor-based Domino's has been around for more than 30 years, but only during the 1980s did it become a national phenomenon. From 1980 to 1985, the company grew by 500 outlets each year. In early 1986 there were 2,800 of them, all claiming to

CARBOHYDRATES (gm.)	ADDED SUGAR[3] (gm.)	FAT[4] (gm.)	FAT % CALORIES	SATURATED FAT (gm.)	CHOLESTEROL (mg.)	SODIUM (mg.)	VITAMIN A (% U.S. RDA)	VITAMIN C (% U.S. RDA)	IRON (% U.S. RDA)	CALCIUM (% U.S. RDA)	GLOOM
38	24	0	0	0	0	80	0	0	4	20	3
43	–	0	0	0	0	80	0	20	4	25	2
49	30	0	0	0	0	100	0	0	6	30	4
35	22	0	0	0	0	70	0	0	4	25	3
48	5	13	28	3	50	1265	67	29	24	8	21
106	21	34	36	14	65	836	6	14	30	19	53
165	73	30	25	11	90	940	8	0	22	64	55
185	62	52	34	24	105	1670	8	0	27	70	90
184	56	68	39	32	170	1530	25	15	46	65	104
183	56	61	43	35	190	1570	30	30	56	65	119

3. To convert grams of sugar to teaspoons of sugar, divide by 4.0.
4. To convert grams of fat to teaspoons of fat, divide by 4.4.

deliver pizzas in 30 minutes or less, all the pizzas kept hot by the special Domino's "hot box" wrapping.

By 1990, more than 5,000 Domino's dotted the landscape, and the company began exploring new ways to get you to eat its pizzas. Domino's began testing actual restaurants, called Pizzazz!, partly to compete with Pizza Hut. Domino's, like Pizza Hut, has also begun delivering pizzas to school cafeterias as part of the school-lunch program. The great thing for the companies is that not only do they make money selling pizzas, they also familiarize kids with their brand of pizza and presumably boost orders outside of school.

Nutritionists—who sometimes feel as if they're always criticizing foods—love to note that pizza, though commonly considered junk food, isn't really junk at all. At its best, pizza is a balanced meal that teenagers actually eat. The crust provides plenty of complex carbohydrates (albeit refined flour), the tomato sauce and some of the toppings count for vegetables, and cheese adds calcium and protein. A typical commercial pizza, however, is hardly a cornucopia of vegetables. Despite the tomato sauce, two slices of Domino's cheese or pepperoni pizza provide only 7 percent of the U.S. Recommended Daily Allowance for vitamin A and even less for vitamin C. Still, two slices of Domino's pizza provide a moderate 16 to 36 Gloom points.

While some upscale pizza parlors may compete by loading their pizzas with tons of cheese and toppings, Domino's dishes up a pizza of more moderate proportions. As a result, the consumer gets a fairly nutritious meal—not as high in calories and fat as those cheese-heavy products. Two slices of a 16-inch cheese pizza contain 376 calories; 24 percent of them are from fat. Top the pizza with mushrooms, green peppers, and onions, and you are doing yourself a good turn. Top it with pepperoni, sausage, anchovies, or extra cheese, and you're adding unnecessary fat, sodium, and calories to the product. Two

DOMINO'S PIZZA

COMPLETE NUTRITIONAL VALUES[*1]	WEIGHT[*2] (gm.)	CALORIES	PROTEIN (gm.)
Cheese Pizza, 2 slices, 16 in. (thin-crust)	—	376	22
Deluxe Pizza, 2 slices, 16 in. (thin-crust)	—	498	27
Double Cheese/Pepperoni, 2 slices, 16 in. (thin-crust)	—	545	32
Ham, 2 slices, 16 in. (thin-crust)	—	417	23

slices of a 16-inch cheese pizza contain under 500 milligrams of sodium; pepperoni alone boosts the sodium by 340 milligrams. Add sausage and olives and anchovies and you'll be adding more sodium than you want to think about. A healthier modification would be to have pizza without any cheese, which Domino's will do upon request. The tomato-sauce–topped pizza—with mushrooms and green pepper—will be extra low in fat and sodium, and you'll finish dinner without that usual stuffed feeling.

In 1989 Domino's expanded its menu for the first time by offering thick, crusty pan pizzas in a 12-inch size. The company does not have nutritional information on these pizzas, nor on its small (12-inch) and medium (14-inch) thin-crust pizzas. We estimate that the small pizzas have about half the calories (and other nutrients) as the 16-inch pizzas; the medium pizzas have about three-quarters the calories, fat, protein, and so on.

You may also be interested in knowing that the small Domino's pizza is about 75 percent larger than the small Pizza Hut pizza. Domino's medium is about 16 percent larger than Pizza Hut's. And Domino's large is more than 10 percent larger than Pizza Hut's. You might want to compare prices in your area.

MSG is added to cooked beef, Italian sausage, and breakfast-sausage toppings. Yellow No. 5 dye colors the regular pizza dough (not pan) and jalapeño and banana-pepper toppings. The company says it is trying to remove the MSG and dye from its foods.

CARBOHYDRATES (gm.)	ADDED SUGAR[3] (gm.)	FAT[4] (gm.)	FAT % CALORIES	SATURATED FAT (gm.)	CHOLESTEROL (mg.)	SODIUM (mg.)	VITAMIN A (% U.S. RDA)	VITAMIN C (% U.S. RDA)	IRON (% U.S. RDA)	CALCIUM (% U.S. RDA)	GLOOM
56	3	10	24	5	19	483	7	2	13	17	16
59	5	20	37	9	40	954	9	4	23	23	31
55	4	25	42	13	48	1042	9	4	22	45	36
58	3	11	24	6	26	805	4	2	19	19	20

DOMINO'S PIZZA

COMPLETE NUTRITIONAL VALUES¹ CONTINUED	WEIGHT² (gm.)	CALORIES	PROTEIN (gm.)	
Pepperoni Pizza, 2 slices, 16 in. (thin-crust)	—	460	24	
Sausage/Mushroom Pizza, 2 slices, 16 in. (thin-crust)	—	430	24	
Veggie Pizza, 2 slices, 16 in. (thin-crust)	—	498	31	

* To estimate nutritional values of the 12-inch pizza multiply the above values by .5. For the 14-inch pizza multiply by .75.
1. A dash means that data not available.
2. To convert grams to ounces (weight), divide by 28.35; to convert grams to fluid ounces (volume), divide by 29.6.

DUNKIN' DONUTS

Well, a health-food store, it ain't. But before we get to the doughnuts, why don't we start with the good news: bran muffins, bagels, orange juice, and milk (sorry, no low-fat milk).

A 3½-ounce oat-bran muffin has 330 calories and 3 grams of fiber (that's a reasonable amount). The bran muffin with raisins provides 310 calories and 4 grams of fiber. About one-fourth of its calories come from fat. Note that these muffins provide more calories than many frozen dinners.

The bagels come in several varieties and are good and chewy. Bagels provide about 250 calories and are low in fat. Surprisingly, even the egg bagel is not high in cholesterol. If you want more than a plain bagel, you can get it with butter, cream cheese, egg and cheese, or egg and cheese plus sausage, bacon, or ham.

Most of the rest of the offerings from this 1,925-store chain are pretty much limited to coffee and dozens of varieties of doughnuts and crullers, all of which are fried—but at least they're fried in vegetable shortening. The smallest, and therefore probably the best, of Dunkin' Donuts fried items is the Glazed French Cruller, which delivers 140 calories and under 2 teaspoons of fat. In contrast, a Glazed

CARBOHYDRATES (gm.)	ADDED SUGAR[3] (gm.)	FAT[4] (gm.)	FAT % CALORIES	SATURATED FAT (gm.)	CHOLESTEROL (mg.)	SODIUM (mg.)	VITAMIN A (% U.S. RDA)	VITAMIN C (% U.S. RDA)	IRON (% U.S. RDA)	CALCIUM (% U.S. RDA)	GLOOM
56	3	17	34	8	28	825	7	2	15	19	28
55	3	16	33	8	28	552	8	2	17	20	23
60	4	18	33	10	36	1035	10	4	26	39	28

3. To convert grams of sugar to teaspoons of sugar, divide by 4.0.
4. To convert grams of fat to teaspoons of fat, divide by 4.4.

Chocolate Ring provides 324 calories and almost 5 teaspoons of fat. You could also pretend you're thinking healthy by choosing a Glazed Whole Wheat Ring, with 330 calories, but why kid yourself? As we said, this ain't no health-food store.

The croissants are the fattiest items, with the chocolate and almond varieties each filled with over 6 teaspoons of grease and 400 calories. Other varieties are probably similar, except for the plain croissant, which is about 25 percent lower in both fat and calories.

The big news out of doughnut-land in 1991 — but don't get too excited — was no-cholesterol doughnuts. That's not as great an advance as it might appear, because the doughnuts only have 1 or 2 milligrams of cholesterol to begin with and they'll all still be fried. Choose a bagel instead.

If you're near Dunkin' Donuts at lunchtime, you can pop in for soup and a croissant sandwich, made with eggs, broccoli, ham, roast beef and cheese, or chicken or tuna salad. No nutrition information is available for any of those products, though it doesn't take a world-class nutritionist to recognize that the soups are salty and the sandwiches fatty.

While this chain is mostly geared to adults, the company seeks to instill "brand recognition" into kids by licensing Ralston Purina to produce "Dunkin' Donuts," a junky, 42-percent-sugar breakfast cereal.

Dunkin' Donuts provides only limited nutrition and no ingredient information for its doughnuts, making it impossible to determine

DUNKIN' DONUTS

COMPLETE NUTRITIONAL VALUES[1]

	WEIGHT[2] (gm.)	CALORIES	PROTEIN (gm.)
Apple Filled Cinnamon Donut	64	190	4
Apple N'Spice Muffin	100	300	6
✔ Bagel, Cinnamon 'N' Raisin	87	250	8
✔ Bagel, Egg	87	250	9
✔ Bagel, Onion	87	230	9
✔ Bagel, Plain	87	240	9
Banana Nut Muffin	103	310	7
Bavarian Filled Donut with Chocolate Frosting	79	240	5
Blueberry Filled Donut	67	210	4
Blueberry Muffin	101	280	6
Boston Kreme Donut	79	240	4
✔ Bran Muffin with Raisins	104	310	6
Chocolate Chunk Cookie	43	200	3
Chocolate Chunk Cookie with Nuts	43	210	3
Chocolate Frosted Yeast Ring	55	200	4
Corn Muffin	96	340	7
Cranberry Nut Muffin	98	290	6
Croissant, Almond	105	420	8
Croissant, Chocolate	94	440	7
Croissant, Plain	72	310	7

accurate sugar contents and Gloom ratings. We would venture to estimate, though, that the various fried products have Gloom ratings of 15 to 30. Bagels have Gloom ratings of around 5.

CARBOHYDRATES (gm.)	ADDED SUGAR[3] (gm.)	FAT[4] (gm.)	FAT % CALORIES	SATURATED FAT (gm.)	CHOLESTEROL (mg.)	SODIUM (mg.)	VITAMIN A (% U.S. RDA)	VITAMIN C (% U.S. RDA)	IRON (% U.S. RDA)	CALCIUM (% U.S. RDA)	GLOOM[5]
25	—	9	43	2	0	220	—	—	—	—	—
52	—	8	24	—	25	360	—	—	—	—	—
49	—	2	7	—	0	370	0	0	15	2	5
47	—	2	7	—	15	380	0	0	15	2	6
46	—	1	4	—	0	480	0	0	15	2	5
47	—	1	4	—	0	450	0	0	15	2	5
49	—	10	29	—	30	410	—	—	—	—	—
32	—	11	41	—	0	260	—	—	—	—	—
29	—	8	34	—	0	240	—	—	—	—	—
46	—	8	26	—	30	340	—	—	—	—	—
30	—	11	41	2	0	250	—	—	—	—	—
51	—	9	26	—	15	560	—	—	—	—	—
25	—	10	45	—	30	110	—	—	—	—	—
23	—	11	47	—	30	100	—	—	—	—	—
25	—	10	45	2	0	190	—	—	—	—	—
51	—	12	32	—	40	560	—	—	—	—	—
44	—	9	28	—	25	360	—	—	—	—	—
38	—	27	58	—	0	280	—	—	—	—	—
38	—	29	59	—	0	220	—	—	—	—	—
27	—	19	55	—	0	240	—	—	—	—	—

DUNKIN' DONUTS

COMPLETE NUTRITIONAL VALUES[1] CONTINUED	WEIGHT[2] (gm.)	CALORIES	PROTEIN (gm.)
Dunkin' Donut (plain cake with handle)	60	240	4
Glazed Buttermilk Ring	74	290	4
Glazed Chocolate Ring	71	324	3
Glazed Coffee Roll	81	280	5
Glazed French Cruller	38	140	2
Glazed Whole Wheat Ring	81	330	4
Glazed Yeast Ring	55	200	4
Honey Dipped Cruller	69	260	4
Honey Dipped Yeast Ring Donut	55	200	4
Jelly Filled Donut	67	220	4
Lemon Filled Donut	79	260	4
Munchkin (average)	—	60	—
✔ Oat Bran Muffins, plain	97	330	7
Oatmeal Pecan Raisin Cookie	46	200	3
Plain Cake Ring Donut	57	262	3
Powdered Cake Ring	62	270	3

1. A dash means that data not available.
2. To convert grams to ounces (weight), divide by 28.35; to convert grams to fluid ounces (volume), divide by 29.6.

HARDEE'S

Hardee's first menu in 1961 listed just eight items, including 15-cent burgers and 10-cent soft drinks. Now Hardee's is the third-largest hamburger chain, and offers everything from roast beef to chicken to salads.

CARBOHYDRATES (gm.)	ADDED SUGAR[3] (gm.)	FAT[4] (gm.)	FAT % CALORIES	SATURATED FAT (gm.)	CHOLESTEROL (mg.)	SODIUM (mg.)	VITAMIN A (% U.S. RDA)	VITAMIN C (% U.S. RDA)	IRON (% U.S. RDA)	CALCIUM (% U.S. RDA)	GLOOM[5]
26	—	14	53	3	0	370	0	4	7	0	21
37	—	14	43	—	10	370	—	—	—	—	—
34	—	21	58	—	2	383	—	—	—	—	—
37	—	12	39	—	0	310	—	—	—	—	—
16	—	8	51	—	30	130	—	—	—	—	—
39	—	18	49	—	5	380	—	—	—	—	—
26	—	9	41	2	0	230	—	—	—	—	—
36	—	11	38	2	0	330	—	—	—	—	—
26	—	9	41	2	0	230	0	4	4	0	14
31	—	9	37	2	0	330	—	—	—	—	—
33	—	12	42	—	0	280	—	—	—	—	—
—	—	—	—	—	—	—	—	—	—	—	—
50	—	11	30	—	0	450	—	—	—	—	—
28	—	9	41	—	25	100	—	—	—	—	—
23	—	18	62	4	0	330	—	—	—	—	—
28	—	16	53	3	0	340	—	—	—	—	—

3. To convert grams of sugar to teaspoons of sugar, divide by 4.0.
4. To convert grams of fat to teaspoons of fat, divide by 4.4.
5. Dunkin' Donuts provides too little information to calculate most Gloom ratings.

And biscuits? Hardee's wasn't first on the block, but when it comes to biscuits, Hardee's can make and sell them. Biscuits with raisins; with bacon, egg, and cheese; with bacon and egg; with ham; with ham and egg; with steak; with steak and egg; with sausage; and with sausage and egg. Even plain!

And all of them loaded with fat and calories. Half of the 320 calories in a plain Rise 'N' Shine Biscuit (Gloom = 29) come from fat. The Bacon & Egg Biscuit provides 410 calories and 42 Gloom points. The worst of all is the Steak & Egg Biscuit, which boasts 550 calories, 1,370 milligrams of sodium, and 57 Gloom points. The least Gloomy biscuits are the ham and Cinnamon 'N' Raisin versions, which rate 27 points.

Your best bet for breakfast is orange juice, a plate of pancakes, and a carton of 2 percent low-fat milk. With syrup and margarine, this meal provides 638 calories and just 22 Gloom points. As at other fast-food restaurants, don't expect to get home-style, crisp-around-the-edge pancakes. They all tend to be a little rubbery.

If you ever wake up in a truly masochistic mood, have Hardee's Big Country Breakfast with Sausage, named in 1989 by *Nutrition Action Healthletter* as one of the five worst fast foods. This meal bulges with 850 calories, 13 teaspoons of fat, a full teaspoon of salt, 340 milligrams of cholesterol, and a whopping 90 Gloom points. The other Big Country Breakfasts aren't quite as bad, though the Country Ham version contains an incredible 2,870 milligrams of sodium (almost 1½ teaspoons of salt).

For lunch or dinner, consider one of Hardee's several packaged salads. The side salad has just 20 calories, reflecting the modest amounts of lettuce, red cabbage, carrots, and tomatoes. The garden salad, which contains egg and cheese, and the chef salad (add ham and turkey to the garden salad) each have under 4 teaspoons of fat. The Chicken Fiesta (also known as Grilled Chicken) Salad is the highest in calories (280) and cholesterol (145), but it's big enough so that if you skip the egg yolk and cheese (to improvise a healthier version), you'll still be full. Reduced-calorie salad dressings are available.

Hardee's Grilled Chicken Sandwich was rated by *Nutrition Action Healthletter* as one of the best fast foods. It has only 2 teaspoons of fat, one-third as much as a McDonald's McChicken. You can cut the fat even further by scraping off the mayonnaise topping. The Grilled Chicken Sandwich provides 310 calories and a moderate 16 Gloom points, including the multigrain bun (which probably has very little whole wheat). Its main defect is the 890 milligrams of sodium. Having the Grilled Chicken Sandwich, regular fries, and orange juice would give you a 620-calorie meal providing 26 Gloom points.

The Chicken Fillet sandwich is one step down on the nutrition ladder, though it's much better than most hamburger sandwiches. Its vital statistics are 370 calories, 3 teaspoons of fat, over 1,000 milligrams of sodium, and 24 Gloom points. The Chicken Stix are fried and breaded fillets, and are less fatty than what McDonald's and KFC (Kentucky Fried Chicken) serve up.

Many Hardee's outlets offer bone-in fried chicken, an item added after Hardee's bought the Roy Rogers restaurant chain. Hardee's, unlike Roy Rogers, fries the chicken in vegetable shortening, not beef fat. (The nutritional listings in the chart are estimates based on Roy Rogers nutritional data.) As in all fried chicken, while the breading and skin are loaded with fat, the meat underneath is excellent. The leg at 140 calories and wing at 192 calories are the lowest-calorie, least-fatty pieces. Adding fries and a shake to any fried-chicken meal, though, easily puts it close to 1,000 calories.

Red-meat lovers should check out the roast-beef sandwiches. Hardee's restaurants in the New York–Washington corridor use the extremely lean roast beef that Roy Rogers restaurants always used — just 1.5 percent fat, according to a CSPI-sponsored study. Other Hardee's, however, probably stayed with a somewhat fattier beef — about 15 percent fat by weight. Hardee's Regular Roast Beef (310 calories and 22 Gloom points) and the Big Roast Beef (360 calories, 28 Gloom points) are the basic sandwiches. The large roast beef sandwich made in the Roy Rogers style provides 373 calories (29 percent from fat) and only 21 Gloom points. Adding cheese adds more fat, sodium, and calories.

The Regular and Big Roast Beef sandwiches rate a bit better than the larger Turkey Club and Chicken Fillet sandwiches. All four get about one-third of their calories from fat (compared to 26 percent for the Grilled Chicken Sandwich). In contrast, the Fisherman's Fillet sandwich gets 43 percent of its 500 calories from fat.

In the hamburger department, where Hardee's started, the regular hamburger has only 270 calories and 490 milligrams of sodium. The regular cheeseburger has 320 calories and 710 milligrams of sodium. But when you hit the Big Deluxe Burger, quarter-pound cheeseburger, Mushroom 'N' Swiss Burger, or bacon cheeseburger, you're hitting grease city. Each of these scores between the 39 and 54 mark on the Gloom scale and provides 6 to 9 teaspoons of fat and about 500 to 600 calories. At least Hardee's uses low-cal mayonnaise in its toppings.

The Real Lean Deluxe has a Gloom rating of 22. Adding a side salad with reduced-calorie Italian dressing and an orange juice gives you a 530-calorie meal with 26 Gloom points.

In 1991 Hardee's came out with its Real Lean Deluxe hamburger sandwich. This sandwich has two-thirds less fat than the Big Deluxe Burger, and one-third more fat than McDonald's McLean Deluxe, which is the lowest-fat burger in the business.

Adding fries and a shake to a large burger sandwich will push you over the 1,000-calorie mark. For instance, a bacon cheeseburger (610 calories), "Big Fry" (500 calories), apple turnover (270 calories), and chocolate shake (460 calories) total a belly-bulging 1,840 calories, 19 teaspoons of fat, and 120 Gloom Points.

Hardee's provides a nutrition brochure at all its outlets, though it does not provide information about vitamins. The company has a free telephone number (see Appendix) to provide consumers with information and respond to problems. Unfortunately, Hardee's won't disclose to customers the ingredients from which its products are made. But they will tell you, if you write them a letter, which specific foods contain ingredients that you say you are sensitive to. The

HARDEE'S

COMPLETE NUTRITIONAL VALUES[1]

	WEIGHT[2] (gm.)	CALORIES	PROTEIN (gm.)	
Apple Turnover	91	270	3	
Bacon Cheeseburger	219	610	34	
Barbecue Sauce, 1 packet (½ oz.)	14	14	0	
Big Cookie	49	250	3	
Big Deluxe Burger	216	500	27	
Big Twin	173	450	23	

company, however, did provide us with ingredient information for most of its products.

Only the apple turnover contains sulfites. MSG is present in Chicken Stix and Chicken Fillet, barbecue sauce, Mushrooms 'N' Sauce, salad croutons, Au Jus seasoning, house dressing, chicken biscuit, and a few other items. Yellow No. 5 dye is in low-fat chocolate milk, coffee creamer, and Cool Twist Cones.

Hardee's, headquartered in the small southern town of Rocky Mount, North Carolina, has built an empire of over 4,000 outlets and has overtaken Wendy's as the number-three burger chain, even though it has no stores in California or the Southwest. It is still about two thousand units and two billion dollars a year behind number-two Burger King. Hardee's, though, seems to have its sights set high. One sign of its ambition is that in early 1990 it paid Ronald Reagan $60,000 to give a speech to 1,400 franchise owners. Next, it bought the 600-unit Roy Rogers fast-food chain and immediately converted hundreds of those shops to Hardee's. If it hooked its ambition to nutrition, Hardee's might attract millions of customers who would never step into a McDonald's, Burger King . . . or current Hardee's. Incidentally, Hardee's is owned by Imasco Ltd., a Canadian cigarette, restaurant, and financial-services conglomerate.

CARBOHYDRATES (gm.)	ADDED SUGAR³ (gm.)	FAT⁴ (gm.)	FAT % CALORIES	SATURATED FAT (gm.)	CHOLESTEROL (mg.)	SODIUM (mg.)	VITAMIN A (% U.S. RDA)	VITAMIN C (% U.S. RDA)	IRON (% U.S. RDA)	CALCIUM (% U.S. RDA)	GLOOM
38	8	12	40	4	0	250	—	—	4	0	18
31	5	39	58	16	80	1030	—	—	30	20	54
4	0	0	0	0	0	140	—	—	0	0	1
31	16	13	47	4	5	240	—	—	4	0	22
32	5	30	54	12	70	760	—	—	30	20	42
34	5	25	50	11	55	580	—	—	20	20	33

HARDEE'S

COMPLETE NUTRITIONAL VALUES[1] CONTINUED	WEIGHT[2] (gm.)	CALORIES	PROTEIN (gm.)	
Biscuit 'N' Gravy	221	440	9	
Biscuit, Bacon	93	360	10	
Biscuit, Bacon & Egg	124	410	15	
Biscuit, Bacon, Egg & Cheese	137	460	17	
Biscuit, Canadian Rise 'N' Shine	161	470	22	
Biscuit, Chicken	146	430	17	
Biscuit, Cinnamon 'N' Raisin	80	320	4	
Biscuit, Country Ham	108	350	11	
Biscuit, Country Ham & Egg	139	400	16	
Biscuit, Ham	106	320	10	
Biscuit, Ham & Egg	138	370	15	
Biscuit, Ham, Egg & Cheese	151	420	18	
Biscuit, Rise 'N' Shine	83	320	5	
Biscuit, Sausage	118	440	13	
Biscuit, Sausage & Egg	150	490	18	
Biscuit, Steak	148	500	15	
Biscuit, Steak & Egg	179	550	20	
Cheeseburger	122	320	16	
Chicken Fillet	173	370	19	
Chicken Stix, 6 pieces	100	210	19	
Cool Twist Cone, chocolate	119	200	4	
Cool Twist Cone, vanilla	119	190	5	
Cool Twist Cone, vanilla/chocolate	119	190	4	
Cool Twist Sundae, caramel	169	330	6	

CARBOHYDRATES (gm.)	ADDED SUGAR[3] (gm.)	FAT* (gm.)	FAT % CALORIES	SATURATED FAT (gm.)	CHOLESTEROL (mg.)	SODIUM (mg.)	VITAMIN A (% U.S. RDA)	VITAMIN C (% U.S. RDA)	IRON (% U.S. RDA)	CALCIUM (% U.S. RDA)	GLOOM
45	0	24	49	6	15	1250	—	—	10	15	42
34	0	21	53	4	10	950	—	—	10	10	34
35	0	24	53	5	155	990	—	—	20	15	42
35	0	28	55	8	165	1220	—	—	20	20	48
35	0	27	52	8	180	1550	—	—	20	20	51
42	0	22	46	4	45	1330	—	—	10	15	38
37	0	17	48	5	0	510	—	—	10	10	27
35	0	18	46	3	25	1550	—	—	15	10	35
35	0	22	50	4	175	1600	—	—	20	15	46
34	0	16	45	2	15	1000	—	—	10	10	27
35	0	19	46	4	160	1050	—	—	20	15	37
35	0	23	49	6	170	1270	—	—	20	20	42
34	0	18	51	3	0	740	—	—	10	12	29
34	0	28	57	7	25	1100	—	—	15	15	44
35	0	31	57	8	170	1150	—	—	20	15	53
46	0	29	52	7	30	1320	—	—	20	15	47
47	0	32	52	8	175	1370	—	—	25	15	57
33	5	14	39	7	30	710	—	—	20	20	22
44	0	13	32	2	55	1060	—	—	15	10	24
13	0	9	39	2	35	680	—	—	4	2	15
31	14	6	27	4	20	65	—	—	10	10	10
28	14	6	28	4	15	100	—	—	0	10	10
29	14	6	28	4	20	80	—	—	10	10	10
54	33	10	27	5	20	290	—	—	4	20	20

HARDEE'S

COMPLETE NUTRITIONAL VALUES¹ CONTINUED	WEIGHT² (gm.)	CALORIES	PROTEIN (gm.)
Cool Twist Sundae, hot fudge	168	320	7
Cool Twist Sundae, stawberry	166	260	5
Crispy Curls (fries)	85	300	4
Dipping Sauce, BBQ, 1 oz.	28	30	0
Dipping Sauce, Honey, ½ oz.	14	45	0
Dipping Sauce, Sweet 'n' Sour, 1 oz.	28	40	0
Dipping Sauce, Sweet Mustard, 1 oz.	28	50	0
Fisherman's Fillet	207	500	23
French Fries, "Big Fry," 5½ oz.	156	500	6
French Fries, large, 4 oz.	113	360	4
French Fries, regular, 2.5 oz.	71	230	3
Fried Chicken Breast	—	412	33
Fried Chicken Breast and Wing	—	604	44
Fried Chicken Leg	—	140	12
Fried Chicken Leg and Thigh	—	436	30
Fried Chicken Thigh	—	296	18
Fried Chicken Wing	—	192	11
✔ Grilled Chicken Sandwich	192	310	24
✔ Hamburger	110	270	13
Hash Rounds	79	230	3
Horseradish, 1 packet (¼ oz.)	7	25	0
Hot Dog, All Beef	120	300	11
Hot Ham 'N' Cheese	149	330	23
Margarine/Butter Blend	5	35	0

CARBOHYDRATES (gm.)	ADDED SUGAR[3] (gm.)	FAT[4] (gm.)	FAT % CALORIES	SATURATED FAT (gm.)	CHOLESTEROL (mg.)	SODIUM (mg.)	VITAMIN A (% U.S. RDA)	VITAMIN C (% U.S. RDA)	IRON (% U.S. RDA)	CALCIUM (% U.S. RDA)	GLOOM
45	33	12	34	6	25	270	—	—	6	20	22
43	33	8	28	5	15	115	—	—	4	15	15
36	0	16	48	3	0	840	—	—	8	2	27
8	0	0	0	0	0	300	—	—	0	9	2
11	11	0	0	0	0	0	0	0	0	0	2
10	9	0	0	0	0	95	—	—	0	6	2
10	9	0	0	0	0	160	—	—	0	15	2
49	5	24	43	6	70	1030	—	—	20	20	38
66	0	23	41	5	0	180	—	—	10	2	30
48	0	17	43	3	0	135	—	—	8	0	21
30	0	11	43	2	0	85	—	—	6	0	14
17	0	24	52	—	0	609	—	—	—	—	28
25	0	37	55	—	165	894	—	—	—	—	52
6	0	8	51	—	40	190	—	—	—	—	11
17	0	28	58	—	125	596	—	—	—	—	39
12	0	20	61	—	85	406	—	—	—	—	28
9	0	13	61	—	47	285	—	—	—	—	18
34	5	9	26	1	60	890	—	—	15	15	16
33	5	10	33	4	20	490	—	—	15	10	16
24	0	14	55	3	0	560	—	—	6	0	22
1	0	2	72	0	5	35	0	0	0	0	4
25	5	17	51	8	25	710	—	—	15	8	28
32	5	12	33	5	65	1420	—	—	15	30	25
0	0	4	103	0	5	40	—	—	0	0	5

HARDEE'S

COMPLETE NUTRITIONAL VALUES[1] CONTINUED

	WEIGHT[2] (gm.)	CALORIES	PROTEIN (gm.)
Mayonnaise, ½ oz.	14	50	0
Milk, 2% Low-fat, 8 fl. oz.	244	121	8
Muffin, Blueberry	106	400	6
Muffin, Oat Bran Raisin	122	440	8
Mushroom 'N' Swiss Burger	186	490	30
✔ Pancakes, 3	137	280	8
Pancakes, 3 with 1 Sausage Patty	176	430	16
Pancakes, 3 with 2 Bacon Strips	150	350	13
Quarter-Pound Cheeseburger	182	500	29
✔ Real Lean Deluxe	205	340	23
✔ Roast Beef Sandwich (RR)	—	350	26
Roast Beef Sandwich with Cheese (RR)	—	403	29
Roast Beef Sandwich with Cheese, large (RR)	—	427	38
✔ Roast Beef Sandwich, large (RR)	—	373	35
Roast Beef, Big	169	360	24
Roast Beef, Regular	141	310	20
Salad Dressing, Blue Cheese, 2 oz.	56	210	1
Salad Dressing, House, 2 oz.	56	290	1
✔ Salad Dressing, Reduced Calorie French, 2 oz.	56	130	1
Salad Dressing, Reduced Calorie Italian, 2 oz.	56	90	0
Salad Dressing, Thousand Island, 2 oz.	56	250	1
Salad, Chef	294	240	22
Salad, Chicken Fiesta/Grilled Chicken	298	280	26
Salad, Garden	241	210	14

CARBOHYDRATES (gm.)	ADDED SUGAR[3] (gm.)	FAT[4] (gm.)	FAT % CALORIES	SATURATED FAT (gm.)	CHOLESTEROL (mg.)	SODIUM (mg.)	VITAMIN A (% U.S. RDA)	VITAMIN C (% U.S. RDA)	IRON (% U.S. RDA)	CALCIUM (% U.S. RDA)	GLOOM
1	1	5	90	1	5	75	0	0	0	0	8
12	0	4	33	3	18	122	10	4	0	30	5
51	24	19	43	4	80	320	—	—	6	4	33
62	15	18	37	3	55	350	—	—	—	8	29
33	5	27	50	13	70	940	—	—	30	30	39
56	0	2	6	1	15	890	—	—	20	6	10
56	0	16	33	6	40	1290	—	—	20	8	32
56	0	9	23	3	25	1110	—	—	20	6	21
34	5	29	52	14	70	1060	—	—	30	25	42
35	5	13	22	—	80	650	—	—	25	10	22
37	5	11	28	—	58	732	—	—	—	—	20
37	5	15	33	—	70	954	—	—	—	—	27
31	5	17	36	—	94	1062	—	—	—	—	28
31	5	12	29	—	82	840	—	—	—	—	21
33	5	15	38	6	65	1150	—	—	30	10	28
32	5	12	35	5	50	930	—	—	25	10	22
10	9	18	77	3	20	790	—	—	0	23	35
6	0	29	90	4	25	510	—	—	0	0	44
21	17	5	35	1	0	480	—	—	0	0	14
5	4	8	80	1	0	310	—	—	0	7	14
9	8	23	83	3	35	540	—	—	0	9	37
5	0	15	56	9	115	930	—	—	10	30	20
4	5	15	48	9	145	640	—	—	10	30	19
3	0	14	60	8	105	270	—	—	6	30	15

HARDEE'S

COMPLETE NUTRITIONAL VALUES¹ CONTINUED	WEIGHT² (gm.)	CALORIES	PROTEIN (gm.)
✔ Salad, Side	112	20	2
Shake, Chocolate	341	460	11
Shake, Strawberry	341	440	11
Shake, Vanilla	341	400	13
Syrup	43	120	0
Tartar Sauce, ⅔ oz.	19	90	0
Turkey Club	208	390	29
Yogurt Cone (only)	5	20	1
Yogurt, Chocolate, Frozen, Soft Serve	113	170	6
Yogurt, Vanilla, Frozen, Soft Serve	113	160	6
Yogurt, NutraSweet Chocolate	113	120	6
Yogurt, NutraSweet Vanilla	113	110	5
Hardee's Meals			
✔ Pancakes (3) with Syrup & Margarine/Butter Blend, Orange Juice, 2% Low-fat Milk	612	638	17
Big Country Breakfast (Ham)	251	620	28
Chicken Biscuit, Hash Rounds, 2% Low-fat Milk	469	781	28
Big Country Breakfast (Bacon)	217	660	24
Big Country Breakfast (Country Ham)	254	670	29
Big Country Breakfast with Sausage, Orange Juice, 2% Low-fat Milk	701	1053	42
Big Country Breakfast (Sausage)	274	850	33
✔ Grilled Chicken Sandwich, French Fries (regular), Orange Juice	446	622	28

CARBOHYDRATES (gm.)	ADDED SUGAR[3] (gm.)	FAT[4] (gm.)	FAT % CALORIES	SATURATED FAT (gm.)	CHOLESTEROL (mg.)	SODIUM (mg.)	VITAMIN A (% U.S. RDA)	VITAMIN C (% U.S. RDA)	IRON (% U.S. RDA)	CALCIUM (% U.S. RDA)	GLOOM
1	0	0	0	0	0	15	—	—	2	2	0
85	55	8	16	5	45	340	—	—	6	50	20
82	54	8	16	5	40	300	—	—	0	50	19
66	41	9	20	6	50	320	—	—	0	50	19
31	31	0	0	0	0	25	0	0	4	0	5
2	0	9	90	1	10	160	—	—	0	3	14
32	5	16	37	4	70	1280	—	—	15	15	28
4	—	0	0	0	0	10	—	—	0	0	—
27	—	4	21	3	10	75	—	—	8	15	—
27	—	4	23	3	10	75	—	—	4	15	—
22	0	0	0	0	0	75	—	—	0	20	—
21	0	1	8	1	0	75	—	—	2	15	—
119	31	11	16	4	38	1079	13	123	24	36	22
51	0	33	48	7	325	1780	—	—	30	15	64
78	0	41	47	10	63	2012	—	—	16	45	65
51	0	40	55	10	305	1540	—	—	30	15	70
52	0	38	51	9	345	2870	—	—	35	15	82
83	0	62	53	19	358	2104	—	—	35	50	88
51	0	57	60	16	340	1980	—	—	35	20	90
84	5	20	29	3	60	977	—	—	21	15	26

HARDEE'S

COMPLETE NUTRITIONAL VALUES¹ CONTINUED

	WEIGHT² (gm.)	CALORIES	PROTEIN (gm.)
✔ Real Lean Deluxe, Side Salad with Reduced Calorie Italian Dressing, Orange Juice	556	530	26
Hamburger, French Fries (regular), Big Cookie, Diet Coke (medium)	890	751	19
Regular Roast Beef Sandwich, Garden Salad with Reduced Calorie French Dressing, French Fries (regular)	509	880	38
Large Roast Beef Sandwich, Crispy Curls, Vanilla Shake	426	1073	52
Fisherman's Fillet Sandwich, French Fries (Big Fry), Vanilla Cool Twist Cone	482	1190	34
Fried Chicken Breast and Wing, French Fries (large), Strawberry Shake	454	1404	59
Bacon Cheeseburger, French Fries (Big Fry), Chocolate Shake, Apple Turnover	807	1840	54

1. A dash means that data not available.
2. To convert grams to ounces (weight), divide by 28.35; to convert grams to fluid ounces (volume), divide by 29.6.

JACK IN THE BOX

The Southwest's 1,000-restaurant Jack in the Box chain offers high fat in just about everything—burgers, breakfast, fish, salads, double fudge cake. Among major menu items, only the Chicken Fajita Pita gets fewer than 30 percent of its calories from

CARBOHYDRATES (gm.)	ADDED SUGAR[3] (gm.)	FAT[4] (gm.)	FAT % CALORIES	SATURATED FAT (gm.)	CHOLESTEROL (mg.)	SODIUM (mg.)	VITAMIN A (% U.S. RDA)	VITAMIN C (% U.S. RDA)	IRON (% U.S. RDA)	CALCIUM (% U.S. RDA)	GLOOM
60	9	21	36	—	80	975	—	—	27	19	26
94	21	34	41	10	25	815	—	—	25	10	48
86	22	42	43	16	155	1765	—	—	37	40	61
133	46	37	31	9	132	2000	—	—	8	52	71
143	19	53	40	15	85	1310	—	—	30	32	74
155	54	62	40	8	205	1329	—	—	8	50	97
220	68	82	40	30	125	1800	—	—	50	72	120

3. To convert grams of sugar to teaspoons of sugar, divide by 4.0.
4. To convert grams of fat to teaspoons of fat, divide by 4.4.

fat. Most of the other offerings rank in the 40- to 50-percent range, with a few—such as the breakfast Crescent (croissant) sandwiches and the Grilled Sourdough Burger—exceeding 60 percent of calories from fat. No wonder the Gloom ratings climb so high. But let's walk bravely through the doors of the fifth-biggest hamburger chain and search for the best of the offerings.

First, a compliment. The chairs in the restaurants we visited actually move! You can sit on a real, adult, wooden chair with a cushioned bottom, instead of those plastic seats that are permanently attached to tables and are used by many fast-food restaurants.

If you find yourself in a Jack in the Box at breakfast time, your best bet is plain pancakes, orange juice, and low-fat milk. Otherwise, it's slim pickin's. The Pancake Platter's 5 teaspoons of fat come from bacon. The Scrambled Egg Platter, which comes with bacon, hash browns, and an English muffin, "costs" you 63 Gloom points, and uses up much of your daily fat and cholesterol quota. The Scrambled Egg Pocket is pita bread crammed full of egg, bacon, ham, and cheese . . . and plenty of fat, cholesterol, and sodium. It rates 45 on the Gloom scale.

The Sausage and Supreme Crescent sandwiches each has at least 9 teaspoons of fat. That's a grim way to start the day! However, the croissant itself is one of the tastiest we've had in a fast-food restaurant; too bad it's the fat that makes it tasty.

A lunch or dinner at Jack in the Box could start with orange juice or 2 percent low-fat milk or a 320-calorie milk shake. Then add a Chicken Fajita Pita, which was named by *Nutrition Action Healthletter* as one of the five best fast-food items in 1989. This pocket sandwich contains lots of grilled chicken strips, onions, lettuce, tomato, and grated Cheddar cheese. It gets only 25 percent of its 292 calories from fat and has a modest Gloom rating of 14. Ordering it without cheese would lower all those figures. Adding orange juice and a side salad to the chicken sandwich would give you quite a decent meal, still with only 15 Gloom points.

The Grilled Chicken Fillet, with 408 calories, features a tasty, marinated piece of white meat with tomato and lettuce, but it is fairly high in fat (4 teaspoons) and sodium. You can trim the fat considerably by asking the clerk to hold the mayonnaise or slice of cheese.

Both the Chicken and Fish Supreme sandwiches belie the low-fat reputation of chicken and fish. They each have about 8 teaspoons of fat and over 500 calories.

Among salads, the side salad is a nice-size portion that has only 51 calories. The other salads pack a fair amount of calories, sodium, and fat. The chef salad gets half its 325 calories from 4 teaspoons of fat. The Taco Salad provides 503 calories, more than half from fat. Jack in the Box really needs to add a low-calorie, full-meal salad. It also needs to improve upon its dressings; a packet of the reduced-calorie French dressing still offers a hefty 176 calories, far more than the "lite" dressings offered by other chains. Jack in the Box says that low-calorie dressings may be available soon.

If you're thinking burger, think small and plain. The regular, 267-calorie burger yields only 18 Gloom points. The Jumbo Jack has 584 calories and 50 Gloom points, but the worst is yet to come.

The larger, specialty burgers head for the Gloom stratosphere. The Bacon Cheeseburger and the Grilled Sourdough Burger each provide over 700 calories, about 10 teaspoons of fat, and over 1,100 milligrams of sodium. They register 63 and 72, respectively, on the Gloom scale. Add onion rings and a Sprite and you're over 1,200 calories.

Much worse is the 942-calorie Ultimate Cheeseburger, with its 16 teaspoons of fat and 88 Gloom points. Truly, this is the Ultimate Artery-clogger and one of the worst sandwiches on the market. It contains all the fat and saturated fat you should have in an entire day. Combine this monster with a jumbo order of fries, milkshake, and hot apple turnover, and you hit 2,006 calories, 26 teaspoons of fat, and 158 Gloom points.

Jack in the Box is a pioneer in the dubious field of finger foods: they offer egg rolls, Taquitos (minced beef in a fried tortilla), and breaded chicken strips (nice piece of chicken in scads of breading). You'd think that the creative geniuses at Jack in the Box, with headquarters in California, could develop at least one waistline-friendly finger food, such as fresh fruit or carrot sticks.

Though we can't praise Jack in the Box for too many of its foods, we can congratulate it for offering at all of its outlets a concise, readable pamphlet that provides nutrition and ingredient information. According to that pamphlet, the "wheat bun" contains very little whole wheat. Sulfite-sensitive individuals should avoid the apple turnovers. At least the chain now fries its potatoes, apple turnovers, and everything else in vegetable shortening. Also, keep an eye out for the lower-fat hamburger, chicken, and turkey sandwiches that Jack in the Box is testing.

JACK IN THE BOX

COMPLETE NUTRITIONAL VALUES

	WEIGHT[1] (gm.)	CALORIES	PROTEIN (gm.)
BBQ Sauce, 1 oz.	28	44	1
Bacon Cheeseburger	242	705	35
Breakfast Jack	126	307	18
Cheeseburger	112	315	15
Cheesecake	99	309	8
✔ Chicken Fajita Pita	189	292	24
Chicken Strips, 4 pieces	112	285	25
Chicken Strips, 6 pieces	177	451	39
Chicken Supreme	245	641	27
Coca-Cola Classic, 32 fl. oz.	960	384	0
Coca-Cola Classic, 12 fl. oz.	360	144	0
Coffee, 8 fl. oz.	240	2	0
Crescent, Sausage	156	584	22
Crescent, Supreme	146	547	20
Diet Coke, 12 fl. oz.	360	1	0
Double Cheeseburger	149	467	21
Double Fudge Cake	85	288	4
Dr Pepper, 12 fl. oz.	360	144	0
Egg Rolls, 3 pieces	165	437	3
Egg Rolls, 5 pieces	285	753	5
Fish Supreme	218	510	24
French Fries, Jumbo	123	396	5
French Fries, regular	109	351	4
French Fries, small	68	219	3

CARBOHYDRATES (gm.)	ADDED SUGAR[2] (gm.)	FAT[3] (gm.)	FAT % CALORIES	SATURATED FAT (gm.)	CHOLESTEROL (mg.)	SODIUM (mg.)	VITAMIN A (% U.S. RDA)	VITAMIN C (% U.S. RDA)	IRON (% U.S. RDA)	CALCIUM (% U.S. RDA)	GLOOM
11	0	0	0	0	0	300	0	0	0	0	3
41	5	45	57	15	113	1240	7	13	28	25	63
30	0	13	38	5	203	871	9	0	17	17	29
33	5	14	40	6	41	746	4	0	15	25	23
29	24	18	51	9	63	208	0	0	3	11	31
29	0	8	25	3	34	703	10	0	15	25	14
18	0	13	41	3	52	695	0	0	4	0	22
28	0	20	40	5	82	1100	0	0	6	0	35
47	5	39	55	10	85	1470	8	10	16	24	58
96	96	0	0	0	0	37	0	0	0	0	15
36	36	0	0	0	0	14	0	0	0	0	6
0	0	0	0	0	0	26	0	0	0	0	0
28	0	43	66	16	187	1012	11	0	16	17	68
27	0	40	66	13	178	1053	11	0	15	15	64
0	0	0	0	0	0	26	0	0	0	0	0
33	5	27	52	12	72	842	8	0	15	40	40
49	25	9	28	0	20	259	4	0	10	4	17
37	37	0	0	0	0	18	0	0	0	0	6
54	0	24	49	7	29	957	0	6	20	8	41
92	0	41	49	12	49	1640	0	11	34	15	71
44	5	27	48	6	55	1040	0	9	15	16	41
51	0	19	43	5	0	219	0	49	8	0	24
45	0	17	44	4	0	194	0	43	7	0	22
28	0	11	45	3	0	121	0	27	4	0	14

JACK IN THE BOX

COMPLETE NUTRITIONAL VALUES CONTINUED	WEIGHT* (gm.)	CALORIES	PROTEIN (gm.)
Fries, Seasoned Curly	109	358	5
Grape Jelly	14	38	0
Grilled Chicken Fillet	205	408	31
Grilled Sourdough Burger	223	712	32
Guacamole	25	55	1
Ham and Turkey Melt	218	592	27
✔ Hamburger	96	267	13
Hash Browns	57	156	1
Hot Apple Turnover	101	348	3
Iced Tea, 12 fl. oz.	360	3	0
Jumbo Jack	222	584	26
Jumbo Jack with Cheese	242	677	32
Milk Shake, Chocolate	322	330	11
Milk Shake, Strawberry	328	320	10
Milk Shake, Vanilla	317	320	10
✔ Milk, 2% Low-fat, 8 fl. oz.	244	122	8
Onion Rings	103	380	5
Old Fashion Patty Melt	215	713	33
✔ Orange Juice, 6 fl. oz.	183	80	1
Pancake Platter	231	612	15
Pancake Syrup, 1.5 oz.	42	121	0
Ramblin' Root Beer, 12 fl. oz.	360	176	0
Salad Dressing, Bleu Cheese, 2.3 fl. oz.	70	262	0
Salad Dressing, Buttermilk House, 2.3 fl. oz.	70	362	0

CARBOHYDRATES (gm.)	ADDED SUGAR[2] (gm.)	FAT[3] (gm.)	FAT % CALORIES	SATURATED FAT (gm.)	CHOLESTEROL (mg.)	SODIUM (mg.)	VITAMIN A (% U.S. RDA)	VITAMIN C (% U.S. RDA)	IRON (% U.S. RDA)	CALCIUM (% U.S. RDA)	GLOOM
39	0	20	50	5	0	1030	0	9	9	3	36
9	8	0	0	0	0	3	0	0	0	0	1
33	5	17	38	4	64	1130	4	13	12	17	28
34	5	50	63	16	109	1140	14	0	24	19	72
2	0	5	82	0	0	130	0	0	0	2	7
40	5	36	55	11	79	1120	30	0	14	43	49
28	5	11	37	4	26	556	0	0	10	15	18
14	0	11	63	3	0	312	0	12	2	0	17
42	18	19	49	6	7	316	0	4	10	0	32
0	0	0	0	0	0	5	0	0	0	0	0
42	5	34	52	11	73	733	0	0	17	14	50
46	5	40	53	14	102	1090	0	0	21	27	59
55	38	7	19	4	25	270	0	0	4	35	15
55	39	7	20	4	25	240	0	0	2	35	15
57	38	6	17	4	25	230	0	0	0	35	14
12	0	5	37	3	18	122	10	4	0	30	6
38	0	23	54	6	0	451	0	5	12	3	35
42	5	46	58	15	92	1360	11	6	21	21	68
20	0	0	0	0	0	0	8	160	2	2	0
87	0	22	32	9	99	888	8	10	10	10	41
30	30	0	0	0	0	6	0	0	0	0	5
46	46	0	0	0	0	20	0	0	0	0	7
14	0	22	76	4	18	918	0	0	0	0	42
8	0	36	90	6	21	694	0	0	0	0	58

JACK IN THE BOX

COMPLETE NUTRITIONAL VALUES CONTINUED

	WEIGHT[1] (gm.)	CALORIES	PROTEIN (gm.)
✔ Salad Dressing, Reduced-Calorie French, 2.3 fl. oz.	70	176	0
Salad Dressing, Thousand Island, 2.3 fl. oz.	70	312	0
Salad, Chef	332	325	30
✔ Salad, Side	111	51	7
Salad, Taco	402	503	34
Scrambled Egg Platter	213	559	18
Scrambled Egg Pocket	183	431	29
Sesame Breadsticks	16	70	2
Sirloin Cheesesteak	231	621	36
Sprite, 12 fl. oz.	360	144	0
Sweet & Sour Sauce, 1 oz.	28	40	0
Taco	78	187	7
Taco, Super	126	281	12
Taquitos, 5 pieces	134	362	15
Taquitos, 7 pieces	189	511	22
Tortilla Chips	28	139	2
Ultimate Cheeseburger	280	942	47
Jack in the Box Meals			
Breakfast Jack, Low-fat Milk, Coffee	610	431	26
Scrambled Egg Pocket, Orange Juice, Low-fat Milk	610	633	38
Sausage Crescent, Hash Browns, Orange Juice, Coffee	641	782	25
✔ Chicken Fajita Pita, Side Salad, Orange Juice	483	423	32
Grilled Chicken Fillet Sandwich, Side Salad, water	316	459	38

CARBOHYDRATES (gm.)	ADDED SUGAR[2] (gm.)	FAT[3] (gm.)	FAT % CALORIES	SATURATED FAT (gm.)	CHOLESTEROL (mg.)	SODIUM (mg.)	VITAMIN A (% U.S. RDA)	VITAMIN C (% U.S. RDA)	IRON (% U.S. RDA)	CALCIUM (% U.S. RDA)	GLOOM
26	0	8	41	1	0	600	0	0	0	0	18
12	0	30	87	5	23	700	0	0	0	0	51
10	0	18	50	8	142	900	73	46	8	44	25
0	0	3	53	2	0	84	0	0	0	6	3
28	0	31	55	13	92	1600	27	15	21	41	45
50	0	32	52	9	378	1060	14	16	27	15	63
31	0	21	44	7	354	1060	21	0	20	21	45
12	0	2	26	0	0	110	0	0	0	0	3
51	0	30	43	9	79	1450	17	0	32	29	44
36	36	0	0	0	0	46	0	0	0	0	6
11	0	0	0	0	0	160	0	0	0	0	2
15	0	11	53	4	18	414	0	0	7	11	17
22	0	17	54	6	29	718	12	4	12	16	25
42	0	15	37	3	24	462	0	3	15	15	22
60	0	21	37	5	34	681	0	4	21	21	31
18	0	6	39	0	0	134	0	0	0	0	9
33	5	69	66	26	127	1176	15	0	35	60	88
42	0	18	38	8	221	1019	19	4	17	47	34
63	0	26	37	10	372	1182	39	164	22	53	43
59	0	50	58	17	187	1249	19	160	20	23	64
49	0	11	23	5	34	787	18	160	17	33	15
33	5	20	39	6	64	1214	4	13	12	23	31

JACK IN THE BOX

COMPLETE NUTRITIONAL VALUES CONTINUED	WEIGHT[1] (gm.)	CALORIES	PROTEIN (gm.)
Chef Salad with Reduced-Calorie French Dressing, Sesame Breadsticks, Diet Coke	778	572	32
Hamburger, French Fries, small, Iced Tea, Cheesecake	623	798	24
Super Taco, Egg Rolls (3 pieces), Chocolate Milk Shake	613	1048	26
Fish Supreme, French Fries, regular, Side Salad with Reduced-Calorie French Dressing, Coca-Cola Classic (12 fl. oz.)	868	1232	35
Taco Salad, Egg Rolls (5 pieces), Coca-Cola Classic (12 fl. oz.)	1047	1400	39
Ultimate Cheeseburger, French Fries, Jumbo, Vanilla Milk Shake, Hot Apple Turnover	821	2006	65

1. *To convert grams to ounces (weight), divide by 28.35; to convert grams to fluid ounces (volume), divide by 29.6.*

KFC (KENTUCKY FRIED CHICKEN)

Unlike most other fast-food giants, KFC (Kentucky Fried Chicken), the third-biggest chain, concentrates on just one thing: fried chicken. But at a time when "fried food" = "bad food," KFC has been forced to make some changes. One change is in the name: Kentucky Fried Chicken has been replaced by a simple KFC.

In 1986 KFC was bought by PepsiCo, which also owns Pizza Hut and Taco Bell, giving PepsiCo more fast-food outlets (19,464 in 1990) than McDonald's. KFC had more than 8,000 outlets—some of which were offering home-delivery—and sales of $5.8 billion in 1990. The chain sells nearly 5 billion pieces of chicken a year. Those chickens start out lean, but they usually don't end up that way.

CARBOHYDRATES (gm.)	ADDED SUGAR[2] (gm.)	FAT[3] (gm.)	FAT % CALORIES	SATURATED FAT (gm.)	CHOLESTEROL (mg.)	SODIUM (mg.)	VITAMIN A (% U.S. RDA)	VITAMIN C (% U.S. RDA)	IRON (% U.S. RDA)	CALCIUM (% U.S. RDA)	GLOOM
48	0	28	44	10	142	1636	73	46	8	44	41
85	29	40	45	16	89	890	0	27	17	26	63
131	38	48	41	17	83	1945	12	10	36	59	81
151	41	55	40	13	55	1932	0	52	22	22	88
156	36	72	46	25	141	3254	27	26	55	56	120
183	61	113	51	40	159	1941	15	53	53	95	158

2. *To convert grams of sugar to teaspoons of sugar, divide by 4.0.*
3. *To convert grams of fat to teaspoons of fat, divide by 4.4.*

The menu is short and sweet—chicken and grease. Original Recipe Chicken, the old standby, is about one-third less fatty than Extra Tasty Crispy Chicken. The Hot & Spicy Chicken offered in some outlets is the same as Extra Tasty Crispy, except that the chicken is marinated in additional spices.

Tests sponsored by the Center for Science in the Public Interest in 1990 found that if you discard the skin and breading from Original Recipe Chicken, you'll also get rid of half the calories and two-thirds of the fat! While 37 percent of the chicken meat's calories come from fat, 70 percent of the skin and breading's come from fat. The meat, released from its greasy, salty covering, actually tastes great.

Dinner packages, which include two or three pieces of chicken, a biscuit, mashed potatoes, and coleslaw, might well carry a surgeon general's warning about fat and sodium. An Original Recipe 2-Piece Dinner with breast and thigh provides 2,482 milligrams of sodium, more than 12 teaspoons of fat, and a Gloom rating of 88, almost your whole day's quota. Add a soft drink to that meal and you'll have well over a thousand calories. You can do worse, though. An Extra Tasty Crispy 3-Piece Dinner can provide as many as 135 Gloom points from its 1,378 calories, 2,951 milligrams of sodium, and 20 teaspoons of fat.

In their own diminutive way, Kentucky Nuggets are as bad as the dinners. An order of six nuggets provides 840 milligrams of sodium and 4 teaspoons of fat; 57 percent of the calories come from fat. The nuggets are made of processed chicken (like McDonald's Chicken McNuggets), rather than whole pieces of chicken. Colonel Sanders, if he were still alive, might grimace if he were served this item.

Another processed chicken diminutive is the Chicken Littles Sandwich. Each one is only 169 calories, compared to 260 in a small McDonald's hamburger, but more than half the calories still come from fat. Frankly, these are pretty disappointing little chicken patties on cute little buns. Plus, you won't be able to scrape the breading off such a tiny amount of chicken.

A larger processed chicken product is the Colonel's Chicken Sandwich, which is a breaded and fried patty served on a bun with a few little pieces of lettuce and a slug of mayo. This sandwich delivers 6 teaspoons of fat and rates 45 Gloom points. The deluxe version is topped with bacon and cheese and rates 54 Gloom points.

It should be no surprise that KFC's Hot Wings are loaded with fat and cholesterol. Fifty-eight percent of the calories come from fat. A 6-piece serving rates 40 Gloom points.

In the beverage department, don't expect to find soft drinks not

KFC

COMPLETE NUTRITIONAL VALUES[1]

	WEIGHT[2] (gm.)	CALORIES	PROTEIN (gm.)
✔ Baked Beans	113	133	6

made by parent PepsiCo and don't be surprised if the only milk offered is whole and not low-fat.

For dessert, go someplace else. The parfaits—strawberry shortcake, lemon cream, or chocolate cream flavors—are among the most skippable items we've seen at any restaurant.

All that said, if you are still interested in eating at KFC, it is possible to have a fairly healthful meal. First, be sure to have a glass of water. Then discard the skin and breading from a breast and drumstick. Mashed potatoes (sorry about the taste!) will save almost 200 calories over the fries. The tasty coleslaw that comes with the dinner, or the corn on the cob (without the "butter," if the clerk is cooperative) that you can have as a side order, will give you fiber and vitamin C. Some KFC outlets carry side orders of rice and red beans.

Though KFC has been slow to recognize America's health consciousness, it is testing a raft of new products, some of which should be more healthful than fried chicken. Unfortunately, none of these have been successful enough to join the regular menu. KFC is also testing batter-coated french fries and other items not destined for the Nutrition Hall of Fame.

Fortunately for KFC fans, the fat used for frying is a vegetable shortening that is much less saturated than the beef fat used by Popeyes and Church's. KFC refuses to disclose other ingredient information, so if you're allergic to MSG, dyes, and other possible ingredients, eat elsewhere. Sulfites may be present in the mashed potatoes. KFC does provide detailed nutrition information, but it's not always in a very helpful form. For instance, it provides information about one Kentucky Nugget, forcing you to do the multiplying to figure out what's in orders of six or nine nuggets.

CARBOHYDRATES (gm.)	ADDED SUGAR[3] (gm.)	FAT[4] (gm.)	FAT % CALORIES	SATURATED FAT (gm.)	CHOLESTEROL (mg.)	SODIUM (mg.)	VITAMIN A (% U.S. RDA)	VITAMIN C (% U.S. RDA)	IRON (% U.S. RDA)	CALCIUM (% U.S. RDA)	GLOOM
—	—	2	11	1	1	492	—	—	—	—	7

KFC

COMPLETE NUTRITIONAL VALUES[1] CONTINUED	WEIGHT[2] (gm.)	CALORIES	PROTEIN (gm.)
Buttermilk Biscuit	65	235	4
Chicken Littles Sandwich	47	169	6
Chocolate Pudding	93	156	2
Cole Slaw	91	119	2
Colonel's Chicken Sandwich	166	482	21
Colonel's Deluxe Chicken Sandwich	187	547	25
✔ Corn on the Cob	143	176	5
Extra Tasty Crispy Center Breast*	135	342	33
Extra Tasty Crispy Drumstick	69	204	14
Extra Tasty Crispy Side Breast*	110	343	22
Extra Tasty Crispy Thigh	119	406	20
Extra Tasty Crispy Wing	65	254	12
French Fries	77	244	3
Hot Wings, 6 pieces	119	376	22
Kentucky Nuggets, 6 pieces	96	276	17
✔ Mashed Potatoes and Gravy	98	71	2
Nugget Sauce, Barbecue	28	35	0
Nugget Sauce, Honey	14	49	0
Nugget Sauce, Mustard	28	36	1
Nugget Sauce, Sweet 'n Sour	28	58	0
Original Recipe Center Breast*	115	283	28
Original Recipe Drumstick	57	146	13
Original Recipe Side Breast*	90	267	19
Original Recipe Thigh	104	294	18

CARBOHYDRATES (gm.)	ADDED SUGAR[3]	FAT[4] (gm.)	FAT % CALORIES	SATURATED FAT (gm.)	CHOLESTEROL (mg.)	SODIUM (mg.)	VITAMIN A (% U.S. RDA)	VITAMIN C (% U.S. RDA)	IRON (% U.S. RDA)	CALCIUM (% U.S. RDA)	GLOOM
28	—	12	45	3	1	655	0	0	9	10	20
14	3	10	54	2	18	331	0	0	10	2	15
20	—	7	43	6	2	127	0	0	3	6	11
13	0	7	50	1	5	197	6	36	0	3	7
39	5	27	51	6	47	1060	0	0	7	5	45
—	0	32	53	8	64	1362	—	—	—	—	54
32	0	3	16	0	0	0	5	4	4	0	3
12	0	20	52	5	114	790	0	0	5	3	32
6	0	14	61	3	71	324	0	0	4	1	22
14	0	22	59	5	81	748	0	0	5	3	36
14	0	30	66	8	129	688	3	0	7	5	48
9	0	19	66	4	67	422	0	0	4	2	30
31	—	12	44	3	2	139	0	26	3	0	16
17	0	24	58	5	148	677	0	0	—	—	40
13	0	17	57	4	71	840	0	0	4	1	31
12	0	2	20	0	0	339	0	0	2	2	5
7	—	1	15	0	0	450	7	0	0	0	4
12	12	0	0	0	0	0	0	0	0	0	2
6	—	1	23	0	0	346	0	0	0	0	4
13	—	1	9	0	0	148	0	0	0	0	2
9	0	15	49	4	93	672	0	0	5	4	25
4	0	8	52	2	67	275	0	0	6	2	14
11	0	16	56	4	77	735	0	0	7	7	27
11	0	20	60	5	123	619	2	0	7	7	33

KFC

COMPLETE NUTRITIONAL VALUES[1] CONTINUED	WEIGHT[2] (gm.)	CALORIES	PROTEIN (gm.)
Original Recipe Wing	55	178	12
Parfait, Apple Shortcake	123	276	2
Parfait, Chocolate Creme	132	360	4
Parfait, Fudge Brownie	120	331	3
Parfait, Lemon Cream	151	513	8
Parfait, Strawberry Shortcake	113	230	2
Potato Salad	113	177	2
Vanilla Pudding	94	159	2
KFC Meals			
Kentucky Nuggets (6 pieces), French Fries, Chocolate Pudding	266	676	22
Original Recipe 2-Piece Dinner with Side Breast & Drumstick	401	838	41
Hot Wings (6 pieces), Cole Slaw, Corn on the Cob, Lemon Cream Parfait	504	1184	38
Original Recipe 2-Piece Dinner with Center Breast & Thigh	473	1002	55
Extra Tasty Crispy 2-Piece Dinner with Side Breast & Drumstick	433	972	44
Original Recipe 3-Piece Dinner with Center Breast, Wing, & Drumstick	481	1032	62
Original Recipe 3-Piece Dinner with Thigh, Side Breast, & Drumstick	505	1132	59
Extra Tasty Crispy 2-Piece Dinner with Center Breast & Thigh	508	1173	62
Extra Tasty Crispy 3-Piece Dinner with Center Breast, Wing, & Drumstick	523	1225	68

CARBOHYDRATES (gm.)	ADDED SUGAR³ (gm.)	FAT⁴ (gm.)	FAT % CALORIES	SATURATED FAT (gm.)	CHOLESTEROL (mg.)	SODIUM (mg.)	VITAMIN A (% U.S. RDA)	VITAMIN C (% U.S. RDA)	IRON (% U.S. RDA)	CALCIUM (% U.S. RDA)	GLOOM
6	0	12	59	3	64	372	0	0	7	5	18
44	—	10	33	5	23	248	0	83	3	3	12
44	30	19	48	11	3	231	0	0	7	5	34
55	—	11	30	5	40	299	0	0	6	4	21
74	—	20	36	9	9	232	4	8	3	22	28
36	24	9	34	5	20	162	1	30	3	3	15
—	0	12	59	2	14	497	—	—	—	—	15
21	—	7	41	6	1	130	0	0	1	6	11
64	0	37	49	13	75	1106	0	26	10	7	59
68	0	45	48	11	150	2201	6	36	24	24	72
136	0	54	41	16	162	1106	15	48	7	25	80
73	0	55	49	14	222	2482	8	36	23	26	88
73	0	56	52	14	158	2263	6	36	20	19	88
72	0	54	48	14	230	2510	6	36	29	26	88
79	0	65	51	16	273	2820	8	36	31	31	104
79	0	69	53	17	249	2669	9	36	23	23	109
80	0	72	53	17	258	2727	6	36	24	21	113

KFC

	WEIGHT[2] (gm.)	CALORIES	PROTEIN (gm.)
COMPLETE NUTRITIONAL VALUES[1] CONTINUED			
Extra Tasty Crispy 3-Piece Dinner with Side Breast, Thigh, & Drumstick	552	1378	64

* Chicken breast is an excellent choice, if the breading and batter is removed.
1. A dash means that data not available.
2. To convert grams to ounces (weight), divide by 28.35; to convert grams to fluid ounces (volume), divide by 29.6.

LONG JOHN SILVER'S

Long John Silver's (LJS) has seen the light! Until recently, the nation's biggest seafood chain (about 1,500 restaurants) fried everything they got their hands on. But the chain has discovered that it can capitalize on the public's growing concern about nutrition by offering baked fish and chicken. LJS is also testing a line of grilled fish and chicken, and a few restaurants offer baked potatoes.

Nice-sized servings of baked fish (cod), shrimp scampi, or chicken can be the centerpiece of a meal that also includes rice, green beans, corn on the cob, or cole slaw. Seasoning the fish and vegetables with lemon juice can help keep the calories down. Having two pieces of fish, rice pilaf, and a small salad would give you an almost fat-free 300-calorie dinner. Three pieces of baked fish with rice pilaf, green beans, slaw, and a breadstick has twice the calories, but still a fairly reasonable Gloom rating of 31. The breadstick is now fried, but will be changed to baked "real soon," according to a company official.

If you're in the mood for a dinner salad, the 150-calorie Ocean Chef Salad or 230-calorie Seafood Salad are good choices. Be aware, though, that the Ocean Chef Salad contains lots of surimi, as well as some little shrimp, shredded carrots, lettuce, and cheese. Surimi is simulated crab, which at Long John Silver's tastes more of salt than crab. It is made from a blend of pollock and other whitefish,

CARBOHYDRATES (gm.)	ADDED SUGAR[3] (gm.)	FAT[4] (gm.)	FAT % CALORIES	SATURATED FAT (gm.)	CHOLESTEROL (mg.)	SODIUM (mg.)	VITAMIN A (% U.S. RDA)	VITAMIN C (% U.S. RDA)	IRON (% U.S. RDA)	CALCIUM (% U.S. RDA)	GLOOM
88	0	86	56	21	287	2951	9	36	27	24	135

3. *To convert grams of sugar to teaspoons of sugar, divide by 4.0.*
4. *To convert grams of fat to teaspoons of fat, divide by 4.4.*

crab meat, and a flock of additives and salt. Surimi is popular in Japan, and probably OK — *if* consumers know they're getting an imitation product and are told the ingredients. Two clerks at a Washington, D.C., restaurant insisted that the stuff was real crab until we got the manager to acknowledge that it was "crab and whitefish." A serving of thick seafood gumbo or seafood chowder with cod can round out your meal.

Long John Silver's still has a vast array of meals based on breaded, battered, and fried fish, chicken, and potatoes. These meals are loaded with calories, fat, and sodium. Consider, for instance, a three-piece Homestyle (that means breaded, not battered) Fish Dinner, which comes with french fries (or "Fryes"), coleslaw, and two hush puppies. It rates 48 on the Gloom scale, provides 830 calories, 9 teaspoons of fat, and half a teaspoon of salt. A three-piece Chicken Plank Dinner (generous-sized chicken tenderloins), with fries, coleslaw, and hush puppies, provides 860 calories, 8 teaspoons of fat, and 64 Gloom points. You'll definitely have to wash your hands after those meals! Still, LJS deserves credit for cutting the fat and sodium of many of its meals by as much as one-third in the past few years.

If you want fried foods, consider ordering a couple of Chicken Planks (130 calories each) or pieces of Homestyle Fish (125 calories each), along with vegetables or Corn Cobbette (without the fake butter sauce) and a breadstick. Remove the greasy coating on the fish or chicken if you're more health-conscious than starving.

Long John Silver's may be the only fast-food restaurant that serves french fries without added salt, even though practically every-

Long John Silver's Complete Nutritional Values[1]	WEIGHT[2] (gm.)	CALORIES	PROTEIN (gm.)
Breadstick	—	—	—
Chicken Plank, 1 piece	56	130	9
✔ Chicken Sandwich, Baked, without sandwich sauce	189	320	33
Chicken Sandwich, Battered-Dipped, 2 pieces, without sandwich sauce	184	440	25
Chowder, Seafood, with Cod	198	140	11
Clams, Breaded	66	240	7
Club Crackers, 1 packet	7	35	0
Cole Slaw, drained on fork	98	140	1
Corn Cobbette, ½ ear	94	140	3
Dijon Herb Sauce, 1 packet (1 oz.)	25	90	0
Fish Sandwich, Battered-Dipped, 1 piece, without sandwich sauce	160	380	19
Fish, Battered, 1 piece	88	210	12
Fish, Homestyle, 1 piece	47	125	7
✔ Fryes	85	170	3
✔ Green Beans	113	30	1
Honey-Mustard Sauce, 1 packet (1 oz.)	25	45	0

thing else is loaded with salt. They do all their frying in vegetable shortening.

Long John Silver's provides nutrition, but no ingredient, information about its products by means of a toll-free consumer hotline (see Appendix).

CARBOHYDRATES (gm.)	ADDED SUGAR[3] (gm.)	FAT[a] (gm.)	FAT % CALORIES	SATURATED FAT (gm.)	CHOLESTEROL (mg.)	SODIUM (mg.)	VITAMIN A (% U.S. RDA)	VITAMIN C (% U.S. RDA)	IRON (% U.S. RDA)	CALCIUM (% U.S. RDA)	GLOOM
—	—	—	—	—	—	—	—	—	—	—	—
10	0	6	42	2	45	490	0	2	4	0	13
29	0	8	23	2	70	900	2	0	10	15	17
47	0	17	35	4	95	1280	0	6	15	15	33
10	0	6	39	2	20	590	15	0	10	20	9
26	0	12	45	3	4	410	0	0	6	4	19
5	0	2	51	0	0	85	0	0	0	0	4
20	0	6	39	1	15	260	4	0	4	6	9
18	0	8	51	0	0	0	6	15	2	0	8
6	0	7	70	1	5	220	0	0	0	0	12
40	0	16	38	4	30	860	0	4	15	15	26
13	0	12	51	3	30	570	0	0	4	0	21
9	0	7	50	2	20	200	2	0	45	0	8
26	0	6	32	2	0	55	0	20	10	0	7
6	0	0	0	0	0	540	0	6	4	2	3
10	0	0	0	0	0	125	0	0	0	0	1

LONG JOHN SILVER'S

COMPLETE NUTRITIONAL VALUES[1] CONTINUED	WEIGHT[2] (gm.)	CALORIES	PROTEIN (gm.)
Hushpuppy, 1 piece	24	70	2
✔ Milk, 2% Low-fat, 8 fl. oz.	244	121	8
Pie, Apple	128	320	3
Pie, Cherry	128	360	4
Pie, Lemon	113	340	7
✔ Rice Pilaf	142	210	5
✔ Salad Dressing, Lite Italian, 1 packet (1 oz.)	28	12	0
Salad Dressing, Ranch, 1 packet (1 oz.)	28	90	0
Salad Dressing, Sea Salad, 1 packet (1 oz.)	28	90	0
Salad, Ocean Chef	234	150	15
✔ Salad, Seafood	278	230	16
✔ Salad, Side	115	25	1
✔ Salad, small	54	11	0
Seafood Gumbo with Cod	198	120	9
Seafood Sauce, 1 packet	25	35	0
Shrimp, battered, 1 piece	17	60	4
Shrimp, Homestyle, 1 piece	16	45	2
Sweet 'n Sour Sauce, 1 packet (1 oz.)	25	40	0
Tartar Sauce, 1 packet (1 oz.)	25	70	0
Long John's Meals			
✔ Light Portion Baked Fish with paprika, 2 pieces with rice pilaf and small salad	284	300	24
Light Portion Baked Fish with lemon crumb, 2 pieces, with rice pilaf and small salad	291	320	24

CARBOHYDRATES (gm.)	ADDED SUGAR[3] (gm.)	FAT[4] (gm.)	FAT % CALORIES	SATURATED FAT (gm.)	CHOLESTEROL (mg.)	SODIUM (mg.)	VITAMIN A (% U.S. RDA)	VITAMIN C (% U.S. RDA)	IRON (% U.S. RDA)	CALCIUM (% U.S. RDA)	GLOOM
10	0	2	26	0	4	25	0	0	4	4	3
12	0	4	33	3	18	122	10	4	0	30	5
45	18	13	37	5	0	420	0	8	6	0	23
55	22	13	33	4	5	200	8	15	6	0	20
60	32	9	24	3	45	120	0	0	4	15	19
43	0	2	9	0	0	570	0	0	6	4	8
1	0	0	0	0	0	410	0	0	0	0	5
18	0	2	20	0	0	230	0	0	0	2	5
13	0	4	40	2	0	160	4	0	0	0	7
13	0	5	30	0	40	860	70	35	20	20	9
31	0	5	20	1	90	580	60	30	25	15	10
5	0	0	0	0	0	15	40	25	4	2	0
2	0	0	0	0	0	5	15	15	0	0	0
4	0	8	60	2	25	740	20	0	10	10	12
6	0	0	0	0	0	380	4	0	2	2	3
4	0	4	60	1	15	180	0	0	0	0	7
4	0	3	60	1	15	70	0	0	0	0	5
10	0	0	0	0	0	95	0	0	0	0	1
10	0	3	39	1	0	70	0	0	0	2	5
45	0	2	6	1	70	650	35	10	10	6	9
49	0	4	11	1	75	900	35	10	10	6	13

LONG JOHN SILVER'S

COMPLETE NUTRITIONAL VALUES[1] CONTINUED

	WEIGHT[2] (gm.)	CALORIES	PROTEIN (gm.)
Baked Fish with paprika, 3 pieces with Fryes, slaw, 2 Hushpuppies	515	610	38
Ocean Chef Salad with Sea Salad Dressing, Seafood Gumbo with Cod, Club Crackers	467	395	24
Baked Fish with lemon crumb, 3 pieces with rice pilaf, green beans, slaw, breadstick	522	640	39
Baked Fish with scampi sauce, 3 pieces with rice pilaf, green beans, slaw, breadstick	529	660	38
Baked Chicken with rice pilaf, green beans, slaw, breadstick	498	630	35
Baked Shrimp Scampi with rice pilaf, green beans, slaw, breadstick	529	610	25
Long John's Homestyle Fish dinner (3 pieces, with Fryes, slaw, 2 Hushpuppies)	372	830	28
Homestyle Shrimp dinner (6 shrimp with Fryes, slaw, 2 Hushpuppies)	327	740	17
Fish (2 pieces) & Fryes with 2 Hushpuppies	309	720	29
Long John's Homestyle Fish dinner (4 pieces, with Fryes, slaw, 2 Hushpuppies)	419	960	35
Batter-dipped Fish Sandwich with tartar sauce, Fryes, Side Salad, Apple Pie	513	965	26
Fish & More (2 Fish, Fryes, slaw, 2 Hushpuppies)	407	860	31
Chicken Planks dinner (3 Planks, Fryes, slaw, 2 Hushpuppies)	399	860	34
Homestyle Shrimp dinner (9 shrimp with Fryes, slaw, 2 Hushpuppies)	375	880	22
Batter-dipped Shrimp dinner (6 shrimp with Fryes, slaw, Hushpuppies)	333	800	19

CARBOHYDRATES (gm.)	ADDED SUGAR[3] (gm.)	FAT[4] (gm.)	FAT % CALORIES	SATURATED FAT (gm.)	CHOLESTEROL (mg.)	SODIUM (mg.)	VITAMIN A (% U.S. RDA)	VITAMIN C (% U.S. RDA)	IRON (% U.S. RDA)	CALCIUM (% U.S. RDA)	GLOOM
85	0	12	18	2	125	1620	45	6	25	20	28
35	0	19	43	4	65	1845	94	35	30	30	28
89	0	14	20	2	125	1870	50	6	25	20	31
87	0	18	25	3	125	1780	45	6	25	20	35
85	0	17	24	3	85	2170	6	6	25	20	38
87	0	18	27	3	220	2120	8	6	25	20	46
93	0	39	42	9	75	980	10	25	160	20	48
90	0	33	40	8	90	780	4	25	25	20	49
72	0	35	44	10	60	1240	0	25	25	10	54
102	0	46	43	11	95	1180	15	25	210	20	55
126	18	38	35	11	30	1420	40	57	35	19	56
92	0	42	44	11	75	1500	4	25	30	20	64
96	0	37	39	9	160	1840	4	30	30	20	64
102	0	41	42	11	130	980	4	25	30	20	64
90	0	40	45	10	95	1420	4	25	25	20	65

LONG JOHN SILVER'S

COMPLETE NUTRITIONAL VALUES[1] CONTINUED

	WEIGHT[2] (gm.)	CALORIES	PROTEIN (gm.)
Fish (1 piece) & Chicken (2 pieces) dinner with Fryes, slaw, 2 Hushpuppies	431	930	37
Fish (3 pieces) & Fryes with 2 Hushpuppies	397	930	41
Chicken Planks dinner (4 Planks, Fryes, slaw, 2 Hushpuppies)	455	990	43
Fish & More (3 Fish with Fryes, slaw, 2 Hushpuppies)	495	1070	43
Batter-dipped Shrimp dinner (9 shrimp with Fryes, slaw, Hushpuppies)	384	970	24

1. A dash means that data not available.
2. To convert grams to ounces (weight), divide by 28.35; to convert grams to fluid ounces (volume), divide by 29.6.

McDONALD'S

There are more and more reasons to go to McDonald's. And they all added up to $18.7 billion in sales in 1990 and a total of over 80 billion hamburgers served over the years.

Twenty-two million people a day can't all be wrong. And they're not. Accessibility has something to do with it. About 8,600 pairs of golden arches adorn American street corners, interstates, and shopping malls. Over 3,200 more pairs of arches have sprouted up in 52 countries overseas, with expansion proceeding much more rapidly than in the United States. A new McDonald's opens its doors somewhere on this globe every 15 hours. Walk through any one of them and you find the standard for the fast-food french fry, a decent cup of coffee, and a hamburger not more than 10 minutes old.

McDonald's has long been the leader among fast-food restaurants. It has the most massive television advertising campaigns,

CARBOHYDRATES (gm.)	ADDED SUGAR[3] (gm.)	FAT[4] (gm.)	FAT % CALORIES	SATURATED FAT (gm.)	CHOLESTEROL (mg.)	SODIUM (mg.)	VITAMIN A (% U.S. RDA)	VITAMIN C (% U.S. RDA)	IRON (% U.S. RDA)	CALCIUM (% U.S. RDA)	GLOOM
99	0	43	42	11	140	1920	4	30	30	20	71
85	0	47	45	13	90	1800	0	25	30	15	74
106	0	44	40	11	205	2330	4	30	35	20	78
105	0	53	45	14	100	2070	4	30	35	20	82
102	0	51	47	13	135	1940	4	25	25	20	86

3. To convert grams of sugar to teaspoons of sugar, divide by 4.0.
4. To convert grams of fat to teaspoons of fat, divide by 4.4.

amounting to over $400 million annually in the United States. It broke open the breakfast market. It showed that burger joints could also fry chicken. And it has made more nutritional breakthroughs in the past couple of years than any other restaurant chain.

McDonald's menu has come a long way from its burger, fries, and shake or soft-drink days. For instance, if you chose a fat-free apple-bran muffin, Cheerios, milk (1% low-fat), and fruit juice for breakfast you'd have a very low-fat, 450-calorie meal with a Gloom of just 7. Congratulations, McDonald's!

On the other hand, die-hard cholesterol addicts can still choose a Biscuit with Sausage & Egg (63 Gloom points) and begin their day with a meal that gets about 60 percent of its 500 calories from fat and contains 1,210 milligrams of sodium. Or they could grab a cinnamon raisin danish (440 calories and 5 teaspoons of fat) and coffee with cream.

The lunch and dinner menus offer some surprises, both good

and bad. While fish and chicken are often touted as healthful foods, McDonald's Filet-O-Fish sandwich and Chicken McNuggets get about 50 percent of their calories from fat, putting them in the same ballpark as the Big Mac. In fact, the Filet-O-Fish sandwich has twice as much fat as the regular hamburger. Hold the tartar sauce!

Though arch-competitors Burger King and Hardee's make chicken nuggets out of strips of chicken, McDonald's nuggets are made of processed chicken. A small order of McNuggets provides 270 calories, more than 3 teaspoons of fat, and 24 Gloom points, that's one-sixth fewer calories and one-fourth less fat than five years ago. Moreover, McNuggets no longer contain ground-up chicken skin.

If you want a chicken sandwich, the McChicken, also made from bits and pieces of chicken, is for you. But ask for it without the mayonnaise, which probably accounts for at least 100 of its 415 calories and about 10 of its 30 Gloom points. You can slice off more calories by removing the batter. And maybe set aside the spongy bun. Oh, well, maybe this sandwich—named one of the worst in 1989 by *Nutrition Action Healthletter*—isn't for you, after all. Burger King's BK Broiler Chicken Sandwich (Gloom = 15) or Carl's Jr.'s Charbroiler BBQ Chicken Sandwich (Gloom = 12) is far better. McDonald's started testing its own low-fat grilled chicken sandwich in 500 stores in 1990, so help may be on the way. Its Gloom rating of 10 beats all of its competitors.

Burgers, though they account for a much smaller percentage of sales than they used to, are still the core of McDonald's business. Eighty billion sold means something.

"Thinking small" is our best advice. The regular McDonald's burger "costs" 255 calories, the Quarter Pounder 410 calories, the double-burger Big Mac 500 calories. Adding cheese will add 50 calories to the small burger, 100 to the Quarter Pounder.

McDonald's made national headlines in November 1990, when it started testing the lowest-fat hamburger available at any restaurant. Together with Auburn University scientists, McDonald's developed a hamburger with half the normal fat content. The addition of 9 percent water, natural flavorings, and a small amount of carrageenan, a harmless additive, makes the remarkably low fat content possible. A McLean Deluxe quarter-pound burger sandwich with ketchup, mustard, tomato, and other toppings gives you 90 fewer calories than a Quarter Pounder, as well as half the fat. The McLean Deluxe's Gloom rating of 18 compares favorably to the Quarter Pounder's Gloom rating of 31. And it is only two points higher than the regular hamburger's rating. A packet of reduced-calorie mayonnaise is available with the sandwich. Adding a slice of cheese to the McLean Deluxe boosts the calories to 370 and the Gloom to 24. Though the reviews of the McLean Deluxe's taste have been mixed, the revolutionary new product is selling well enough throughout the nation that McDonald's has eliminated the fatty McD.L.T. sandwich from its menu.

McDonald's has long courted the kiddie market with Happy Meals, which usually consist of a hamburger, fries, and a Coke in a cartoon-festooned bag. That meal's 22 grams of fat leads to a Gloom rating of 38. You could create your own "Happier Meal" by choosing a hamburger (McLean Deluxe for bigger appetites), carrot sticks, and low-fat milk. The clerk will even sell you the official bag. Such a meal has only 11 grams of fat, two days' worth of vitamin A, and a Gloom rating of just 14.

McDonald's offers several prepackaged salads, which account for about 7 percent of the company's total sales. Probably even more people would choose salads if they were less expensive. At several McDonald's that we visited, the chicken and chef salads were the most expensive items on the menu, costing as much as 50 percent more than the next-highest-price item, a McLean Deluxe with Cheese.

The salads, which weigh 4 to 10 ounces, are a wonderful alternative to greasy burgers. The Chunky Chicken Salad is your best nutritional bet, with under 1 teaspoon of saturated fat and 150 calories. The chef salad provides 170 calories and 3 teaspoons of fat (from the cheese, ham, and egg). The lowest-calorie salads are the garden salad (50 calories) and the side salad (30 calories). Several regular and reduced-calorie dressings are available; the Lite Vinaigrette is lowest in calories.

Whatever entrée you choose, you can boost the nutrients and hold down the calories by drinking water, orange or grapefruit juice, or low-fat milk (McDonald's was the first chain to offer milk with just 1 percent fat). The shakes, reformulated in 1990 with a remarkable 80 percent less fat, are the healthiest in the business, but they still add about 300 calories to a meal.

In another amazing development, McDonald's outlets now offer carrot and celery sticks. Those items, which come in sealed plastic bags, are a rare source of dietary fiber in fast-food land and make great take-out meals.

For dessert or a snack, McDonald's now offers a low-fat frozen yogurt cone (105 calories) and a yogurt-based sundae (210 to 270 calories). Some outlets offer nonfat Orange Sorbet Ice. (This product is made largely from water, sugar, and orange juice, and is not true sorbet, which contains real fruit.) Those products put McDonald's way ahead of other burger joints. On the other hand, the fried apple pie and Chocolaty Chip Cookies are still high in calories and fat. Of course, you could also take a nice walk for dessert and burn up a few of the calories from the rest of your meal.

Pizza Hut and Domino's: Beware! McDonald's is testing a 12-inch pizza (made from frozen dough) in at least 200 outlets. Early reviews of its taste are mixed, at best. A few outlets are also testing bagels, turkey and other subs, huevos rancheros, catfish sandwiches, grilled

McDonald's

Complete Nutritional Values[1]

	WEIGHT[2] (gm.)	CALORIES	PROTEIN (gm.)
Apple Juice, 6 fl. oz.	180	91	0
Apple Pie	85	260	2
Bacon Bits	3	15	1
Barbeque Sauce, 1 fl. oz.	32	50	0
Big Mac	215	500	25
Biscuit with Biscuit Spread	75	260	5
Biscuit with Bacon, Egg & Cheese	153	430	15
Biscuit with Sausage	118	420	12
Biscuit with Sausage & Egg	175	500	19

chicken sandwiches, bone-in fried and skinless broiled chicken, and dinner meals, including pasta dishes and fried shrimp with coleslaw and curly fries.

Many McDonald's meals are still fairly high in the baddies (fat, cholesterol, sodium, and sugar) and low in the goodies (vitamins A and C and fiber), but the company deserves great credit for not having disastrously large and fatty sandwiches (like Burger King's Double Whopper with Cheese), for developing a variety of innovative and healthful new items, and for making numerous improvements in old items. For instance, compared to several years ago, an Egg McMuffin has one-third less fat, the Hotcakes have 40 percent less sodium, and the Biscuit with Bacon, Egg & Cheese is down 40 calories and 6 grams of fat. In 1990 McDonald's switched to vegetable shortening from beef fat for all of its frying. However, the Filet-O-Fish has much more sodium than it used to.

McDonald's has long published nutrition information about its products. Many outlets will give you a free copy of a nutrition and ingredient brochure. And all outlets display a large poster with the same information.

CARBOHYDRATES (gm.)	ADDED SUGAR[3] (gm.)	FAT[4] (gm.)	FAT % CALORIES	SATURATED FAT (gm.)	CHOLESTEROL (mg.)	SODIUM (mg.)	VITAMIN A (% U.S. RDA)	VITAMIN C (% U.S. RDA)	IRON (% U.S. RDA)	CALCIUM (% U.S. RDA)	GLOOM
50	0	0	0	0	0	5	0	2	4	0	0
30	12	15	52	4	6	240	0	20	4	0	23
0	0	1	60	1	1	95	0	0	0	0	2
12	11	1	9	0	0	340	4	4	2	0	4
42	5	26	47	9	100	890	6	2	20	25	41
32	1	13	45	3	1	730	0	0	8	8	23
33	1	26	54	8	248	1190	10	0	15	20	52
32	1	28	60	8	44	1040	0	0	10	8	47
33	1	33	59	10	270	1210	6	0	20	10	63

McDONALD'S

COMPLETE NUTRITIONAL VALUES[1] CONTINUED	WEIGHT[2] (gm.)	CALORIES	PROTEIN (gm.)
Breakfast Burrito	105	280	12
✔ Carrot Sticks	85	37	0
✔ Celery Sticks	85	14	0
✔ Cheerios, ¾ cup	19	80	3
Cheeseburger	116	305	15
Chicken Fajitas	82	185	11
Chicken McNuggets, 6 pieces	113	270	20
Chocolaty Chip Cookies	56	330	4
Coca-Cola Classic,* 12 fl. oz.	360	140	0
Coca-Cola Classic, 32 fl. oz.	960	380	0
Coke, Diet, 12 fl. oz.	360	1	0
Croutons	11	50	1
Danish, Apple	115	390	6
Danish, Cinnamon Raisin	110	440	6
Danish, Iced Cheese	110	390	7
Danish, Raspberry	117	410	6
English Muffin with Margarine	58	170	5
Filet-O-Fish	141	370	14
French Fries, large	122	400	6
French Fries, medium	97	320	4
French Fries, small	68	220	3
✔ Frozen Yogurt Cone, Vanilla, Low-fat	85	105	4
Frozen Yogurt Sundae, Hot Caramel, Low-fat	174	270	7
Frozen Yogurt Sundae, Hot Fudge, Low-fat	169	240	7

CARBOHYDRATES (gm.)	ADDED SUGAR[3] (gm.)	FAT[4] (gm.)	FAT % CALORIES	SATURATED FAT (gm.)	CHOLESTEROL (mg.)	SODIUM (mg.)	VITAMIN A (% U.S. RDA)	VITAMIN C (% U.S. RDA)	IRON (% U.S. RDA)	CALCIUM (% U.S. RDA)	GLOOM
21	0	17	55	6	135	580	10	10	8	10	29
9	0	0	0	0	0	40	240	10	2	2	0
3	0	0	0	0	0	100	0	10	0	2	1
14	1	1	11	0	0	210	15	15	30	2	2
30	5	13	38	5	50	710	8	4	15	20	22
20	0	8	39	3	35	310	2	8	4	8	12
17	0	15	50	4	56	580	0	0	6	0	24
42	18	16	43	5	4	280	0	0	10	2	27
38	38	0	0	0	0	15	0	0	0	0	6
101	101	0	0	0	0	40	0	0	0	0	16
0	0	0	0	0	0	30	0	0	0	0	0
7	0	2	36	1	0	140	0	0	0	0	4
51	30	17	39	4	25	370	0	25	8	0	29
58	32	21	43	5	34	430	0	6	10	4	38
42	24	21	48	6	47	420	4	0	8	4	38
62	36	16	35	3	26	310	0	6	8	0	30
26	0	5	26	1	0	230	2	0	8	15	7
38	5	18	44	4	50	930	2	0	10	15	32
46	0	22	50	5	0	200	0	25	6	0	29
36	0	17	48	4	0	150	0	20	4	0	23
26	0	12	49	3	0	110	0	15	2	0	16
22	6	1	7	0	3	80	2	0	0	10	2
59	30	3	9	2	13	180	6	0	0	20	9
50	29	3	12	2	6	170	4	0	2	25	8

McDONALD'S

COMPLETE NUTRITIONAL VALUES[1] CONTINUED

	WEIGHT[2] (gm.)	CALORIES	PROTEIN (gm.)
Frozen Yogurt Sundae, Strawberry, Low-fat	171	210	6
✔ Grapefruit Juice, 6 fl. oz.	183	80	1
✔ Grilled Chicken Breast Sandwich (test product)	177	252	24
✔ Hamburger	102	255	12
Hashbrown Potatoes	53	130	1
Honey, ½ oz.	14	45	0
Hot Mustard Sauce, 1 fl. oz.	30	70	0
✔ Hotcakes with Margarine and Syrup	176	410	8
McChicken	187	415	19
McDonaldland Cookies	56	290	4
✔ McLean Deluxe	206	320	22
McLean Deluxe with Cheese	219	370	24
McMuffin, Egg	135	280	18
McMuffin, Sausage	135	345	15
McMuffin, Sausage with Egg	159	415	21
McRib Sandwich	184	445	24
Milk Shake, Chocolate Low-fat	293	320	11
Milk Shake, Strawberry Low-fat	293	320	11
Milk Shake, Vanilla Low-fat	293	290	11
✔ Milk, 1% Low-fat, 8 fl. oz.	240	110	9
✔ Muffin, Fat-free Apple Bran	75	180	5
Muffin, Fat-free Blueberry	75	170	3
Orange Drink, 12 fl. oz.	360	130	0
✔ Orange Juice, 6 fl. oz.	183	80	1

CARBOHYDRATES (gm.)	ADDED SUGAR[3] (gm.)	FAT[4] (gm.)	FAT % CALORIES	SATURATED FAT (gm.)	CHOLESTEROL (mg.)	SODIUM (mg.)	VITAMIN A (% U.S. RDA)	VITAMIN C (% U.S. RDA)	IRON (% U.S. RDA)	CALCIUM (% U.S. RDA)	GLOOM
49	30	1	5	1	5	95	4	2	0	20	5
19	0	0	0	0	0	0	0	100	0	0	0
30	5	4	14	1	50	740	8	8	15	15	11
30	5	9	32	3	37	490	4	4	15	10	16
15	0	7	48	1	0	330	0	2	0	0	13
12	12	0	0	0	0	0	0	0	0	0	2
8	8	4	46	1	5	250	0	0	0	2	9
74	17	9	20	1	8	640	4	0	10	10	18
39	5	20	43	4	42	770	2	4	15	15	30
47	13	9	29	2	0	300	0	0	10	0	17
35	5	10	28	4	60	670	10	10	20	15	18
35	5	14	34	5	75	890	15	10	20	20	24
28	0	11	35	4	224	710	10	0	15	25	26
27	0	20	52	7	57	770	4	0	15	20	31
27	0	25	54	8	256	915	10	0	20	25	45
48	0	22	44	0	75	972	—	—	—	—	40
66	35	2	5	1	10	240	6	0	0	35	8
67	35	1	4	1	10	170	6	0	0	35	7
60	33	1	4	1	10	170	6	0	0	35	6
12	0	2	16	2	10	130	10	4	0	30	3
40	9	0	0	0	0	200	0	0	6	4	3
40	0	0	0	0	0	220	0	2	4	8	2
33	33	0	0	0	0	10	0	0	0	0	5
19	0	0	0	0	0	0	0	120	0	0	0

McDONALD'S

COMPLETE NUTRITIONAL VALUES[1] CONTINUED	WEIGHT[2] (gm.)	CALORIES	PROTEIN (gm.)
Orange Sorbet Ice, cone (4 oz.)	—	106	0
Orange Sorbet Ice, sundae (6½ oz.), no topping	—	142	0
Orange Sorbet Ice/Low-fat Frozen Yogurt Twist Cone	—	104	2
Orange Sorbet Ice/Low-fat Frozen Yogurt Twist Sundae	—	138	3
Pork Sausage	48	180	8
Quarter Pounder	166	410	23
Quarter Pounder with Cheese	194	510	28
Salad Dressing, Bleu Cheese, 2.5 fl. oz.	75	250	2
✔ Salad Dressing, Lite Vinaigrette, 2 fl. oz.	60	48	0
Salad Dressing, Ranch, 2 fl. oz.	60	220	0
Salad Dressing, Red French Reduced-Calorie, 2 fl. oz.	60	160	0
Salad Dressing, Thousand Island, 2.5 fl. oz.	75	390	0
Salad, Chef	265	170	17
✔ Salad, Chunky Chicken	255	150	25
✔ Salad, Garden	189	50	4
✔ Salad, Side	106	30	2
Sausage	43	160	7
Scrambled Eggs	100	140	12
Sprite, 12 fl. oz.	360	140	0
Sweet and Sour Sauce, 1 fl. oz.	32	60	0
Wheaties, ¾ cup	23	90	2

CARBOHYDRATES (gm.)	ADDED SUGAR[3] (gm.)	FAT[2] (gm.)	FAT % CALORIES	SATURATED FAT (gm.)	CHOLESTEROL (mg.)	SODIUM (mg.)	VITAMIN A (% U.S. RDA)	VITAMIN C (% U.S. RDA)	IRON (% U.S. RDA)	CALCIUM (% U.S. RDA)	GLOOM
27	23	0	2	0	0	25	0	30	0	0	2
38	30	0	0	0	0	0	0	55	0	0	3
25	18	1	4	0	1	50	0	15	0	6	3
34	23	1	3	0	2	45	0	25	0	10	3
0	0	16	82	6	48	350	0	0	4	0	26
34	5	20	44	8	85	650	4	6	20	15	31
34	5	28	49	11	115	1090	15	6	20	30	43
5	3	20	72	5	35	750	0	0	0	0	39
8	7	2	38	0	0	240	0	0	0	0	7
4	3	20	82	4	20	520	0	0	0	0	35
20	18	8	45	1	0	460	0	0	0	0	19
10	8	40	92	5	40	500	0	0	0	0	64
8	0	9	48	4	111	400	100	35	8	15	12
7	0	4	24	1	78	230	170	45	6	4	6
6	0	2	36	1	65	70	90	35	8	4	3
4	0	1	30	0	33	35	80	20	4	2	2
0	0	15	84	5	43	310	0	0	4	0	24
1	0	10	64	3	399	290	10	0	10	6	28
36	36	0	0	0	0	15	0	0	0	0	6
14	12	0	3	0	0	190	6	0	0	0	3
19	3	1	10	0	0	220	20	20	20	2	2

McDONALD'S

COMPLETE NUTRITIONAL VALUES¹ CONTINUED

	WEIGHT² (gm.)	CALORIES	PROTEIN (gm.)
McDonald's Meals			
✔ Cheerios, 1% Low-fat Milk, Orange Juice, Fat-Free Apple Bran Muffin	517	450	18
✔ Hotcakes with Margarine & Syrup, Orange Juice, 1% Low-fat Milk	597	600	18
Cinnamon Raisin Danish, Orange Juice	293	520	8
Scrambled Eggs, English Muffin with Margarine, black coffee	158	310	17
Biscuit with Sausage & Egg, 1% Low-fat Milk, Orange Juice, Hash Brown Potatoes	651	820	30
✔ Chunky Chicken Salad, Lite Vinaigrette Dressing, Orange Juice	498	278	26
✔ Happier Meal: Hamburger, Carrot Sticks, Low-fat Milk	427	402	21
McLean Deluxe, Side Salad, 1% Low-fat Milk	552	460	33
Filet-O-Fish, Side Salad, Lite Vinaigrette Dressing, Diet Coke	667	449	16
Happy Meal: Hamburger, French Fries, small, Coca-Cola, 12 fl. oz.	530	620	16
Chef Salad, Croutons, Bacon Bits, Ranch Dressing, Diet Coke	699	456	19
Hamburger, French Fries, medium, Strawberry Low-fat Frozen Yogurt Sundae, Coca-Cola Classic, 12 fl. oz.	730	925	22
Chicken McNuggets (6 pieces), Barbeque Sauce, French Fries, medium, Strawberry Low-fat Milk Shake	535	960	35

CARBOHYDRATES (gm.)	ADDED SUGAR[3] (gm.)	FAT[4] (gm.)	FAT % CALORIES	SATURATED FAT (gm.)	CHOLESTEROL (mg.)	SODIUM (mg.)	VITAMIN A (% U.S. RDA)	VITAMIN C (% U.S. RDA)	IRON (% U.S. RDA)	CALCIUM (% U.S. RDA)	GLOOM
85	10	3	6	2	10	540	25	139	36	36	7
105	17	11	17	3	18	770	14	124	10	40	17
76	32	21	36	4	34	430	0	126	10	24	27
27	0	15	44	4	399	520	12	0	18	21	37
79	1	42	46	13	280	1670	16	126	20	40	65
34	7	6	19	1	78	470	170	165	6	4	9
51	5	11	25	5	47	660	254	18	17	42	14
51	5	13	25	6	103	835	100	34	24	47	20
50	12	21	42	5	83	1235	82	20	14	17	32
94	43	22	31	7	37	585	4	19	17	10	38
19	3	32	63	9	132	1185	100	35	8	15	41
153	73	27	26	7	42	750	8	26	19	30	49
132	46	34	32	8	66	1240	10	24	12	35	58

McDONALD'S

COMPLETE NUTRITIONAL VALUES¹ CONTINUED	WEIGHT² (gm.)	CALORIES	PROTEIN (gm.)
McChicken, French Fries, medium, Coca-Cola Classic, 32 fl. oz., Low-fat Frozen Yogurt Hot Fudge Sundae	1413	1355	30
Quarter Pounder, French Fries, medium, Side Salad, Blue Cheese Dressing, Low-fat Frozen Yogurt	529	1115	35
Quarter Pounder with Cheese, French Fries, large, Chocolate Low-fat Milk Shake, Apple Pie	669	1490	47

* Also available in 16 fl. oz. and 22 fl. oz.
1. A dash means that data not available.
2. To convert grams to ounces (weight), divide by 28.35; to convert grams to fluid ounces (volume), divide by 29.6.

PIZZA HUT

Considering that Pizza Hut takes 5 minutes to bake lunch-time pizzas and between 15 and 20 minutes to serve up a meal at other times, the 6,600-restaurant chain (plus another 1,400 shops abroad) is not a standard fast-food chain. But Pizza Hut can get you a pizza fast, if you call in your order in advance, and over 1,500 outlets now offer home delivery. As one executive told *Nation's Restaurant News,* "We want to be there for every occasion when a customer thinks pizza, be it delivery, take-out, drive-through, lunch, eat-in, dinner, all of it." In any case, because Pizza Hut is so huge (its $4 billion sales represent one-quarter of the entire pizza market) and has the fast-food look and a standardized menu, we thought Pizza Hut deserved a close look.

Pizza Hut does offer a few non-pizza dishes, but pizza is king. You have four choices: Thin 'n Crispy, Traditional Hand-Tossed

CARBOHYDRATES (gm.)	ADDED SUGAR³ (gm.)	FAT⁴ (gm.)	FAT % CALORIES	SATURATED FAT (gm.)	CHOLESTEROL (mg.)	SODIUM (mg.)	VITAMIN A (% U.S. RDA)	VITAMIN C (% U.S. RDA)	IRON (% U.S. RDA)	CALCIUM (% U.S. RDA)	GLOOM
226	135	40	27	10	48	1130	6	24	21	40	77
101	14	59	47	17	156	1665	86	46	28	27	82
176	52	67	40	21	131	1770	21	51	30	65	103

3. To convert grams of sugar to teaspoons of sugar, divide by 4.0.
4. To convert grams of fat to teaspoons of fat, divide by 4.4.

(medium-thick crust), and Pan (deep-dish) pizzas; all cost the same, but generally increase in calories. The fourth choice, Personal Pan Pizzas (available only during lunch hours), contain up to 675 calories, so consider taking half of one home for dinner. The traditional Hand-Tossed variety is said to be best for home delivery, because it is least prone to getting soggy.

Toppings range from pepperoni, beef, pork, and extra cheese to mushrooms, green pepper, black olives, and onions. Obviously, the second group of toppings will be kinder to your cholesterol level than the first. The Super Supreme offers all nine toppings at once; the Meat Lover's Pizza has six different toppings (Pizza Hut does not provide nutrition information for Meat Lover's Pizza). If you were at one of the several Pizza Huts in Moscow, you could try a salmon and sardine pizza.

Pizza, with its tomato sauce, dough, vegetable toppings, and cheese, provides a variety of nutrients. A typical serving of two slices of a medium pizza at Pizza Hut provides substantial amounts of protein, iron, and several B vitamins; moderate amounts of vitamin C and fiber (about 6 grams); and about half your daily requirement for calcium. (Pizza Hut provides nutrition information only for its medium pizzas. Small pizzas are about half as big. Large pizzas are about one-third bigger. You can adjust the numbers in the chart accordingly.)

But pizza is not always quite as healthful as pizza-industry publicists would have you believe. For example, that typical two-slice serving of a medium hand-tossed pizza offers upwards of 500 calories, with 40 percent coming from fat. It provides hefty amounts of sodium (over 1,200 milligrams), but only 10 percent of the U.S. RDA for vitamin A.

Your best bets are probably the medium Thin 'n Crispy and pan cheese pizzas. They have about 4 teaspoons of fat and less than 1,000 milligrams of sodium per serving. Their Gloom ratings of 22 and 24 are just over half that of a Big Mac. The thin-crusted Thin 'n Crispy pizzas are a bit lower in calories then the other types.

You can create your own light pizza by not having cheese on the pizza; you'll also leave the restaurant without that terrible stuffed feeling that often comes after eating several slices of pizza. (We bet you're thinking, "No cheese? You've got to be kidding!" In fact cheeseless pizza is delicious. Try it! You may become a believer.) We tested hand-tossed pizzas with and without cheese and found we cut the fat by 85 percent and calories by 40 percent, though most of the calcium was lost also.

Pizza Hut offers a salad bar, which gives you some opportunity to round out your meal, though it's a bit paltry and tired compared to, say, Wendy's. It offers the usual cottage cheese, sliced fresh mushrooms, canned garbanzos, and canned fruit cocktails and

Pizza Hut

Complete Nutritional Values[1]

	WEIGHT[2] (gm.)	CALORIES	PROTEIN (gm.)
✔ Pan Pizza, Cheese, medium, 2 slices	205	492	30

peaches. But it's all-you-can-eat, so for $2.49 you can skip the pizza and fill yourself up. If Coke is your drink, you won't find it at Pizza Hut—the chain is owned by PepsiCo.

In order to squeeze into crowded shopping centers, train stations, and other locales that aren't big enough for a full-size restaurant, Pizza Hut has created Pizza Hut Express units. The company is installing these little operations in convenience stores, supermarkets, and wherever else they can. The Express in New York's Penn Station offers only the Personal Pan Pizza and the Big Topper, which is one-third bigger and comes with sausage, beef, pork, pepperoni, green pepper, mushroom, and onion.

Pizza Huts in Chicago and California have offered whole-wheat crust on a limited-time basis. If you are a whole-wheat fan, tell the manager of the Pizza Hut you patronize that you wish they'd offer whole-wheat pizzas, the way some independent pizzerias do. The company says that if there's enough demand, it will roll out the whole wheat.

Pizza Hut has joined the battle among fast-food giants for the children's market. Along with their "Kid's Night" program that allows kids 12 and under to eat free on Tuesday nights, Pizza Hut has mounted promotional tie-ins with Nickelodeon (the child-oriented cable network) and a variety of child-oriented movies. It is also advertising on "Teenage Mutant Ninja Turtles" video cassettes and delivering pizzas to hundreds of schools at lunchtime.

Pizza Hut does not provide nutritional information on salad-bar items, breadsticks, soft drinks, and other menu items. You can get certain ingredient information by contacting Pizza Hut's Consumer Affairs Department (see Appendix). A Pizza Hut spokesperson told us that MSG is present only in cooked breakfast sausage, and Yellow No. 5 dye is found only in hot peppers.

CARBOHYDRATES (gm.)	ADDED SUGAR[3] (gm.)	FAT[4] (gm.)	FAT % CALORIES	SATURATED FAT (gm.)	CHOLESTEROL (mg.)	SODIUM (mg.)	VITAMIN A (% U.S. RDA)	VITAMIN C (% U.S. RDA)	IRON (% U.S. RDA)	CALCIUM (% U.S. RDA)	GLOOM
57	0	18	33	9	34	940	9	12	30	63	24

PIZZA HUT

COMPLETE NUTRITIONAL VALUES[1] CONTINUED	WEIGHT[2] (gm.)	CALORIES	PROTEIN (gm.)
Pan Pizza, Pepperoni, medium, 2 slices	211	540	29
Pan Pizza, Super Supreme, medium, 2 slices	257	563	33
Pan Pizza, Supreme, medium, 2 slices	255	589	32
Personal Pan Pizza, Supreme, whole	264	647	33
Personal Pan Pizza, Pepperoni, whole	256	675	37
✔ Thin 'n Crispy Pizza, Cheese, medium, 2 slices	148	398	28
Thin 'n Crispy Pizza, Pepperoni, medium, 2 slices	146	413	26
Thin 'n Crispy Pizza, Super Supreme, medium, 2 slices	203	463	29
Thin 'n Crispy Pizza, Supreme, medium, 2 slices	200	459	28
Traditional Hand-Tossed Pizza, Cheese, medium, 2 slices	220	518	34
Traditional Hand-Tossed Pizza, Pepperoni, medium, 2 slices	197	500	28
Traditional Hand-Tossed Pizza, Super Supreme, medium, 2 slices	243	556	33
Traditional Hand-Tossed Pizza, Supreme, medium, 2 slices	239	540	32

1. Small pizzas are about half as big as the portions shown. Large pizzas are about one-third bigger.
2. To convert grams to ounces (weight), divide by 28.35; to convert grams to fluid ounces (volume), divide by 29.6.

CARBOHYDRATES (gm.)	ADDED SUGAR[3] (gm.)	FAT[4] (gm.)	FAT % CALORIES	SATURATED FAT (gm.)	CHOLESTEROL (mg.)	SODIUM (mg.)	VITAMIN A (% U.S. RDA)	VITAMIN C (% U.S. RDA)	IRON (% U.S. RDA)	CALCIUM (% U.S. RDA)	GLOOM
62	0	22	37	9	42	1127	10	14	35	52	31
53	0	26	42	12	55	1447	12	18	37	54	36
53	0	30	46	14	48	1363	12	16	28	50	41
76	0	28	39	11	49	1313	12	18	37	52	39
76	0	29	39	12	53	1335	12	17	32	73	39
37	0	17	38	10	33	867	7	8	18	66	22
36	0	20	44	11	46	986	7	10	18	45	28
44	0	21	41	10	56	1336	10	14	27	46	31
41	0	22	43	11	42	1328	10	16	33	43	31
55	0	20	35	14	55	1276	10	16	30	75	29
50	0	23	41	13	50	1267	10	12	28	44	33
54	0	25	40	13	54	1648	11	20	38	44	37
50	0	26	43	14	55	1470	11	20	45	48	36

3. To convert grams of sugar to teaspoons of sugar, divide by 4.0.
4. To convert grams of fat to teaspoons of fat, divide by 4.4.

POPEYES

Popeyes, a chain with about 750 outlets, hails from Louisiana and provides a slightly different experience for the eater. What a surprise it is to encounter tastes at a fast-food restaurant other than (or, rather, in addition to) greasy and salty.

Popeyes is a cross between authentic Cajun cooking and KFC. Don't ask us what Popeye has to do with this chain; canned spinach is certainly not offered. Popeyes does not reveal (or doesn't know) the nutrient content or ingredients of its products, so our discussion of its foods is necessarily limited.

Popeyes serves fried chicken, regular or spicy. Don't be put off by Popeyes' catch phrase, "raging cajun flavor"—the spices ain't so ragin'. You can order chicken by the piece, or get a meal consisting of several pieces of chicken, a salty biscuit, and fries or rice or mashed potatoes. The chicken and nuggets seem at least as greasy as KFC's, possibly more so. And, this chain is one of the few that still fries in beef fat instead of vegetable oil. By the way, don't bother searching the menu board for a grilled chicken sandwich or salad or frozen yogurt or other such product. As we said, Popeyes is in the fried-chicken business.

Popeyes' offerings share certain of the defects of traditional Southern cooking, often containing lard and loads of salt. The mashed potatoes would be fine were they not so salty and served with a gravy that contains bits of ground beef ("Hold the gravy!"). The Cajun rice is a nice, spicy item, but vegetarians should be warned that it is mixed with ground beef. The red beans and rice provides some dietary fiber, but this dish is flavored with ham and/or lard.

The corn-on-the-cob we sampled in Washington, D.C., in the middle of winter was excellent, and the clerk didn't blink an eye when we asked for it without butter. The coleslaw was good, not drenched in mayonnaise. Some outlets offer vitamin-packed collard greens. Others serve breakfasts, including such Southern favorites as grits (the starchy part of corn), sausage, bacon, and hash browns. They also offer Sausage and Egg Biscuit, waffles, and toast.

The beverage line is thin: no orange juice, only whole milk, and plenty of soda.

In 1989 Popeyes spawned several Super Popeyes, which offer

grilled chicken, salads, and alcoholic beverages, in addition to Popeyes standard fare. In 1989, Al Copeland, the owner of Popeyes, bought Church's Fried Chicken restaurants. In mid 1991, though, Copeland was in deep financial trouble. Don't expect him to drop everything and rush you a complete nutrition brochure.

ROY ROGERS

Roy Rogers, once boasting 600 restaurants in the Washington—New York corridor, was bought out by Hardee's in 1990. Many of those restaurants have been converted to Hardee's, though some in New York and New Jersey may still continue as Roy Rogers restaurants.

The fried chicken featured by Roy Rogers will probably be incorporated into Hardee's national menu. However, Roy Rogers extremely lean (1.5 percent fat by weight), but more expensive, roast beef may only be used in Hardee's restaurants in the Washington—New York region. For further information, see the Hardee's section (page 182).

SUBWAY

Subway sandwich shops are multiplying faster than just about any other fast-food chain in the country. By early 1991 the chain had 5,100 shops in the United States and abroad, and was growing at the rate of 100 per month. One reason for the rapid growth is that Subway outlets are much cheaper to build and furnish than a hamburger operation.

Subway's subs (or grinders or heroes, depending upon what region of the country you're in) are made with bread baked fresh in their stores. The sandwiches contain tuna, "seafood" (made with surimi, a salty, additive-laden concoction) and crab, roast beef, meatballs, turkey (actually, turkey roll), sausages, or ham. Toppings include cheese, lettuce, tomato, pickles, green peppers, black olives, and onions. And unlike most fast-food outlets, you're not overcome by the smell of hot grease when you walk past. Traditionally, subs are cold sandwiches, but Subway microwaves its meatball, steak-and-cheese, and barbecued-beef sandwiches.

According to the company, "Each sandwich is custom made according to the customer's preference," so you can make your meal as (un)healthy as you like. You can skip the pickles, cheese, salt, mayonnaise, and oil, if you wish. Subway clerks and literature sometimes say they offer whole-wheat bread, but when we obtained a list of ingredients we discovered that white flour was the major ingredient. (After the Center for Science in the Public Interest filed a complaint against the Subway company in 1991 about mislabeled bread, Subway promised to stop the deceptive labeling.)

Subway offers 6-inch and 12-inch sandwiches and salads (sandwiches minus the bread). The sandwiches and salads contain about the same amount of fat, but because the bread adds almost fat-free calories, fat comprises a much smaller percentage of calories in the sandwiches than in the salads. The sandwiches average about 30 percent calories from fat and compare favorably to fast-food hamburgers and most roast-beef sandwiches. The 6-inch sandwich high-

SUBWAY

COMPLETE NUTRITIONAL VALUES*1

	WEIGHT[2] (gm.)	CALORIES	PROTEIN (gm.)
Salad, Chef, small	—	189	19
✔ Salad, Garden, large	—	46	2
Salad, Ham, small	—	170	14
Salad, Roast Beef, small	—	185	18
Salad, Seafood and Crab, small	—	198	12
Salad, Tuna, small	—	212	20
Salad, Turkey, small	—	167	15
✔ Sandwich, Ham, 6 in.	—	360	20

est in fat (4 teaspoons) and calories (429) is the meatball sub, which still has only half as much fat as a Big Mac. The turkey and club subs are lowest in fat. As in most fast foods, all the sandwiches and salads have more sodium than they should.

If you need a conversation stopper, or starter, at a party, Subway's has a Giant Party Sub. At a cost of about $12 per foot, they'll sell you sandwiches that are from 3 to 100 feet long, or even longer.

Subway provides partial nutrition information for its small sandwiches and salads (without dressings). The figures shown in the chart are for products containing 2 to 2½ ounces of meat or fish, plus some onion, lettuce, tomatoes, dill pickles, green peppers, half a black olive, teaspoon of oil, and slice of cheese. The Gloom ratings range from 3 for a small garden salad to 21 for a 6-inch, 429-calorie meatball sub. Of course, a large version of the meatball sandwich, with over 800 calories, would be a fine meal for two people. The company doesn't provide ingredient information, but says none of its products contains MSG, Yellow No. 5 dye, or sulfites.

CARBOHYDRATES (gm.)	ADDED SUGAR[3] (gm.)	FAT[4] (gm.)	FAT % CALORIES	SATURATED FAT (gm.)	CHOLESTEROL (mg.)	SODIUM (mg.)	VITAMIN A (% U.S. RDA)	VITAMIN C (% U.S. RDA)	IRON (% U.S. RDA)	CALCIUM (% U.S. RDA)	GLOOM
6	0	10	47	—	—	479	10	42	0	12	11
10	0	0	10	—	0	634	10	84	6	4	3
6	0	10	53	—	—	479	10	42	0	12	11
6	0	10	49	—	—	479	10	42	0	12	11
13	0	11	51	—	—	946	10	42	7	12	14
8	0	12	49	—	—	545	10	42	9	13	12
4	0	9	50	—	—	479	10	42	0	12	10
45	0	11	28	—	—	839	10	42	0	12	16

SUBWAY

COMPLETE NUTRITIONAL VALUES¹ CONTINUED	WEIGHT² (gm.)	CALORIES	PROTEIN (gm.)
Sandwich, Meatball, 6 in.	—	429	26
✔ Sandwich, Roast Beef, 6 in.	—	375	24
Sandwich, Seafood and Crab, 6 in.	—	388	18
Sandwich, Steak, 6 in.	—	423	28
✔ Sandwich, Subway Club, 6 in.	—	379	25
Sandwich, Tuna, 6 in.	—	402	26
✔ Sandwich, Turkey, 6 in.	—	357	21

* *Subway provides nutrition information only for its 6-inch sandwiches. Double the values shown for 12-inch sandwiches.*
1. A dash means that data not available.
2. To convert grams to ounces (weight), divide by 28.35; to convert grams to fluid ounces (volume), divide by 29.6.

TACO BELL

Among the fastest growing limited-menu restaurants are Mexican-style restaurants, and Taco Bell accounts for about two-thirds of all sales. Glen Bell opened the first Taco Bell in 1962 in Downey, California. Taco Bell, now owned by PepsiCo, has grown to over 3,300 restaurants with total sales of $2.5 billion in 1990. It's aiming for 10,000 stores in 2001. Its major strategy for growth has been to cut prices of some of its popular items—which has spurred competitors to do the same.

Visiting a Mexican-style restaurant offers a break from the hamburger and fried-chicken joints, even if it's not quite like visiting Mexico. The menu, like those of a hamburger or roast-beef restaurant, is built on ground beef, cheese, tomatoes, and lettuce. But instead of a big fluffy bun, you get a soft-flour or fried-corn tortilla; instead of ketchup, you get a spicy Mexican-style tomato sauce.

CARBOHYDRATES (gm.)	ADDED SUGAR[3] (gm.)	FAT[4] (gm.)	FAT % CALORIES	SATURATED FAT (gm.)	CHOLESTEROL (mg.)	SODIUM (mg.)	VITAMIN A (% U.S. RDA)	VITAMIN C (% U.S. RDA)	IRON (% U.S. RDA)	CALCIUM (% U.S. RDA)	GLOOM
45	—	16	34	—	—	876	10	42	21	13	21
45	—	11	26	—	—	839	10	42	0	12	16
52	—	12	29	—	—	1306	10	42	13	12	20
46	—	14	30	—	—	883	10	42	22	13	19
45	—	11	26	—	—	839	10	42	0	12	16
45	—	13	28	—	—	905	10	42	15	13	17
46	—	10	26	—	—	839	0	42	0	12	16

3. To convert grams of sugar to teaspoons of sugar, divide by 4.0.
4. To convert grams of fat to teaspoons of fat, divide by 4.4.

Taco Bell has one thing that most hamburger chains don't: beans. Beans are one of the best sources of dietary fiber and protein, with very little fat. For your fill of fiber, choose the bean burrito (a flour tortilla, not fried, wrapped around beans and other ingredients). Only one-fourth of its 447 calories come from the 3 teaspoons of fat. Getting the Red Sauce adds about 125 milligrams more sodium than the hotter Green Sauce, but either way, the bean burrito is unfortunately high in sodium.

The bean burrito has a Gloom rating of 21, compared to a rating of 32 for the Burrito Supreme and 36 for the regular beef burrito, each of which provide about 450 to 500 calories. The "Supreme" burrito contains ground beef plus beans, tomatoes, and a couple of other ingredients not in the regular beef burrito.

A way to get even more beans (along with more salt than you probably want) is to order a side dish of Pintos 'n Cheese. A serving rates just under 200 calories and 11 on the Gloom scale.

Taco Bell's new chicken and steak tacos are among the best choices on the menu. They are made with several chunks of grilled meat, along with the standard shredded lettuce, cheese, and sauce, in a soft (nonfried) flour tortilla. These tacos weigh in at a bit over 200 calories, 2 teaspoons of fat, and about 15 Gloom points. Skipping the cheese will save you, and adding guacamole or sour cream will cost you, a few dozen more calories and several Gloom points.

Taco Bell's best-seller is still the standard hard-shell taco, which flies out of restaurants at the rate of 2 million a day. This 183-calorie item has about 2½ teaspoons of fat and a Gloom of 15, three points worse than the chicken taco. The much larger 335-calorie Taco Bellgrande has twice the fat and twice the Gloom. The inexpensive Fiesta Taco is two-thirds the size — and price — of the regular tacos. The Fiesta Burrito, tostada, and soft taco are similarly priced. The ingredients are the same as their larger counterparts.

Far and away the fattiest food is the Taco Salad, proving that "salad" doesn't necessarily mean low-cal. More than half of its 905 calories come from the 14 teaspoons of fat, yielding a Gloom rating of 68. You can cut the calories, fat, and Gloom in half by skipping the shell. The Taco Salad was named by *Nutrition Action Healthletter* in 1989 as one of the worst fast foods.

Taco Bell features a special children's combination called the Kids' Fiesta Meal. It comes with either a taco or bean burrito, Cinnamon Twists, and small soft drink in a colorful bag. A cheap little toy is thrown in, too.

A few notes on other items: Nachos (cheese sauce on corn tortilla chips) are loaded with fat and calories. Worse is the Bellgrande version, with its 649 calories and 8 teaspoons of fat from the added beef, sour cream, and other ingredients; if you must have it, take half of it home or share it with a couple of friends. The Mexican

TACO BELL

COMPLETE NUTRITIONAL VALUES[1]

	WEIGHT[2] (gm.)	CALORIES	PROTEIN (gm.)
Burrito Supreme with Red Sauce	255	503	20

Pizza, made on a fried flour tortilla, gets more than half of its 575 calories from fat. Enchiritos are very salty.

In the beverage department, you can get 2% low-fat milk and various sizes of (what else?) Pepsi-brand soft drinks, and in some stores orange juice. In a first-ever experience, the manager of a Taco Bell in northern California voluntarily offered water, after we said we didn't want anything to drink!

Taco Bell has begun testing a breakfast menu, so soon maybe taco-lovers will be able to get egg, cheese, or potato-filled versions of tacos and burritos, along with bacon or sausage. You can't yet read about the nutritional values of those and other products by asking a clerk for a pamphlet, but you can write to Taco Bell's headquarters or call them (see Appendix).

Taco Bell has a terrific automated cash register that permits easy substitutions. Thus, the clerk can easily modify any food to suit your taste. We asked one clerk to leave the cheese off of, and add guacamole to, the Pintos 'n Cheese. He was patient with our other requests, but he finally did throw up his hands in despair when we asked for the Kids' Fiesta Meal without the taco, soft drink, and Cinnamon Twists. He just gave us the bag and toy gratis and was probably pleased to see us leave.

Taco Bell has come a long, long way since 1986, when it refused to disclose its ingredients, did not know its products' nutritional values, and fried everything in highly saturated coconut oil. Noisy criticism by consumer advocates and cardiologists persuaded Taco Bell to fry in a less harmful corn-oil shortening. It also removed lard from beans and tortillas. The company said it was also removing MSG and HVP from all its menu items and reducing sodium levels. Good work!

CARBOHYDRATES (gm.)	ADDED SUGAR[3]	FAT[4] (gm.)	FAT % CALORIES	SATURATED FAT (gm.)	CHOLESTEROL (mg.)	SODIUM (mg.)	VITAMIN A (% U.S. RDA)	VITAMIN C (% U.S. RDA)	IRON (% U.S. RDA)	CALCIUM (% U.S. RDA)	GLOOM
55	0	22	39	8	33	1181	18	43	22	19	32

TACO BELL

COMPLETE NUTRITIONAL VALUES[1] CONTINUED	WEIGHT[2] (gm.)	CALORIES	PROTEIN (gm.)	
✔ Burrito, Bean, with Red Sauce	206	447	15	
Burrito, Beef, with Red Sauce	206	493	25	
✔ Burrito, Chicken, no Red Sauce	171	334	17	
Burrito, combination	198	407	18	
Burrito, Fiesta	114	226	8	
Chilito	156	383	18	
Cinnamon Twists		–no data–		
Enchirito with Red Sauce	213	382	20	
Green Sauce, 1 oz.	28	4	0	
Guacamole, ¾ oz.	21	34	0	
Hot Taco Sauce, 1 packet (⅓ fl. oz.)	11	3	0	
Jalapeno Peppers	100	20	1	
Mexican Pizza	223	575	21	
Meximelt, Beef	106	266	13	
Meximelt, Chicken	107	257	14	
Nacho Cheese	56	103	4	
Nachos	106	346	7	
Nachos Bellgrande	287	649	22	
Nachos Supreme	145	367	12	
Pico de Gallo	28	8	0	
Pintos 'n Cheese with Red Sauce	128	190	9	
Red Sauce, 1 oz.	28	10	0	
Salad, Chicken	153	125	8	
Salad Dressing, Ranch, 2½ fl. oz.	74	236	2	

CARBOHYDRATES (gm.)	ADDED SUGAR[3] (gm.)	FAT[4] (gm.)	FAT % CALORIES	SATURATED FAT (gm.)	CHOLESTEROL (mg.)	SODIUM (mg.)	VITAMIN A (% U.S. RDA)	VITAMIN C (% U.S. RDA)	IRON (% U.S. RDA)	CALCIUM (% U.S. RDA)	GLOOM
63	0	14	28	2	9	1148	7	88	21	19	21
48	0	21	38	8	57	1311	10	3	23	15	36
38	0	12	32	4	52	880	9	19	43	11	20
46	0	16	35	5	33	1136	9	45	19	15	25
29	–	9	36	3	9	652	5	57	15	15	12
36	0	18	42	8	47	893	17	0	14	27	28
–no data–											
31	0	20	47	9	54	1243	19	47	16	27	29
1	0	0	0	0	0	136	2	0	0	0	1
3	0	2	53	0	0	113	2	5	1	1	3
0	0	0	0	0	0	82	3	0	1	0	0
4	0	0	0	0	0	1370	5	4	2	4	7
40	0	37	58	11	52	1031	20	51	21	26	46
19	0	15	51	8	38	689	16	3	11	25	22
19	0	15	53	7	48	779	10	4	20	22	22
5	0	8	70	3	9	393	3	2	2	11	12
37	0	18	47	6	9	399	11	3	5	19	24
61	0	35	49	12	36	997	23	96	19	30	43
41	0	27	66	5	18	471	14	50	2	26	30
1	0	0	0	0	1	88	11	3	1	1	0
19	0	9	43	4	16	642	9	86	8	16	11
2	0	0	0	0	0	261	5	0	0	1	2
5	0	8	58	4	32	252	16	17	17	9	9
1	0	25	95	5	35	571	5	0	3	3	38

TACO BELL

COMPLETE NUTRITIONAL VALUES[1] CONTINUED	WEIGHT[2] (gm.)	CALORIES	PROTEIN (gm.)
Salsa, ⅓ fl. oz.	10	18	1
Sour Cream, ⅔ fl. oz.	21	46	1
Taco, hard shell, beef	78	183	10
Taco Bellgrande	163	335	18
Taco, hard shell, chicken	86	171	12
Taco, Fiesta (beef)	57	127	6
Taco, Soft, Fiesta	68	147	7
Taco Salad, without shell	520	484	28
Taco Salad, with shell	575	905	34
Taco Sauce, 1 packet (⅓ fl. oz.)	11	2	0
Taco, Soft, Beef	92	225	12
✔ Taco, Soft, Chicken	107	213	14
Taco, Soft, Steak	100	218	14
Taco Supreme	92	230	11
Taco Supreme, Soft	124	272	13
Tostada, Beef, with Red Sauce	156	243	9
Tostada, Chicken, with Red Sauce	164	264	12
✔ Tostada, Fiesta	93	167	6
Taco Bell Meals			
Chicken Soft Taco, Pintos 'n Cheese with Red Sauce	228	400	22
Taco, Bean Burrito with Red Sauce	284	630	25
Kids' Fiesta Meal with Beef Taco, Cinnamon Twist, Soft Drink, small	473	494	12
Enchirito with Red Sauce, Beef Meximelt	319	648	33

CARBOHYDRATES (gm.)	ADDED SUGAR[3] (gm.)	FAT[4] (gm.)	FAT % CALORIES	SATURATED FAT (gm.)	CHOLESTEROL (mg.)	SODIUM (mg.)	VITAMIN A (% U.S. RDA)	VITAMIN C (% U.S. RDA)	IRON (% U.S. RDA)	CALCIUM (% U.S. RDA)	GLOOM
4	0	0	0	0	0	376	5	0	3	4	2
1	0	4	78	2	0	0	3	0	0	2	4
11	0	11	54	5	32	276	7	2	6	8	15
18	0	23	62	11	56	472	17	9	11	18	30
11	0	9	47	3	52	337	6	5	31	8	12
10	—	7	50	3	16	139	5	1	4	6	9
15	—	7	43	3	16	361	3	1	6	5	11
22	0	31	58	14	80	680	33	124	22	29	34
55	0	61	61	19	80	910	33	125	33	32	68
0	0	0	0	0	0	126	4	0	0	0	1
18	0	12	48	5	32	554	4	2	13	12	18
19	0	10	42	4	52	615	4	4	35	8	16
18	0	11	45	5	14	456	3	2	16	11	15
12	0	15	59	8	32	276	11	5	6	11	20
19	0	16	53	8	32	554	9	5	13	14	22
27	0	11	41	4	16	596	13	75	9	18	14
20	0	15	51	7	37	454	18	36	19	18	18
17	—	7	38	2	9	324	8	45	5	11	8
37	0	19	43	8	60	1232	15	90	18	27	26
74	0	25	36	7	41	1424	14	90	27	27	35
73	46	19	35	8	32	525	7	2	8	8	37
50	0	35	49	17	92	1932	35	50	27	52	51

TACO BELL

COMPLETE NUTRITIONAL VALUES[1] CONTINUED	WEIGHT[2] (gm.)	CALORIES	PROTEIN (gm.)
Nachos Bellgrande, Guacamole, Cinnamon Twists	343	854	24
Mexican Pizza, Nachos, Cinnamon Twists	364	1092	30
Taco Salad with Shell and Ranch Dressing, Guacamole	670	1175	36

1. *A dash means that data not available.*
2. *To convert grams to ounces (weight), divide by 28.35; to convert grams to fluid ounces (volume), divide by 29.6.*

TCBY (THE COUNTRY'S BEST YOGURT)

TCBY had 1,800 stores in 1990 and is growing so fast — sales almost tripled between 1987 and 1989 — that we'll surely have more and more opportunities to try it out.

TCBY offers coffee, yogurt-based pies, yogurt with M&M's and crumbled-up candy bars, and a few other foods, but frozen yogurt in a cup or cone is its basic item. Flavors change from day to day, and many outlets offer low-fat and nonfat versions, as well as sugar-free (made with NutraSweet/aspartame) frozen yogurt. Small cups of TCBY provide between 118 calories (nonfat, sugar-free) and 192 (regular) calories and have stellar Gloom ratings of 0 to 8. There's less than a teaspoon of fat in the small (regular yogurt) serving and very little sodium. Having larger sizes of yogurt in a cone will add more of everything. For instance, a super size (15.2 ounce) cup of regular yogurt has 494 calories, 3 teaspoons of fat, and 22 Gloom points.

All yogurt, of course, is a great source of calcium. A small cup of nonfat or sugar-free TCBY yogurt provides 12 percent of the U.S.

CARBOHYDRATES (gm.)	ADDED SUGAR³ (gm.)	FAT⁴ (gm.)	FAT % CALORIES	SATURATED FAT (gm.)	CHOLESTEROL (mg.)	SODIUM (mg.)	VITAMIN A (% U.S. RDA)	VITAMIN C (% U.S. RDA)	IRON (% U.S. RDA)	CALCIUM (% U.S. RDA)	GLOOM
88	8	45	47	15	36	1344	25	101	22	31	59
101	8	63	52	20	61	1664	31	54	28	45	84
59	0	88	67	24	115	1594	40	130	37	36	102

3. To convert grams of sugar to teaspoons of sugar, divide by 4.0.
4. To convert grams of fat to teaspoons of fat, divide by 4.4.

RDA of calcium; the large size of those yogurts provides 21 percent. TCBY says its regular yogurt is slightly lower in calcium. TCBY also says that its regular yogurt, but not the other yogurts, contains iron, but it's unclear where the iron comes from.

All the TCBY shops we've visited offer pamphlets listing nutrition information, but they are terribly confusing. The pamphlet lists the nutrients in a 4-ounce serving, but there is no 4-ounce serving. The kiddie portion is 3.2 fluid ounces, and the small is 5.9 fluid ounces. Got your calculator? Other sizes are 8.2, 10.5, 15.2, and 31.6 fluid ounces. The brochure notes that calorie content varies with the flavor, but doesn't say how much. Chocolate is moderately higher in calories than the vanilla and other of its flavors.

TCBY won't disclose the nutritional value or ingredients of its cones, nor will it disclose the amount of sugar in any of its products. The figures shown in the chart are based on tests sponsored by the Center for Science in the Public Interest, which found that sugar makes up 10 percent of the weight of TCBY's vanilla yogurt. A small cup of regular or nonfat frozen yogurt contains 3½ teaspoons of sugar; a large cup contains just over 6 teaspoons of sugar.

TCBY

COMPLETE NUTRITIONAL VALUES[1]

	WEIGHT (gm.)	CALORIES	PROTEIN (gm.)
Nonfat* Frozen Yogurt, giant (31.6 fl. oz. cup)	758	869	32
✔ Nonfat Frozen Yogurt, kiddie (3.2 fl. oz. cup)	77	88	3
Nonfat Frozen Yogurt, large (10.5 fl. oz. cup)	252	289	10
Nonfat Frozen Yogurt, regular (8.2 fl. oz. cup)	197	226	8
✔ Nonfat Frozen Yogurt, small (5.9 fl. oz. cup)	142	162	6
Nonfat Frozen Yogurt, super (15.2 fl. oz. cup)	365	418	15
Regular Frozen Yogurt, giant (31.6 fl. oz. cup)	758	1027	32
Regular Frozen Yogurt, kiddie (3.2 fl. oz. cup)	77	104	3
Regular Frozen Yogurt, large (10.5 fl. oz. cup)	252	342	10
Regular Frozen Yogurt, regular (8.2 fl. oz. cup)	197	267	8
Regular Frozen Yogurt, small (5.9 fl. oz. cup)	142	192	6
Regular Frozen Yogurt, super (15.2 fl. oz. cup)	365	494	15
Sugar-Free* Frozen Yogurt, giant (31.6 fl. oz. cup)	758	632	32
Sugar-Free Frozen Yogurt, kiddie (3.2 fl. oz. cup)	77	64	3
Sugar-Free Frozen Yogurt, large (10.5 fl. oz. cup)	252	210	10
Sugar-Free Frozen Yogurt, regular (8.2 fl. oz. cup)	197	164	8
Sugar-Free Frozen Yogurt, small (5.9 fl. oz. cup)	142	118	6
Sugar-Free Frozen Yogurt, super (15.2 fl. oz. cup)	365	304	15

* *Nonfat yogurt contains no fat but does contain added sugar. Sugar-free yogurt contains no fat or added sugar.*
1. *Values vary somewhat with flavors.*

CARBOHYDRATES (gm.)	ADDED SUGAR[2] (gm.)	FAT[3] (gm.)	FAT % CALORIES	SATURATED FAT (gm.)	CHOLESTEROL (mg.)	SODIUM (mg.)	VITAMIN A (% U.S. RDA)	VITAMIN C (% U.S. RDA)	IRON (% U.S. RDA)	CALCIUM (% U.S. RDA)	GLOOM
182	78	0	0	0	0	356	0	0	0	63	11
18	8	0	0	0	0	36	0	0	0	6	1
60	26	0	0	0	0	118	0	0	0	21	4
47	20	0	0	0	0	92	0	0	0	16	3
34	15	0	0	0	0	66	0	0	0	12	2
87	38	0	0	0	0	171	0	0	0	30	6
182	78	24	21	16	79	474	0	0	32	47	46
18	8	2	17	2	8	48	0	0	3	5	4
60	26	8	21	5	26	156	0	0	11	16	15
47	20	6	20	4	20	126	0	0	8	12	12
34	15	4	19	3	15	90	0	0	6	9	8
87	38	11	20	8	38	228	0	0	15	23	22
142	0	0	0	0	0	316	32	0	0	63	2
14	0	0	0	0	0	32	3	0	0	6	0
47	0	0	0	0	0	105	11	0	0	21	1
37	0	0	0	0	0	82	8	0	0	16	1
27	0	0	0	0	0	59	6	0	0	12	0
68	0	0	0	0	0	152	15	0	0	30	1

2. To convert grams of sugar to teaspoons of sugar, divide by 4.0.
3. To convert grams of fat to teaspoons of fat, divide by 4.4.

WENDY'S

Wendy's, the fourth biggest burger chain, has the top-rated burgers in the business, according to *Restaurants & Institutions*. But you should think of Wendy's as a salad-bar chain, not a hamburger chain.

Wendy's salad bar is the best in the fast-food business. While Burger King and others replaced their all-you-can-eat salad bars with prepackaged salads, Wendy's expanded some of its salad bars into 50-item (hot and cold) SuperBars. The SuperBar offers everything from broccoli and fresh fruit to tacos and pasta. Wendy's salad bar has attracted many a person who would otherwise never set foot in a fast-food outlet.

Load up on the low-calorie, vitamin-packed vegetables, topped with reduced-calorie Italian dressing, and make a dessert from the cantaloupe, watermelon, and other seasonal fresh fruit. If you're hungrier than that, have some spaghetti and meatless sauce at the pasta section. Or try refried beans and Spanish rice at the Mexican Fiesta section.

Of course, the SuperBar has some fat traps, like the potato salad, regular dressings, cheese, Cheddar chips, and pepperoni. But if you're smart enough to head for the salad bar, you're smart enough to avoid most of the less healthful offerings. For take out or drive-thru customers, Wendy's does offer prepackaged salads.

If the SuperBar isn't enough for you, Wendy's also offers a plain baked potato that's big, hot, filling, and good for you. It provides 50 percent of the U.S. RDA for vitamin C and 20 percent of the iron allowance in just 270 calories. At your request, Wendy's will heap the potatoes with all sorts of things, including vitamin- and mineral-rich cheese and broccoli. However, the stuffed spuds provide as many as 520 calories and ¾ teaspoon of salt (the bacon and cheese version) and 5 teaspoons of fat (sour cream and chives). The bacon and cheese potato has a Gloom rating of 29 compared to a Gloom of 0 for the plain potato. You can improve a stuffed potato by scraping off some of the calorie-rich topping or getting the topping on the side.

A plain baked potato, garden salad with reduced-calorie Italian dressing, and low-fat milk would be a tasty, nutritious meal with 550 calories and costing about $3 and 18 Gloom points.

Another decent offering is Chili, which is made with beef and beans. The regular, 9-ounce serving provides 6 grams of dietary fiber, only 220 calories, and less than 2 teaspoons of fat (but it is fairly high in sodium: 750 milligrams). This chili (Gloom = 12), other chains' roast-beef sandwiches, and the new low-fat hamburgers are just about the only red-meat dishes that get 30 percent or less of their calories from fat.

Once you venture away from the salad bar, baked potatoes, and chili, our advice is to eat small. Wendy's is famous for having tasty burgers, but your waistline may pay the price. The Jr. Hamburger, made with a 2-ounce patty, is only 260 calories and has a Gloom rating of 15. Adding a slice of cheese brings it to just over 300 calories and 21 Gloom points. The Kids' Meal sandwiches are identical to the Jr. sandwiches, except they don't have onion. (In addition to the burger, the Kids' Meal includes fries and a small drink.)

The Single Hamburger (340 calories and 23 Gloom points) and Big Classic (570 calories and 48 Gloom points) sandwiches are made with 4-ounce patties. The Big Classic comes with, the Single without, all the toppings: mayonnaise, ketchup, pickles, onion, tomato, and lettuce. The Big Classic is similar to the Single with Everything, except that it is served on a Kaiser Bun and gets 150 extra calories from an extra dose of mayonnaise.

Many people order double-burger versions of the various sandwiches. Adding a quarter-pound patty to either the Single or Big Classic tacks on another 180 calories and 3 teaspoons of fat. The monster sandwich is the Big Classic Double with Cheese, which provides over 800 calories, 12 teaspoons of fat, and 72 Gloom points.

One of the most grotesque meals you could have (we're doing this just for fun; don't you dare try it!) at a fast-food restaurant would be a Big Classic Double with Cheese, a Biggie order of fries, a large Frosty, and a cookie. That spells 2,224 calories, 25 teaspoons of fat, 2,456 milligrams of sodium, and a diet-busting 168 Gloom points. Compared to that, a Jr. Hamburger, small order of fries, and small Coke looks like a miniature, doll-house-size dinner, with "just" 600 calories, 5 teaspoons of fat, and 36 Gloom points.

Of Wendy's three chicken sandwiches, all made with a breast fillet, the Grilled Chicken is the winner, at 320 calories and 2 teaspoons of fat (Gloom = 16). Next best is the 430-calorie Chicken

Sandwich, which is made with fried chicken (Gloom = 25). The Chicken Club is identical, except that it has bacon, so it has 506 calories, almost 6 teaspoons of fat, and 33 Gloom points.

Adding a garden salad with reduced-calorie dressing and plain baked potato to the Grilled Chicken Sandwich would give you an excellent — and filling — 760-calorie dinner with 22 Gloom points. Less concerned diners might stuff themselves with a Chicken Club Sandwich, Biggie Fries, and medium Frosty, which provide almost twice as many calories, more than twice as much fat, and three times as many Gloom points (94).

The Crispy Chicken Nuggets are the usual greasy fare with a 31 Gloom score. They are made of minced chicken, chicken skin, and additives and are even fattier than McDonald's. The nuggets are

WENDY'S

COMPLETE NUTRITIONAL VALUES[1]

	WEIGHT[2] (gm.)	CALORIES	PROTEIN (gm.)
Alfalfa Sprouts, fresh, 1 oz.	28	8	1
Alfredo Sauce, 2 oz.	56	35	1
Applesauce, Chunky, 1 oz.	28	22	0
Bacon Bits, ½ oz.	14	40	5
Baked Potato, Hot Stuffed Bacon & Cheese	362	520	20
Baked Potato, Hot Stuffed Broccoli & Cheese	350	400	8
Baked Potato, Hot Stuffed Cheese	318	420	8
Baked Potato, Hot Stuffed Chili & Cheese	403	500	15
Baked Potato, Hot Stuffed Sour Cream & Chives	323	500	8
✔ Baked Potato, Plain	250	270	6
Bananas, 1 oz.	28	26	0
Big Classic	260	570	27
Big Classic with Cheese	278	640	31

fried in vegetable shortening, as are Wendy's other fried items.

A few of Wendy's outlets offer a multi-grain bun, but bleached flour is its main ingredient, and most of its dark color comes from caramel color and molasses. Perhaps one of these days Wendy's or another chain will dare to give its customers a real whole-grain bun.

Various Wendy's outlets are open for breakfast with scrambled egg platters and egg sandwiches on toast, English muffins, or biscuits, available with the usual fast-food cheese, sausage, and bacon fillers. Your best breakfast bet is an order of pancakes with orange juice and 2 percent low-fat milk. Some outlets may have blueberry muffins. Wendy's does not provide nutritional information on these products.

CARBOHYDRATES (gm.)	ADDED SUGAR[3] (gm.)	FAT[4] (gm.)	FAT % CALORIES	SATURATED FAT (gm.)	CHOLESTEROL (mg.)	SODIUM (mg.)	VITAMIN A (% U.S. RDA)	VITAMIN C (% U.S. RDA)	IRON (% U.S. RDA)	CALCIUM (% U.S. RDA)	GLOOM
0	0	0	0	0	0	0	0	4	0	0	0
5	0	1	26	1	0	300	0	0	0	6	3
6	0	0	0	0	0	0	0	0	0	0	0
0	0	2	45	1	10	400	0	2	2	0	5
70	0	18	31	5	20	1460	10	60	25	8	29
58	0	16	36	3	0	455	14	60	15	10	18
66	0	15	32	4	10	310	10	50	20	6	18
71	0	18	32	4	25	630	15	60	28	8	23
67	0	23	41	9	25	135	50	75	20	10	24
63	0	0	0	0	0	20	0	50	20	2	0
7	0	0	0	0	0	0	0	4	0	0	0
47	5	33	52	6	90	1085	10	20	35	15	48
47	5	39	55	9	105	1345	16	20	35	27	56

WENDY'S

COMPLETE NUTRITIONAL VALUES[1] CONTINUED	WEIGHT[2] (gm.)	CALORIES	PROTEIN (gm.)
Big Classic, Double	334	750	46
Big Classic, Double, with Cheese	352	820	50
Breadsticks, 2	7	30	1
Broccoli, fresh, ½ cup	43	12	1
Butterscotch Pudding, ¼ cup	57	90	1
Cantaloupe, fresh, 2 oz.	57	20	0
Carrots, fresh, ¼ cup	27	12	0
Cauliflower, fresh, ½ cup	57	14	1
Cheddar Cheese, shredded, 1 oz.	28	110	7
Cheddar Chips, 1 oz.	28	160	3
Cheese Sauce, 2 oz.	56	39	1
Cheese, Parmesan, grated, 1 oz.	28	130	12
Cheese, Parmesan, imitation, 1 oz.	28	80	9
Cheese, Shredded, Salad Bar, Imitation, 1 oz.	28	90	6
Cheeseburger	144	410	28
Cheeseburger, Double	218	590	47
Chicken Club Sandwich	205	506	30
Chicken Salad, 2 oz.	56	120	7
Chicken Sandwich, fried	219	430	26
✔ Chili, 9 oz.	255	220	21
Chives, 1 oz.	28	71	6
Chocolate Chip Cookie	64	275	3
Chocolate Milk, 8 fl. oz.	240	160	7
Chocolate Pudding, ¼ cup	57	90	0

CARBOHYDRATES (gm.)	ADDED SUGAR[3] (gm.)	FAT[4] (gm.)	FAT % CALORIES	SATURATED FAT (gm.)	CHOLESTEROL (mg.)	SODIUM (mg.)	VITAMIN A (% U.S. RDA)	VITAMIN C (% U.S. RDA)	IRON (% U.S. RDA)	CALCIUM (% U.S. RDA)	GLOOM
47	5	45	54	11	155	1295	10	20	55	15	64
47	5	51	56	14	170	1555	16	20	55	27	72
5	0	1	30	0	0	30	0	0	2	2	1
2	0	0	0	0	0	10	6	65	2	2	0
11	9	4	40	—	0	85	0	0	2	6	6
5	0	0	0	0	0	5	20	30	0	0	0
2	0	0	0	0	0	10	80	4	0	0	0
3	0	0	0	0	0	10	0	70	2	2	0
1	0	10	82	6	30	175	10	0	0	20	12
12	0	12	68	—	5	445	0	0	2	6	18
5	0	2	46	1	0	305	0	0	0	6	5
1	0	9	62	5	20	525	6	0	0	40	12
4	0	3	34	3	0	410	20	0	0	50	4
1	0	6	60	4	0	125	4	0	0	20	6
30	5	21	46	9	80	760	6	0	30	22	31
30	5	33	50	14	145	970	6	0	50	22	47
42	5	25	44	5	70	930	2	15	80	10	33
4	0	8	60	1	0	215	0	4	2	0	10
41	5	19	40	3	60	725	2	8	80	10	25
23	1	7	29	3	45	750	15	15	35	8	12
18	0	1	13	0	0	20	195	313	30	25	1
40	20	13	43	4	15	256	2	0	8	2	23
24	11	5	28	3	15	140	15	4	4	25	7
12	9	4	40	—	0	70	0	0	2	15	5

WENDY'S

COMPLETE NUTRITIONAL VALUES[1] CONTINUED	WEIGHT[2] (gm.)	CALORIES	PROTEIN (gm.)
Chow Mein Noodles, ½ oz.	14	74	1
Coca-Cola, Biggie (28 fl. oz.)	840	350	0
Coca-Cola, large (16 fl. oz.)	480	200	0
Coca-Cola, medium (12 fl. oz.)	360	150	0
Coca-Cola, small (8 fl. oz.)	240	100	0
Coke, Diet, small (8 fl. oz.)	240	1	0
Cole Slaw, 2 oz.	57	70	0
Cottage Cheese, ½ cup	105	108	13
Crispy Chicken Nuggets, 6 pieces	93	280	14
Croutons, ½ oz.	14	60	2
Cucumber, fresh, 4 slices	14	2	0
Dr Pepper, small (8 fl. oz.)	240	100	0
Eggs, hard-cooked, 1 Tbsp.	20	30	3
Fettucini, 2 oz.	56	190	4
Fish Filet Sandwich	170	460	18
Flour Tortilla, 1	37	110	3
French Fries, Biggie (6 oz.)	170	449	6
French Fries, large (4.2 oz.)	118	312	4
French Fries, small (3.2 oz.)	91	240	3
Frosty Dairy Dessert, large (14.5 oz.)	413	680	14
Frosty Dairy Dessert, medium (11 oz.)	316	520	10
Frosty Dairy Dessert, small (8.5 oz.)	243	400	8
Garbanzo Beans	28	46	3
Garlic Toast	18	70	2

CARBOHYDRATES (gm.)	ADDED SUGAR[3] (gm.)	FAT[4] (gm.)	FAT % CALORIES	SATURATED FAT (gm.)	CHOLESTEROL (mg.)	SODIUM (mg.)	VITAMIN A (% U.S. RDA)	VITAMIN C (% U.S. RDA)	IRON (% U.S. RDA)	CALCIUM (% U.S. RDA)	GLOOM
8	0	4	49	1	0	60	0	0	4	0	5
88	88	0	0	0	0	35	0	0	0	0	14
50	50	0	0	0	0	20	0	0	0	0	8
38	38	0	0	0	0	15	0	0	0	0	6
25	27	0	0	0	0	10	0	0	0	0	4
0	0	0	0	0	0	20	0	0	0	0	0
8	0	5	64	1	5	130	4	25	0	2	5
3	0	4	33	3	15	425	6	0	0	6	7
12	0	20	64	4	50	600	0	0	4	4	31
8	0	3	45	—	—	155	0	0	4	0	4
0	0	0	0	0	0	0	0	0	0	0	0
26	26	0	0	0	0	5	0	0	0	0	4
0	0	2	60	1	90	25	4	0	0	0	6
27	0	3	14	1	10	3	0	0	6	0	4
42	5	25	49	5	55	780	2	2	15	10	38
19	0	3	25	0	—	220	0	0	2	8	5
62	0	22	45	5	0	271	0	19	7	0	32
43	0	16	45	3	0	189	0	13	5	0	22
33	0	12	45	2	0	145	0	10	4	0	17
100	34	24	32	8	85	374	17	0	10	51	40
77	26	18	32	6	65	286	13	0	9	39	30
59	20	14	32	5	50	220	10	0	6	30	23
8	0	1	20	0	0	5	0	0	6	0	1
9	0	3	39	1	0	65	4	0	2	2	4

WENDY'S

COMPLETE NUTRITIONAL VALUES[1] CONTINUED	WEIGHT[2] (gm.)	CALORIES	PROTEIN (gm.)
Green Peas, frozen, 1 oz.	28	21	1
Green Peppers, fresh, ¼ cup	37	10	0
✔ Grilled Chicken Sandwich	175	320	24
Hamburger, Double	200	520	43
Hamburger, Single, plain	126	340	24
Hamburger, Single, with Everything	210	420	25
Honeydew Melon, 2 oz.	57	20	0
Honey Mustard, ½ oz.	14	50	0
Hot Chocolate, 6 fl. oz.	180	110	2
Jalapeño Peppers, canned, 1 Tbsp.	14	2	0
Jr. Bacon Cheeseburger	155	430	22
Jr. Cheeseburger	125	310	18
✔ Jr. Hamburger	111	260	15
Jr. Swiss Deluxe	163	360	18
Lemon-Lime soft drink, small (8 fl. oz.)	240	100	0
Lemonade, 8 fl. oz.	240	90	0
Lettuce, iceberg, 1 cup	55	8	0
Lettuce, romaine, 1 cup	55	9	0
Milk, 2% Low-fat, 8 fl. oz.	240	110	8
Mushrooms, fresh, ¼ cup	17	4	0
Olives, Black, 1 oz.	28	35	0
Oranges, fresh, 2 oz.	56	26	0
Pasta Medley, 2 oz.	56	60	2
Pasta Salad, ¼ cup	57	35	2

CARBOHYDRATES (gm.)	ADDED SUGAR[3] (gm.)	FAT[4] (gm.)	FAT % CALORIES	SATURATED FAT (gm.)	CHOLESTEROL (mg.)	SODIUM (mg.)	VITAMIN A (% U.S. RDA)	VITAMIN C (% U.S. RDA)	IRON (% U.S. RDA)	CALCIUM (% U.S. RDA)	GLOOM
4	0	0	0	0	0	30	4	8	2	0	0
2	0	0	0	0	0	0	4	60	0	0	0
37	5	9	25	2	60	715	2	8	20	10	16
30	5	27	47	11	130	710	0	0	50	10	39
30	5	15	40	6	65	500	0	0	30	10	23
35	5	21	45	7	70	890	5	15	30	10	31
5	0	0	0	0	0	5	0	25	0	0	0
5	—	3	54	1	5	170	0	0	0	0	7
22	13	1	8	9	0	115	0	0	2	6	4
0	0	0	0	0	0	0	190	0	0	0	0
32	5	25	52	5	50	835	2	15	20	10	36
33	5	13	38	5	34	770	2	4	20	10	21
33	5	9	31	3	34	570	2	4	20	10	15
34	5	18	45	3	40	765	4	10	20	20	26
24	24	0	0	0	0	20	0	0	0	0	4
24	40	0	0	0	0	0	0	15	2	0	4
1	0	0	0	0	0	5	2	4	2	0	0
1	0	0	0	0	0	5	15	20	4	2	0
11	0	4	33	3	20	115	10	4	0	30	5
0	0	0	0	0	0	0	0	0	0	0	0
2	0	3	77	0	0	245	0	0	4	2	5
7	0	0	0	0	0	0	0	50	0	2	0
9	0	2	30	0	0	5	6	15	4	0	2
6	0	0	0	—	0	120	0	0	2	0	1

WENDY'S

COMPLETE NUTRITIONAL VALUES¹ CONTINUED	WEIGHT² (gm.)	CALORIES	PROTEIN (gm.)
Peaches, in syrup, 2 pieces, 2 oz.	57	31	0
Pepperoni, Sliced, 1 oz.	28	140	5
Picante Sauce, 2 oz.	56	18	0
Pineapple Chunks in natural juice, ½ cup	100	60	0
Potato Salad, ¼ cup	57	125	0
Red Onions, fresh, 3 rings	9	2	0
Red Peppers, Crushed, 1 oz.	28	120	5
Refried Beans, 2 oz.	56	70	4
Rotini, 2 oz.	56	90	3
Salad Dressing, Blue Cheese, 4 Tbsp.	60	360	0
Salad Dressing, Celery Seed, 4 Tbsp.	60	280	0
Salad Dressing, French, 4 Tbsp.	60	240	0
Salad Dressing, Golden Italian, 4 Tbsp.	60	180	0
Salad Dressing, Hidden Valley Ranch, 4 Tbsp.	60	200	0
Salad Dressing, Italian Caesar, 4 Tbsp.	60	320	0
✔ Salad Dressing, Reduced Calorie Bacon and Tomato, 4 Tbsp.	60	180	0
Salad Dressing, Reduced Calorie Italian, 4 Tbsp.	57	100	0
Salad Dressing, Sweet Red French, 4 Tbsp.	60	280	0
Salad Dressing, Thousand Island, 4 Tbsp.	60	280	0
Salad, Chef	331	180	15
✔ Salad, Garden	277	70	4
Salad, Taco	791	660	40
Sauce for Chicken Nuggets, Barbecue, 1 oz.	28	50	0

CARBOHYDRATES (gm.)	ADDED SUGAR[3] (gm.)	FAT[4] (gm.)	FAT % CALORIES	SATURATED FAT (gm.)	CHOLESTEROL (mg.)	SODIUM (mg.)	VITAMIN A (% U.S. RDA)	VITAMIN C (% U.S. RDA)	IRON (% U.S. RDA)	CALCIUM (% U.S. RDA)	GLOOM
8	0	0	0	0	0	5	2	2	0	0	0
2	0	12	77	4	35	435	0	0	2	0	22
4	0	0	1	0	—	5	10	30	2	0	0
16	0	0	0	0	0	0	0	15	2	0	0
6	0	11	79	1	10	90	0	10	2	0	13
0	0	0	0	0	0	0	0	0	0	0	0
15	0	4	30	—	0	5	200	15	15	2	2
10	0	3	39	1	0	215	0	0	6	2	5
15	0	2	20	0	0	0	0	0	4	0	2
0	0	40	100	8	40	420	0	0	0	0	63
12	0	18	58	4	20	260	0	0	0	0	29
16	14	24	90	3	0	712	0	0	0	0	43
12	11	16	80	2	0	1000	0	0	0	0	35
0	0	24	108	1	20	380	0	0	0	0	38
0	0	36	101	6	20	560	0	0	0	0	57
12	11	16	80	2	0	760	0	0	0	0	32
6	5	9	77	1	0	700	0	0	0	0	20
20	18	24	77	3	0	500	0	0	0	0	41
8	4	28	90	4	20	420	0	0	0	0	45
10	0	9	45	—	120	140	110	110	15	25	10
9	0	2	26	0	0	60	110	70	8	10	1
46	0	37	50	—	35	1110	80	80	35	80	39
11	0	0	0	0	0	100	6	0	4	0	1

WENDY'S

COMPLETE NUTRITIONAL VALUES[1] CONTINUED	WEIGHT[2] (gm.)	CALORIES	PROTEIN (gm.)
Sauce for Chicken Nuggets, Barbecue, 1 oz.	28	50	0
Sauce for Chicken Nuggets, Honey, ½ oz.	14	45	0
Sauce for Chicken Nuggets, Sweet and Sour, 1 oz.	28	45	0
Sauce for Chicken Nuggets, Sweet Mustard, 1 oz.	28	50	0
Seafood Salad, 2 oz.	56	110	4
Sour Cream, 1 oz.	28	60	1
Sour Topping, 1 oz.	28	58	0
Spaghetti Meat Sauce, 2 oz.	56	60	4
Spaghetti Sauce, 2 oz.	56	28	0
Spanish Rice, 2 oz.	56	70	2
Strawberries, fresh, 2 oz.	56	17	0
Sunflower Seeds and Raisins, 1 oz.	28	140	5
Taco Chips, 1½ oz.	40	260	4
Taco Meat, 2 oz.	56	110	10
Taco Sauce, 1 oz.	28	16	0
Taco Shells, 1	11	45	0
Tartar Sauce, 1 Tbsp.	21	120	0
Three Bean Salad, ¼ cup	57	60	1
Tomatoes, fresh, 1 oz.	28	6	0
Tuna Salad, 2 oz.	56	100	8
Turkey Ham, ¼ cup	28	35	5
Watermelon, ¼ cup	57	18	0

Wendy's Meals

✔ Salad (Romaine Lettuce, Tomatoes, Mushrooms, Cucumbers, Broccoli), 2% Low-fat Milk	397	143	9

CARBOHYDRATES (gm.)	ADDED SUGAR[3] (gm.)	FAT[4] (gm.)	FAT % CALORIES	SATURATED FAT (gm.)	CHOLESTEROL (mg.)	SODIUM (mg.)	VITAMIN A (% U.S. RDA)	VITAMIN C (% U.S. RDA)	IRON (% U.S. RDA)	CALCIUM (% U.S. RDA)	GLOOM
11	0	0	0	0	0	100	6	0	4	0	1
12	12	0	0	0	0	0	0	0	0	0	2
11	0	0	0	0	0	55	0	0	2	0	1
9	0	1	18	0	0	140	0	0	0	0	3
7	0	7	57	0	0	455	0	2	2	20	9
1	0	6	90	4	10	15	6	0	0	20	6
2	0	5	78	5	0	30	0	0	0	0	9
8	0	2	30	1	10	315	4	4	4	0	5
7	0	0	1	0	0	345	0	0	0	0	4
13	0	1	13	0	0	440	6	0	10	4	4
4	0	0	0	0	0	0	0	50	0	0	0
6	0	10	64	7	0	5	0	0	10	2	12
40	0	10	35	1	0	20	0	0	4	8	11
4	0	7	57	2	25	300	0	0	10	4	10
3	0	0	1	0	0	140	4	2	0	0	1
6	0	3	60	1	0	45	0	0	0	0	5
0	0	14	105	2	15	115	0	0	0	0	21
13	1	0	0	0	—	15	4	0	2	0	0
1	0	0	0	0	0	5	2	10	0	0	0
4	0	6	54	1	0	290	0	4	2	0	8
0	0	1	26	0	15	275	0	0	4	0	3
4	0	0	0	0	0	0	2	10	0	0	0
15	0	4	25	3	20	135	33	99	6	34	4

WENDY'S

COMPLETE NUTRITIONAL VALUES[1] CONTINUED	WEIGHT[2] (gm.)	CALORIES	PROTEIN (gm.)
✔ Baked Potato (plain), Garden Salad with Reduced Calorie Italian Dressing, 2% Low-fat Milk	778	550	18
✔ Grilled Chicken Sandwich, Baked Potato (plain), Garden Salad with Reduced Calorie Italian Dressing	713	760	34
✔ Chili (9 oz.), Garden Salad with Reduced Calorie Italian Dressing, 2% Low-fat Milk	783	500	33
Kids' Meal: Jr. Hamburger, French Fries, small, Coca-Cola, small	442	600	18
Single Hamburger, French Fries, small, Coca-Cola, large	697	780	27
Baked Potato with Chili and Cheese, Chef Salad with Reduced Calorie Italian Dressing, 2% Low-fat Milk	1031	890	38
Big Classic, French Fries, large, Coca-Cola, small	618	982	31
Taco Salad with Hidden Valley Ranch Dressing, Baked Potato with Cheese, Coca-Cola, large	1649	1480	48
Chicken Club Sandwich, French Fries, Biggie, Frosty, medium	691	1475	46
Big Classic Double, with Cheese, French Fries, Biggie, Frosty, large, Chocolate Chip Cookie	999	2224	73

1. A dash means that data not available.
2. To convert grams to ounces (weight), divide by 28.35; to convert grams to fluid ounces (volume), divide by 29.6.

CARBOHYDRATES (gm.)	ADDED SUGAR[3] (gm.)	FAT[4] (gm.)	FAT % CALORIES	SATURATED FAT (gm.)	CHOLESTEROL (mg.)	SODIUM (mg.)	VITAMIN A (% U.S. RDA)	VITAMIN C (% U.S. RDA)	IRON (% U.S. RDA)	CALCIUM (% U.S. RDA)	GLOOM
89	5	15	24	4	20	895	120	124	28	42	18
115	10	20	23	3	60	1495	114	128	48	22	27
49	6	22	39	7	65	1625	135	89	43	48	28
91	32	21	32	6	34	725	2	14	24	10	36
113	55	27	31	8	65	665	0	10	34	10	48
98	6	40	40	8	165	1585	135	174	43	63	48
115	32	49	45	9	90	1284	10	33	40	15	75
162	50	76	46	5	65	1820	90	130	55	86	82
181	31	66	40	16	135	1487	15	34	96	49	94
249	59	110	45	31	270	2456	35	39	80	80	168

3. To convert grams of sugar to teaspoons of sugar, divide by 4.0.
4. To convert grams of fat to teaspoons of fat, divide by 4.4.

INGREDIENTS

T his chapter lists the ingredients of fast foods for those companies that disclose such information. Some companies, such as KFC and Popeyes, provide little or no information about ingredients. Thus, anyone with severe allergies should avoid eating at those chains.

Be aware that companies change product formulations from time to time so that you cannot be absolutely sure that the ingredient information that we had in mid 1991, is still current. You may be able to obtain more recent information from the various companies, but their brochures always include a notice that changes may have occurred.

Some chains provide information on the components of their sandwiches and other products. For instance, Arby's indicates that its Junior Roast Beef Sandwich consists of roast beef, a small bun, and bun-toasting oil. Our listings will provide those components, followed later by the ingredients of each of those components.

Popular soft drinks that are sold by many restaurants, such as Coca-Cola and Pepsi-Cola, are listed separately. Common natural foods, such as honey, chives, and broccoli, are generally not included, unless they include preservatives or other unexpected ingredients. Enriched flour contains added vitamin B1 (thiamin), vitamin B2 (riboflavin), vitamin B3 (niacin), and iron; those nutrients are not shown in our listings. Bromated flour is treated with potassium bromate.

We found numerous typographical and other errors in company ingredient listings. We've corrected as many as we could decipher, but have left others. In some cases (e.g., Baskin-Robbins), we obtained ingredient listings from packaging, which was not always legible. We've indicated missing ingredients. Occasionally, companies will list a chemical company's name for an ingredient, rather than the common name (e.g., Arby's lists "color (BSM) double spice" as a bun ingredient, and Hardee's lists an enigmatic PD321); it's unclear what that ingredient really is.

PURPOSES OF COMMON INGREDIENTS

Alpha-tocopherol (vitamin E): preservative to prevent rancidity

Annatto extract: artificial coloring

Aspartame (NutraSweet): artificial sweetener

Azodicarbonamide: flour bleach

Beta-carotene or apo-carotenal: vegetable form of vitamin A that is used as a nutrient or artificial coloring

BHA, BHT: preservatives to prevent rancidity

Brominated vegetable oil: emulsifier in soft drinks keeps flavor oils in suspension and gives cloudy appearance

Caffeine: stimulant drug added to soft drinks

Calcium disodium EDTA, disodium EDTA, EDTA: preservative

Calcium or sodium propionate: anti-mold preservative

Carob bean: thickening agent

Carrageenan: thickening agent

Casein, sodium caseinate: milk protein

Cellulose gum: thickening agent

Citric acid: preservative, acidulant

Dextrose: sugar

Dimethylpolysiloxane: anti-foaming agent

Disodium dihydrogen pyrophosphate: color retention

Disodium guanylate and disodium inosinate: two flavor enhancers used together

Fumaric acid: acidulant

Guar gum: thickening agent

Gum arabic: thickening agent

High-fructose corn syrup: sweetener made from corn syrup

Hydrolyzed plant (or vegetable) protein: flavor enhancer

Lecithin: emulsifier

Locust bean gum: thickening agent

Microcrystalline or powdered cellulose: prevents caking

Modified starch: thickening agent

Mono- and diglycerides: emulsifier, bread softener

MSG (monosodium glutamate): flavor enhancer

Polysorbate 60, 65, 80: emulsifier

Potassium bromate: flour bleach

Propyl gallate: preservative to prevent rancidity

Propyl paraben: preservative

Propylene glycol alginate: thickening agent

Silicon dioxide, silica gel: anti-caking agent

Sodium benzoate: preservative

Sodium erythorbate, erythorbic acid: preservative

Sodium nitrite: preservative, flavoring, and coloring

Sodium silicoaluminate: anti-caking agent

Sorbic acid, potassium sorbate: preservative

Sulfites: preservative

TBHQ: preservative to prevent rancidity

Tripolyphosphate: helps retain moisture

Vanillin, ethyl vanillin: artificial vanilla flavoring

Xanthan gum: thickening agent

Ingredients are generally listed in order of predominance, with the first ingredient being present in the greatest amount. We have sometimes changed the order of minor ingredients to conserve space.

We have not listed the function of ingredients after the chemical name. The following list describes the function of some frequently encountered ingredients. Some additives, such as sodium phosphates, serve a wide variety of technical functions; one would have to contact the company directly to learn what purpose such an additive serves in a given food.

ARBY'S

COMPONENTS
Sandwiches

Bac'N Cheddar Deluxe: roast beef, poppy-seed bun, Cheddar cheese sauce, tomato, lettuce, mayonnaise, bacon, bun-toasting oil.

Beef 'N Cheddar: roast beef, onion bun, Cheddar cheese sauce, ranch dressing, bun-toasting oil.

Chicken Breast Sandwich: breaded skinless chicken breast fried in vegetable oil, poppy-seed bun, lettuce, mayonnaise, bun-toasting oil.

Chicken Cordon Bleu: breaded chicken breast fried in vegetable oil, poppy-seed bun, ham, mayonnaise, pasteurized process Swiss cheese, bun-toasting oil.

Chicken Fajita Pita: roast chicken meat, lettuce, tomato, onion/salsa mix, shredded Cheddar cheese, pita bread pocket.

Fish Fillet: breaded cod fillet fried in vegetable oil, poppy-seed bun, lettuce, tartar sauce, bun-toasting oil.

French Dip: roast beef, deli bun, bun-toasting oil. Served with au jus.

French Dip 'N Swiss: roast beef, deli bun, pasteurized process Swiss cheese, onion, bun-toasting oil. Served with au jus.

Giant Roast Beef: roast beef, super bun, bun-toasting oil.

Grilled Chicken Barbeque: grilled chicken breast, poppy-seed bun, barbecue sauce, dill pickle, onion, bun-toasting oil.

Grilled Chicken Deluxe: grilled chicken breast, poppy-seed bun, shredded lettuce, tomato, reduced-calorie honey mayonnaise, bun-toasting oil.

Hot Ham 'N Cheese: ham, regular bun, pasteurized process Swiss cheese, bun-toasting oil.

Junior Roast Beef: roast beef, junior bun, bun-toasting oil.

Light (chicken, roast beef, turkey) Deluxe: meat, multi-grain bun, lettuce, tomato, reduced-fat and -calorie mayonnaise.

Philly Beef 'N Swiss: roast beef, poppy-seed bun, peppers, onion, pasteurized process Swiss cheese, bun-toasting oil.

Regular Roast Beef: roast beef, regular bun, bun-toasting oil.

Roast Chicken Club: roasted chicken, poppy-seed bun, lettuce, tomato, bacon, pasteurized process Swiss cheese, mayonnaise, bun-toasting oil.

Roast Chicken Deluxe: roasted chicken, poppy-seed bun, lettuce, tomato, mayonnaise, bun-toasting oil.

Sub Deluxe: roast beef, ham, poppy-seed bun, pasteurized process Swiss cheese, lettuce, onion, sub dressing, mayonnaise, bun-toasting oil.

Super Roast Beef: roast beef, super bun, tomato, lettuce, ranch dressing, bun-toasting oil.

Turkey Deluxe: roasted boneless turkey breast, regular bun, tomato, lettuce, mayonnaise, bun-toasting oil.

Baked Potatoes

Broccoli & Cheddar: potato, broccoli, butter or margarine/butter blend, Cheddar cheese sauce, shredded Cheddar cheese.

Deluxe: potato, butter or margarine/butter blend, sour cream, shredded Cheddar cheese, bacon bits, chives.

Mushroom 'N Cheese: potato, pasteurized process Swiss cheese, mushrooms, liquid margarine, pepper seasoning, Parmesan cheese.

Breakfasts

Bacon biscuit: biscuit, liquid margarine, bacon.

Bacon and egg croissant: croissant, bacon, eggs.

Bacon platter: scrambled eggs, bacon strips, potato cake, blueberry muffin.

Egg platter: scrambled eggs, potato cake, blueberry muffin.

Ham and cheese croissant: croissant, ham, pasteurized process Swiss cheese.

Ham biscuit: biscuit, liquid margarine, ham.

Ham platter: scrambled eggs, grilled ham, potato cakes, blueberry muffin.

Mushroom and cheese croissant: croissant, mushrooms, pasteurized process Swiss cheese, liquid margarine, pepper seasoning, Parmesan cheese.

Plain biscuit: biscuit, liquid margarine with butter or margarine/butter blend, jelly.

Plain croissant: croissant, with butter or margarine/butter blend, jelly.

Sausage and egg croissant: croissant, sausage, eggs.

Sausage biscuit: biscuit, liquid margarine, sausage.

Sausage platter: scrambled eggs, sausage patty, potato cake, blueberry muffin.

Toastix: Toastix, powdered sugar, optional meats: sausage, bacon, ham. Served with "maple" syrup.

INGREDIENTS
Sandwich Fillings and Buns

Bacon: pork belly cured with water, salt, sugar, sodium phosphate, sodium erythorbate, sodium nitrite.

Base mix for rolls: sugar and/or dextrose, salt, partially hydrogenated soybean oil, soy flour, corn starch, dough conditioners (vegetable mono-and diglycerides, potassium bromate, ascorbic acid, L-cysteine, enzyme).

Deli bun: enriched flour, water, base mix (see above), sodium stearoyl lactylate, whole egg powder, soybean shortening, color (BSM) double spice.

Poppy-seed bun: flour (high gluten), water, sugar, base mix (see above), vegetable shortening, yeast, egg yolks (dehydrated whole eggs), vegetable emulsifier, spice mix, vinegar, potassium bromate, calcium propionate, poppy seeds.

Onion bun: flour (high gluten), water, sugar, base mix (see above), vegetable shortening, yeast, egg yolk (pasteurized spray-dried eggs), vegetable emulsifier, spice mix, vinegar, potassium bromate, calcium propionate, dehydrated onion, poppy seeds, vegetable oil.

Breaded chicken breast fillet: boneless skinless chicken breast with rib meat, water, and salt; breaded with flour (bleached flour, salt, spices, dehydrated garlic, MSG, soybean oil, hydrolyzed vegetable protein), leavening (sodium bicarbonate, sodium aluminum phosphate, monocalcium phosphate), dextrose; battered with water and starch.

Breaded fish fillet: cod, bleached flour, water, modified starch, corn flour, salt, sugar, yeast, natural flavors, leavening (sodium bicarbonate, sodium aluminum phosphate), garlic powder, MSG, cellulose gum, spice, pepper, guar gum.

Buns, Junior, Regular, Super: flour, vegetable shortening, sugar, salt, water, yeast, sesame seeds.

Cheddar cheese sauce: water, aged Cheddar cheese (cultured milk, salt, enzymes, annatto color), partially hydrogenated soybean oil, modified starch, modified buttermilk solids, Monterey Jack cheese (cultured milk, salt, enzymes), sodium phosphate, salt, yeast extracts, annatto color, yellow #6.

Ham: ham cured with water, salt, sugar (corn syrup), sodium phosphate, smoke flavor, sodium ascorbate (erythorbate), sodium nitrite.

Mozzarella cheese: part skim milk, mozzarella culture, salt, enzymes, sodium propionate, natural flavor.

Pasteurized process Swiss cheese: Swiss cheese (cultured milk, salt, enzymes), water, cream, sodium citrate, salt, sodium phosphate, sorbic acid, citric acid, lecithin.

Roast beef: contains up to 9% of a self-basting solution of water, salt, sodium phosphates. Chunked and formed. Trimmed boneless beef chunks (minimum 70%) combined with chopped beef for a maximum of 12% fat.

Roasted boneless turkey breast: turkey breast, turkey broth, salt, dextrose, sodium phosphate.

Roasted chicken breast with rib meat: chicken breast with rib meat, water, maltodextrin, salt, sugar, onion powder, dextrose, garlic powder, spices, paprika, MSG, seasoning (hydrolyzed plant protein, dextrose, vitamin B1, disodium inosinate and guanylate), powdered lemon juice, corn syrup, citric acid.

Shredded Cheddar cheese: cultured pasteurized milk, salt, enzymes, microcrystalline cellulose, calcium chloride, apo-carotenal (artificial color), characteristic milk Cheddar artificial flavor.

Shredded mild Cheddar cheese: pasteurized milk, cheese culture, microcrystalline cellulose, salt, enzymes, artificial color.

Condiments

Arby's sauce: water, glucose, tomato paste, white vinegar, fructose, modified starch, salt, karaya gum, vegetable oil, citric acid, spices, sodium benzoate.

Au jus beef seasoning dip: salt, dextrose, corn syrup solids, caramel, hydrolyzed plant protein, MSG, spices, gelatin, disodium guanylate and inosinate, natural beef flavor, sugar, citric acid.

Bacon bits: bacon, water, salt, sugar, sodium phosphate, sodium erythorbate, hickory smoke flavor, sodium nitrite.

Barbecue sauce: tomato purée, sugar, vinegar, molasses, hickory smoke flavor, salt, food starch, mustard flour, dried onion, spices, dried garlic.

Cholesterol-free mayonnaise: soybean oil, water, vinegar, salt, egg whites, mustard flour, xanthan gum, cellulose gel (microcrystalline cellulose), polysorbate 60, cellulose gum, lemon juice concentrate, natural flavor, paprika, dried garlic, calcium disodium EDTA, dried onion, beta-carotene.

Croutons: enriched unbleached flour, water, corn syrup, partially hydrogenated soybean oil, yeast, salt, whey, soya flour, nonfat milk, butter, butter flavor.

Dill pickles: pickles, water, distilled vinegar, salt alum, sodium benzoate, natural spice flavors, yellow #5, polysorbate 80.

Horsey sauce: vegetable oil, water, sugar, white vinegar, modified starch, frozen egg yolks, salt, glucose, spices, locust bean gum, artificial flavor, calcium disodium EDTA.

Ketchup: tomatoes, vinegar, sugar, glucose, fructose, salt, onion powder, spices.

Mayonnaise: vegetable oil, whole eggs, water, white vinegar, egg yolks, salt, sugar, lemon juice concentrate, spices and seasonings, calcium disodium EDTA.

Parmesan cheese: part-skim milk, cheese culture, salt, enzymes.

Ranch dressing: soybean oil, corn syrup, vinegar, water, tomato paste, salt, spices, gums (arabic, xanthan, guar), onion, garlic, beet powder.

Reduced-calorie honey mayonnaise: soybean oil, water, high-fructose corn syrup, distilled vinegar, egg yolks, sugar, cider vinegar, modified starch, honey, salt, dehydrated garlic, MSG, maltodextrin, dehydrated onion, spices, mustard flour, dehydrated red pepper, sugar, parsley flakes, xanthan gum, caramel color, natural flavor, yellow #5.

Sour cream: pasteurized cream, lactic culture, modified starch, gelatin, mono- and diglycerides, citric acid, sodium caseinate, lactic acid, propylene glycol monoester, guar gum, artificial flavor, salt, carrageenan, monopotassium phosphate, potassium sorbate.

Sub dressing: soybean oil, water, distilled vinegar, salt, sugar, garlic juice, onion juice, dehydrated garlic, spices, xanthan gum, dehydrated onion, dehydrated red bell pepper, propylene glycol alginate, artificial color, hydrolyzed vegetable protein, calcium disodium EDTA.

Tartar sauce: soybean oil, pickle relish, water, eggs, vinegar, sugar, salt, spices and spice extractives, onion, gums (arabic, guar, xanthan), calcium disodium EDTA.

Side Orders

Curly Fries: potatoes, partially hydrogenated soybean and/or canola oils, flour, salt, cornstarch, modified food starch, spices, garlic powder, nonfat milk, cornmeal, onion powder, yellow corn flour, cellulose gum, dextrose and/or fructose, resin paprika, sodium acid pyrophosphate, natural flavor.

French fries: potatoes, partially hydrogenated soybean oil, dextrose, disodium dihydrogen pyrophosphate.

Potato cakes: potatoes, partially hydrogenated soybean and/or canola oils, salt, corn flour, dehydrated potato flakes, natural flavor, disodium dihydrogen pyrophosphate, dextrose.

Salads

Chef Salad: lettuce and red cabbage, tomato, ⅓ hard-cooked egg, cucumber, broccoli florets, shredded carrot, turkey and ham, shredded cheddar cheese.

Garden Salad: lettuce and red cabbage, shredded carrot, broccoli florets, tomato, cucumber, and shredded cheddar cheese.

Roast Chicken Salad: lettuce and red cabbage, roast chicken strips, shredded carrot, broccoli florets, tomato, cucumber, and shredded cheddar cheese.

Side Salad: lettuce and red cabbage, tomato, and shredded carrot.

SALAD DRESSINGS

Blue cheese: soybean oil, water, blue cheese, eggs, distilled vinegar, sugar, salt, high-fructose corn syrup, xanthan gum, sour cream solids, natural flavor, dehydrated garlic.

Buttermilk ranch: soybean oil, fresh cultured buttermilk, distilled vinegar, high-fructose corn syrup, salt, garlic juice, eggs, MSG, natural flavor, dehydrated onion, polysorbate 60, xanthan gum, sodium benzoate, spice, lactic acid, calcium disodium EDTA.

Honey french: high-fructose corn syrup, soybean oil, corn-cider vinegar, tomato paste, distilled vinegar, salt, paprika, spices, honey, beet juice concentrate, onion, natural flavors, xanthan gum, propylene glycol, alginate, garlic.

Lite Italian: water, distilled vinegar, sugar, salt, soybean oil, garlic, xanthan gum, spices, onion, red bell peppers, calcium disodium EDTA, yellow #5 and #6.

Thousand island: soybean oil, water, high-fructose corn syrup, pickle relish, distilled vinegar, eggs, sugar, tomato paste, salt, spice, propylene glycol, alginate, paprika, dehydrated onion, natural flavor, oleoresin paprika, calcium disodium EDTA, beet juice concentrate.

Soups

Boston clam chowder: dehydrated potatoes, reconstituted whey, pollock, clams and juice, water, vegetable oil margarine (may contain palm oil), modified starch.

Chicken noodle: chicken broth, egg noodles, chicken meat, chicken seasoning (maltodextrin, salt, hydrolyzed plant protein, MSG, chicken fat, sugar, torula yeast, partially hydrogenated soybean and cottonseed oils, spice, caramel color, disodium inosinate and guanylate), carrots, celery, modified starch, dehydrated onions, dehydrated parsley, spices.

Corn chowder: water, corn, cream-style corn, celery, carrots, modified starch, dehydrated potatoes, bacon fat, onions, bacon bits, salt, potato flakes, sugar, green peppers, garlic salt, MSG, spices, annatto.

Cream of broccoli: milk, reconstituted whey, broccoli, processed Cheddar cheese (milk solids, bacterial culture, water, sodium citrate, sodium phosphate, rennet and/or pepsin, calcium chloride, salt, citric acid, sorbic acid, color), modified starch, salt, mechanically deboned chicken, MSG, spices, chicken fat, sugar, chicken seasoning (hydrolyzed plant protein, MSG, salt, sugar, dehydrated chicken meat, soy flour, torula yeast, partially hydrogenated soybean and cottonseed oils, spice), dehydrated onions.

French onion: onions, water, vegetable oil, margarine (may contain palm oil), seasoning, hydrolyzed plant protein, salt, gelatin, onion powder, autolyzed yeast, caramel lecithin, spices, garlic powder.

Lumberjack mixed vegetable: tomatoes, carrots, celery, dehydrated potatoes, barley, water, lima beans, green beans, seasoning (salt, hydrolyzed plant protein, dextrose, MSG, beef fat, onion powder, modified starch,

tomato powder, sugar, caramel, disodium guanylate and inosinate, beef extract, maltodextrin, nonfat milk, chicken fat, garlic powder, carrot powder, natural and artificial flavors, paprika), corn, beef fat, modified starch, dehydrated onions, salt, Worcestershire sauce, sugar, dehydrated parsley, spice.

Split pea with ham: water, split peas, ham (cured with water, salt, sugar, sodium phosphates, corn syrup, sodium erythorbate, sodium nitrite), carrots, lard, flour, modified starch, MSG, salt, garlic powder, sugar, dehydrated onions, spices, onion powder, smoke flavor, dehydrated parsley.

Tomato Florentine: tomatoes, water, noodles, tomato paste, spinach, seasoning (salt, hydrolyzed plant protein).

Vegetable, beef, and barley: water, tomatoes, barley, beef, carrots, celery, seasoning salt, hydrolyzed plant protein, modified starch, MSG, beef fat, dextrose, onion powder, tomato powder, sugar, caramel, beef extract, maltodextrin, nonfat milk, disodium guanylate and inosinate, chicken fat, garlic powder, carrot powder, natural beans, green beans, modified starch, corn, dehydrated onions, salt, beef stock, sugar, caramel, Worcestershire sauce, emulsifier (propylene glycol monoesters, monoglycerides), dehydrated parsley.

Breakfast Items

Biscuit: water, enriched bleached flour, partially hydrogenated cottonseed and/or soybean oils, buttermilk, sodium bicarbonate, sodium aluminum phosphate, sugar, salt, natural flavor, monocalcium phosphate, citric acid.

Blueberry muffin: water, enriched bromated bleached flour, sugar, partially hydrogenated soybean and cottonseed oils, citric acid, modified corn starch, nonfat milk, egg yolk, leavening (baking soda).

Cinnamon nut danish: flour, water, sugar, partially hydrogenated soybean and cottonseed oils, eggs, yeast, nonfat dry milk, raisins, pecans, salt, corn syrup, sodium stearoyl lactylate, cinnamon, natural and artificial flavors, mono- and diglycerides, calcium propionate, yellow color (annatto and turmeric), potassium sorbate, yeast food, agar.

Croissant: enriched flour, butter, vegetable oil shortening (contains palm and cottonseed oils), sugar, yeast, skim milk powder, salt, dough conditioner (flour diacetyl tartaric esters of mono- and diglycerides, ascorbic acid, potassium bromate, fungal amylase).

Toastix: French bread (enriched bleached flour, water, high-fructose corn syrup, yeast, salt, soybean oil, calcium sulfate, corn starch, salt, ammonium sulfate, potassium bromate), dough conditioner (contains mono- and diglycerides, ethoxylated mono- and diglycerides), calcium propionate, water, bleached flour, soybean oil, sugar, yellow corn flour, whole egg solids, modified starch, nonfat milk, salt, soy flour, dextrose, gum arabic, leavening (monocalcium phosphate, sodium bicarbonate), natural and artificial flavors, yeast, polysorbate.

"Maple" syrup: corn syrup, sugar syrup, butter, algin derivative, natural and artificial flavors, salt, sodium benzoate, sorbic acid, sodium citrate, citric acid, caramel color.

Sausage: whole boned hog, salt, spices, sugar, flavoring, propyl gallate, citric acid, BHT.

Desserts

Apple turnover: unbleached flour, water, partially hydrogenated soybean and cottonseed oils, sugar, diced apples, corn syrup, modified starch, raisins, salt, dextrin, citric acid, cinnamon.

Blueberry turnover: enriched flour, water, vegetable shortening, blueberries, sugar, starch, apple powder, salt, citric acid.

Cherry turnover: cherries, unbleached flour, partially hydrogenated soybean and cottonseed oils, sugar, modified starch, corn syrup, water, salt, dextrin.

Cheesecake: cream cheese, sugar, sour cream, whole eggs, flour, vanilla, salt, cookie crust (flour, butter, sugar, egg white).

Chocolate chip cookie: chocolate chips (chocolate with lecithin, vanillin), unbleached flour, sugar, partially hydrogenated soybean and/or palm oils, whole eggs, butter, salt, vanilla, soda.

Beverages

Shake mix: whole milk, sugar, corn sweeteners, sweet whey, cream, nonfat milk solids, artificial vanilla flavor, mono- and diglycerides, cellulose gum, guar gum, carrageenan, dipotassium phosphate, sodium citrate, salt, sodium carbonate monohydrate.

Shake Syrups added to mix:

Vanilla: high-fructose corn syrup and/or sugar, water, caramel color, natural and artificial flavors, sodium benzoate, citric acid.

Chocolate: corn syrup, water, cocoa powder, caramel color, sodium phosphate, vanillin, potassium sorbate.

Jamocha: corn syrup, water, instant coffee, cocoa powder, partially hydrogenated coconut oil, salt, potassium sorbate, carrageenan, tetra-sodium pyrophosphate, vanillin.

Hot chocolate drink base: syrup, sugar and/or corn sweeteners, water, sweetened condensed skim milk, partially hydrogenated soybean, coconut, palm, palm kernel, and/or cottonseed oils, cocoa, whey solids, salt, disodium phosphate, natural and artificial flavors, chocolate liquor, soy lecithin, cream.

POLAR SWIRLS

Butterfinger (candy added to vanilla shake): sugar, corn syrup, ground roasted peanuts, hydrogenated palm kernel and/or soybean oils, cocoa, molasses, buttermilk solids, confectioners corn flakes, skim milk, salt, emulsifiers (glyceryl-lacto esters of fatty acids and soy lecithin), glycerin, artificial flavor, yellow #5, TBHQ, citric acid.

Heath (candy added to jamocha shake): milk chocolate (sugar, cocoa butter, milk, chocolate, lecithin, salt, vanillin), sugar, butter, partially hydrogenated soybean oil, almonds, salt, nonfat milk, natural and artificial flavors, lecithin.

Oreo (cookie added to vanilla shake): sugar, enriched flour, vegetable and animal shortening (partially hydrogenated soybean oil and lard), cocoa processed with alkali, high-fructose corn syrup, corn flour, whey, chocolate, baking soda, salt, soy lecithin, vanillin.

Reese's Peanut Butter cup (candy added to chocolate shake): milk chocolate (sugar, cocoa, butter, milk, chocolate, soy lecithin), peanuts, sugar, dextrose, salt, TBHQ, citric acid.

Snickers (candy added to vanilla shake): milk chocolate (sugar, milk, cocoa, butter, chocolate, soy lecithin, vanillin), peanuts, corn syrup, sugar, cellulose gel, milk, butter, salt, flour, egg whites, soy protein, artificial flavor, almonds, coconut.

Fats

Bun-toasting oil: partially hydrogenated soybean oil, lecithin, artificial flavor, artificial color, TBHQ, citric acid, dimethylpolysiloxane.

Butter: milkfat, nonfat milk solids (curd), water, salt, color.

Liquid margarine: partially hydrogenated soybean oil, TBHQ, citric acid, dimethylpolysiloxane.

Margarine/butter blend: liquid soybean oil, water, butter, salt, whey, vegetable mono- and diglycerides, vegetable lecithin, sodium benzoate, artificial flavor, colored with carotene and vitamin A palmitate.

Vegetable frying oil: partially hydrogenated corn oil, TBHQ, citric acid, dimethylpolysiloxane.

Vegetable frying shortening: partially hydrogenated corn oil, monoglyceride citrate, propyl gallate, propylene glycol, dimethylpolysiloxane.

BASKIN-ROBBINS

The company does not provide ingredient information, but you can ask the clerk to see the lid tops when you are in a store. Aspartame is added to various flavors of sugar-free nonfat frozen dairy desserts.

INGREDIENTS

Chocolate chip ice cream: Fresh cream, nonfat milk, sugar, bittersweet chocolate chips (chocolate liquor processed with alkali, sugar, milk fat, cocoa butter, lecithin), corn sweetener, whey, vanilla extract, carob bean gum, guar gum, lecithin, carrageenan, annatto.

Fruit Whip (soft-serve sorbet): water, sucrose, fruit purée, corn sweetener, natural flavors, citric acid, guar gum, carob bean gum, natural colors.

Jamoca almond fudge ice-cream sundae bar: Ice cream: fresh cream, nonfat milk, sugar, corn sweetener, jamoca extract (coffee infusion, sugar), caramel color, polysorbate 80, cellulose gum, mono- and diglycerides, carrageenan, locust bean gum, guar gum. *Coating* (milk chocolate and nuts): sugar, whole milk solids, chocolate liquor, roasted almonds, cocoa butter, lecithin, butter oil, vanilla extract, salt and other natural flavors, coconut oil, soybean oil. *Ribbon:* corn syrup, water, cocoa processed with alkali, modified starch, artificial flavor, potassium sorbate, carrageenan, citric acid, salt, propyl paraben.

Low-fat frozen yogurt: nonfat milk, sucrose, fruit purée (in fruit flavors), corn sweetener, cream, cultured pasteurized nonfat milk, natural flavors (e.g., jamoca extract, Kahlua, Grand Marnier), cocoa (in chocolate), carob bean gum, guar gum, lecithin, carrageenan, natural colors, viable yogurt culture.

Mint chocolate-chip ice cream: fresh cream, nonfat milk, sugar, bittersweet chocolate chips (chocolate li-

SOFT DRINKS

Either beverage concentrates or final beverages include the following ingredients:

Cherry Coke mix: high-fructose corn syrup, sucrose, water, caramel color, phosphoric acid, natural flavors, caffeine.

Coca-Cola Classic mix: high-fructose corn syrup and/or sugar, water, caramel color, phosphoric acid, natural flavors, caffeine.

Diet Coke: carbonated water, caramel color, phosphoric acid, sodium saccharin, potassium benzoate, natural flavors, citric acid, caffeine, potassium citrate, aspartame, dimethylpolysiloxane.

Diet Dr. Pepper: carbonated water, caramel color, natural flavors, phosphoric acid, sodium saccharin, aspartame, caffeine, sodium benzoate, monosodium phosphate, lactic acid, sodium citrate.

Diet Pepsi: carbonated water, caramel color, phosphoric acid, sodium saccharin, potassium benzoate, citric acid, aspartame, caffeine, natural flavor, dimethylpolysiloxane.

Diet Seven-Up: carbonated water, citric acid, aspartame, sodium benzoate, sodium citrate, natural lemon and lime flavors.

Diet Sprite: carbonated water, citric acid, natural flavors, potassium citrate, potassium benzoate, sodium saccharin, aspartame.

Dr. Nehi mix: sugar, water, caramel color, artificial and natural flavors, phosphoric acid, sodium benzoate, caffeine, monosodium phosphate, lactic acid.

Dr. Pepper: carbonated water, sugar, high-fructose corn syrup, caramel color, artificial and natural flavors, phosphoric acid, caffeine, sodium benzoate, monosodium phosphate, lactic acid.

Fanta Orange mix: high-fructose corn syrup, sucrose, water, citric acid, sodium benzoate, tocopherols, modified starch, natural and artificial flavors, yellow #6, glycerol ester of wood rosin, brominated vegetable oil.

Fanta Root Beer mix: high fructose corn syrup, sucrose, water, caramel color, natural and artificial flavors, sodium benzoate, acacia, quillaia, citric acid.

quor processed with alkali, sugar, milk fat, cocoa butter, lecithin), corn sweetener, whey, natural flavor, cellulose gum, mono- and diglycerides, artificial colors including yellow #5, carob bean gum, guar gum, lecithin, carrageenan.

Nonfat frozen yogurt: nonfat milk, sucrose, fruit purée (in fruit flavors), cultured pasteurized nonfat milk, corn sweetener, maltodextrin, natural flavors, carob bean gum, guar gum, lecithin, carrageenan, natural colors, viable yogurt culture.

Peanut butter and chocolate ice-cream sundae bar: *Ice cream:* nonfat milk, fresh cream, sugar, corn sweetener, cocoa and chocolate liquor processed with alkali, cellulose gum, mono- and diglycerides, carrageenan, polysorbate 80, guar gum, carob bean gum. *Coating* (milk chocolate and nuts): sugar, whole milk solids, chocolate liquor, peanuts, cocoa butter, lecithin,

SOFT DRINKS CONTINUED

Lemon-Lime Slice: carbonated water, high-fructose corn syrup and/or sugar, concentrated fruit juices (apple, white grape, pear, lemon, lime), natural lemon and lime flavors, citric acid, malic acid, potassium citrate, potassium benzoate, salt, dimethylpolysiloxane, phosphoric acid.

Mandarin Orange Slice: carbonated water, high-fructose corn syrup and/or sugar, concentrated fruit juices (apple, pear, mandarin, orange), citric acid, potassium citrate, gum arabic, potassium benzoate, salt, glycerol ester of wood rosin, natural flavors, yellow #5, brominated vegetable oil, ascorbic acid, BHA, artificial color.

Mello Yello mix: high-fructose corn syrup, sucrose, water, concentrated orange juice, citric acid, natural flavors, sodium benzoate, erythorbic acid, EDTA, sodium citrate, caffeine, yellow #5, carob bean gum.

Mountain Dew: carbonated water, high-fructose corn syrup, sugar, fruit juice, natural flavors, citric acid, caffeine, propylene glycol, sodium citrate, gum arabic, yellow #5, brominated vegetable oil, sodium benzoate, erythorbic acid, calcium disodium EDTA.

Mr. Pibb mix: high-fructose corn syrup, sucrose, water, caramel color, phosphoric acid, sodium benzoate, natural and artificial flavors, caffeine, monosodium phosphate, lactic acid.

Pepsi-Cola: carbonated water, high-fructose corn syrup and/or sugar, caramel color, phosphoric acid, caffeine, citric acid, natural flavors.

R.C. Cola mix: sugar, water, caramel color, phosphoric and citric acids, cola extractives, caffeine, natural flavors, gum acacia, sodium benzoate.

R.C. Diet Rite mix: water, caramel color, phosphoric and citric acids, gum arabic, sodium saccharin, natural and artificial flavors, sodium benzoate.

R.C. Root Beer mix: sugar and/or corn sweetener, water, caramel color, natural flavors, gum arabic, vanillin.

Seven-Up mix: corn syrup and/or sugar, water, citric acid, sodium benzoate, sodium citrate, natural lemon and lime flavors.

Sprite: carbonated water, high-fructose corn syrup and/or sucrose, citric acid, natural flavors, sodium citrate, sodium benzoate.

butter oil, vanilla extract, salt and other natural flavors, coconut oil, soybean oil. *Ribbon:* peanuts, peanut oil, salt, mono- and diglycerides.

Pralines and cream ice-cream sundae bar: *Ice cream:* fresh cream, nonfat milk, sugar, corn sweetener, vanilla-vanillin extract, polysorbate 80, cellulose gum, mono- and diglycerides, carrageenan, locust bean gum, guar gum, artificial color including yellow #5 and #6. *Coating* (milk chocolate and nuts): sugar, whole milk solids, chocolate liquor, praline pecans (sugar, pecans, butter, corn syrup, salt), cocoa butter, lecithin, butter oil, vanilla extract, salt and other natural flavors, coconut oil, soybean oil. *Ribbon:* corn syrup, sweetened condensed whole milk, water, butter, stabilizer (algin, sodium phosphate, dextrin), salt, artificial vanilla flavor and other natural extractives, potassium sorbate, lecithin, annatto color, sodium bicarbonate, propyl paraben.

Rocky road ice cream: Fresh cream, nonfat milk, sugar, marshmallows (sugar, corn syrup, corn starch, gelatin), corn sweetener, almonds, cocoa and chocolate liquor processed with alkali, whey, carob bean gum, guar gum, lecithin, carrageenan.

Strawberry shortcake ice cream: Fresh cream, nonfat milk, strawberry ribbon (corn syrup, strawberries, water, citric acid, malic acid, sodium citrate, artificial color, cellulose gum, potassium sorbate, sodium sorbate, artificial flavor, and several other ingredients), cake flakes (white flour, sugar, water, coconut oil, hydrogenated soybean and/or palm oils, mono- and diglycerides, whole egg powder, sodium bicarbonate, monocalcium phosphate, whey powder, salt, vanilla extract, natural color, and several other ingredients), sugar, corn sweetener, whey, artificial flavor, carob bean gum, guar gum, lecithin, carrageenan.

BURGER KING

COMPONENTS
Sandwiches

Bacon Double Cheeseburger: beef patties, sesame-seed bun, bacon, cheese. ***Deluxe (version):*** add tomato, lettuce, and mayonnaise.

BK Broiler Chicken Sandwich: broiled chicken patty, oat-bran bun, lettuce, tomato, BK broiler sauce.

Burger Buddies: beef patties, bun, cheese, ketchup, pickle.

Chicken Sandwich: breaded chicken patty, specialty bun, lettuce, mayonnaise.

Double Cheeseburger: beef patties, sesame-seed bun, cheese, ketchup, pickles, onion.

Hamburger (Cheeseburger): beef patty, sesame-seed bun, ketchup, pickle, mustard. (Add cheese in cheeseburger.) ***Deluxe (version):*** add tomato, lettuce, and mayonnaise.

Ocean Catch Fish Filet: breaded fish filet, oat-bran bun, lettuce, tartar sauce.

Sausage Breakfast Buddy: Buddies bun, cheese, scrambled egg mix, sausage.

Whopper or Double Whopper: beef patty (patties for the Double Whopper), sesame-seed bun, tomato, lettuce, mayonnaise, ketchup, pickle, onion. Also with cheese.

INGREDIENTS
Sandwich Fillings and Buns

Bacon: bacon cured with water, salt, sugar, hickory smoke flavor, sodium phosphate, sodium erythorbate, sodium nitrite, BHA, BHT.

Bacon bits: bacon cured with water, salt, smoke flavor, sodium nitrite, sugar, sodium phosphate, sodium erythorbate, dextrose, sodium ascorbate.

BK Broiler Chicken Breast Patty: breaded chicken breast with rib meat, water, natural flavorings, vegetable oil, salt, sodium phosphates, dehydrated butter (butter, sweetcream buttermilk), maltodextrin, MSG, lecithin, modified corn starch, onion powder, extractives of annatto. Glazed with water, vegetable oil, natural flavors (maltodextrin, salt), salt, brown sugar, modified starch, carbohydrate gum, onion powder, garlic powder.

Burger buddies bun: enriched flour, water, sugar (high-fructose corn syrup, liquid sucrose, or granulated sugar), soybean and/or cottonseed shortenings, yeast, wheat gluten, milk replacer (dry whey, corn flour, milk protein, calcium and sodium propionate, dry buttermilk), salt, powdered artificial butter flavor, spice blend mix (yellow corn flour, extractives of paprika, extractives of turmeric, lecithin, oil of orange, oil of lemon), dough strengthener (sodium stearoyl-2-lactylate, calcium stearoyl-2-lactylate, and/ or ethoxylated mono- and diglycerides), mono- and diglycerides, calcium propionate, dough conditioners [contains one or more of: brew buffer or yeast nutrients (calcium carbonate, ammonium chloride, corn starch, calcium sulfate, salt), monocalcium phosphate, ammonium sulfate, calcium iodate, potassium bromate, azodicarbonamide].

Chicken (for sandwich): chicken, marinade (water, salt, sodium phosphate, MSG, modified starch); battered and breaded with bleached flour, salt, spices, partially hydrogenated vegetable oil, whey, MSG, yeast, sweet peppers, onion, garlic, dextrose, leavening (monocalcium phosphate, sodium acid pyrophosphate, sodium bicarbonate), corn starch, oat flour, natural flavor. Fried in vegetable shortening.

Chicken Tenders: breaded chicken breast with rib meat, marinated with approximately 8% solution of water, salt, sodium phosphate, modified starch and flavoring; battered and breaded with bleached flour, water, corn flour, salt, spices, modified starch, dextrose, garlic powder, MSG, monocalcium phosphate, buttermilk, natural flavors, yeast, sugar. Breading set in vegetable oil; fried in vegetable oil.

Cream cheese: pasteurized milk and cream, cheese culture, salt, stabilizers (xanthan or carob bean and/or guar gums).

Diced chicken: fully cooked chicken, water, natural flavor, salt, vegetable oil, butter powder, buttermilk powder, maltodextrin, sodium phosphates, MSG, lecithin, modified corn starch, onion, annatto extract.

Ham: smoked ham cured with water, salt, dextrose, corn syrup, sodium phosphate, sodium erythorbate, sodium nitrite.

Oat Bran Bun: flour, water, sugar (sucrose or high-fructose corn syrup), oat bran premix (oat bran, milk replacer, dairy flavor, caramel color), oat bran topping, vegetable shortening, salt, gluten, yeast, yeast food, dough conditioners (calcium salts, sulfates, phosphates, and ammonium salts), 0.5% dough conditioners (sodium and/or calcium-2-stearoyl lactylate or ethoxylated mono- and diglycerides), dough softeners

(mono- and diglycerides), potassium sorbate, oxidation/reduction additives (potassium/calcium bromate, ascorbic acid, potassium/calcium iodate, alpha-amylase, azodicarbonamide).

Ocean Catch fish fillet: cod fillet, water, bromated flour, enriched bleached flour, bleached modified starch, yellow corn flour, partially hydrogenated soybean oil, salt, yeast, sugar, onion powder, garlic powder, spice, oleoresin paprika. Fried in vegetable shortening.

Processed American cheese: pasteurized American cheese (cultured milk, salt, enzymes, calcium chloride, annatto color), cream, water, enzyme modified cheese, sodium citrate, salt, sodium aluminum phosphate, sodium phosphate, sorbic acid, lactic acid, acetic acid, phosphoric acid, lecithin, oleoresin paprika, carotenal color.

Processed Swiss: Swiss cheese (cultured milk, salt, enzymes), water, cream, sodium citrate, salt, sodium phosphate, sorbic acid, citric acid, lecithin.

Sausage: pure pork sausage patty with salt, spices, dextrose, sugar, MSG, hydrolyzed vegetable protein. Fried in vegetable oil.

Sesame-seed bun: enriched flour, water, sugar (sucrose or high-fructose corn syrup), sesame seeds, animal and/or vegetable shortenings, salt, wheat gluten, yeast, yeast food (calcium sulfate, potassium iodate, potassium bromate, and/or ammonium sulfate), dough softeners (mono- and diglycerides, protease enzyme), dough conditioners (sodium and/or calcium stearoyl lactylate), calcium propionate, whiteners (calcium peroxide, azodicarbonamide), potassium sorbate, monocalcium phosphate (leavening agent).

Shredded Cheddar: cultured pasteurized milk, salt, enzymes, powdered cellulose or microcrystalline cellulose, artificial color.

Turkey: cooked white turkey meat, water, modified starch, salt, dextrose, sodium phosphates.

Whopper Patty/Burger Buddies

Patties: ground beef (broiled).

Condiments

Barbecue dipping sauce: water, sugar, distilled vinegar, tomato paste, salt, modified starch, hickory smoke flavor, brown sugar, soybean oil, spices, xanthan gum, sodium benzoate, garlic.

BK Broiler Sauce: water, soybean oil, dijon mustard (water, mustard seed, distilled vinegar, salt, white wine, tartaric acid, citric acid, spices), egg yolks, modified food starch, egg white, vinegar, microcrystalline cellulose and sodium carboxymethylcellulose, dehydrated buttermilk, salt, sugar, spices, dehydrated onion and garlic, worcestershire sauce [water, vinegar, corn sweetener, worcesterchire sauce base (water, salt, molasses, vinegar, garlic, MSG, sugar, chili pepper, spices, natural smoke flavor, mustard, onion, polysorbate 80, natural flavors, HVP), sodium benzoate], xanthan gum, MSG, natural citrus and natural flavors, potassium sorbate, calcium disodium EDTA.

Bull's Eye Barbecue Sauce (for sandwiches): tomato purée, sugar, vinegar, molasses, hickory smoke flavor, salt, food starch, mustard flour, dried onion, blend of spices and dried garlic.

Burger King A.M. Express Dip (for French Toast Sticks): water, sugar, corn syrup, modified starch, natural and artificial flavors, xanthan gum, cellulose gum, phosphoric acid, potassium sorbate and sorbic acid, caramel color.

Honey dipping sauce: high-fructose corn syrup, honey, sugar, natural honey flavor, caramel color.

Ranch dipping sauce: soybean oil, cultured lowfat buttermilk, vinegar, garlic juice, water, sugar, salt, natural flavor (with buttermilk), MSG, xanthan gum, sorbic acid, calcium disodium EDTA, polysorbate 60, spice, lemon juice concentrate.

Sweet and sour dipping sauce: water, high-fructose corn syrup, brown sugar, vinegar, pineapple juice concentrate, starch, crushed pineapple, apricot concentrate, soybean oil, Worcestershire sauce, salt, xanthan gum, red bell pepper, garlic powder, spice, red #40.

Tartar sauce: soybean oil, water, dill relish with onion (cucumber, vinegar, onion, salt, natural flavors, mustard flour, alum), corn sweetener, distilled vinegar, egg yolk, spices, salt, sodium benzoate, polysorbate 80, xanthan gum, calcium disodium EDTA.

Side Orders

French fries: potatoes, vegetable shortening, dextrose, sodium acid or disodium dihydrogen pyrophosphate. Fried in vegetable shortening.

Onion rings: rehydrated onion, partially hydrogenated soy, cottonseed, and/or palm oils, bleached flour, gelatinized starch, water, sugar, yellow corn flour, methylcellulose gum, sodium tripolyphosphate,, sodium alginate, soy flour, dextrose, carbohydrate gum, calcium chloride, salt, garlic powder, natural flavor, leavening (monocalcium phosphate, sodium bicarbonate), dried whey, polysorbate 80, silicon dioxide, onion extractives, BHA. Fried in vegetable shortening.

Tater Tenders: potatoes, dehydrated potato flakes, salt, corn flour, enriched flour, natural flavor, disodium dihydrogen pyrophosphate, dextrose. Fried in vegetable shortening.

Salads

Chef: lettuce, ham, turkey, tomatoes, cucumbers, whole eggs, cheese, carrots, celery, radishes.

Chicken: lettuce, diced baked chicken, tomatoes, cucumbers, carrots, celery, radishes.

Croutons: enriched flour, vegetable shortening, salt, whey, corn syrup, yeast, spices, Romano cheese, onion, MSG, distilled vinegar, garlic, beta-carotene.

Garden: lettuce, tomatoes, broccoli, cucumbers, cheese, carrots, celery, radishes.

Side: lettuce, tomatoes, cucumbers,

carrots, celery, radishes.

SALAD DRESSINGS

Blue cheese: soybean oil, blue cheese, vinegar, water, egg yolk, corn syrup, salt, sugar, sour cream solids, xanthan gum, garlic, black pepper, spice, buttermilk.

French: soybean oil, corn syrup, sugar, cider vinegar, tomato paste, water, salt, paprika, spices, natural flavor, xanthan gum, onion, beet powder, garlic.

Olive oil and vinegar: Olive oil, soybean oil, red wine vinegar, water, lemon juice, onion, salt, spices, garlic.

Ranch: soybean oil, water, vinegar, buttermilk, egg yolk, salt, garlic, sugar, onion, natural flavor, spices, xanthan gum, lemon juice.

Reduced-calorie Italian: water, vinegar, soybean oil, olive oil, sugar, salt, spices, xanthan gum, lemon juice, red bell pepper, turmeric.

Thousand island: soybean oil, cider vinegar, sugar, sweet pickle relish, tomato paste, egg yolk, water, salt, nonfat milk, onion, garlic, spices, xanthan gum, beet powder, natural flavor.

Breakfast Items

Blueberry mini muffin: water, bleached and unbleached flour (enriched flour, malted barley flour), sugar, blueberries, soybean oil, whole egg, corn starch, milk powder (sweet whey solids, sodium caseinate, nonfat dry milk, lecithin, calcium phosphate, calcium oxide), dextrose, mono- and diglycerides, salt, sodium aluminum phosphate, bicarbonate of soda, xanthan gum, potassium sorbate, vegetable lecithin.

Croissant: enriched flour (wheat and barley), vegetable shortening, water, sugar, skim milk, yeast, whole eggs, salt, mono- and diglycerides, annatto extract, calcium propionate, sodium phosphate, lecithin, carrageenan, spice oils.

Egg Mix: eggs mixed with 2% milk and salt. Grilled in vegetable shortening.

French toast sticks: enriched egg bread flour (bleached, malted, barley), water, sugar, vegetable shortening, yeast, gluten, salt, corn flour, dextrose, egg yolk, dough conditioners (sodium and/or calcium stearoyl-2-lactylate, ethoxylated mono- and diglycerides, calcium sulfate, ammonium sulfate, ammonium chloride), beta-carotene, potassium bromate; breaded and battered with water, flour, nonfat dry milk, modified starch, salt, soy flour, natural and artificial flavors, yeast, spices, leavening (monocalcium phosphate, sodium bicarbonate), dextrose, polysorbate 80. Fried in vegetable oil.

Desserts

Apple pie: apples, flour, vegetable shortening, corn sweetener, sugar, water, modified starch, salt, potassium sorbate, cinnamon, casein, lecithin, sodium phosphate, carrageenan, extract of turmeric and paprika.

Breyers Chocolate Frozen Yogurt: skim milk, cream, corn syrup, sugar, cocoa, yogurt culture.

Breyers Vanilla Frozen Yogurt: skim milk, cream, corn syrup, sugar, natural flavor, yogurt culture.

Snickers Ice Cream Bar: ice cream (cream, skim milk, whole milk, sugar, corn sweetener, peanut butter, cocoa powder, gelatin, mono- and diglycerides, cellulose gel, cellulose gum, guar gum, polysorbate 80, carrageenan, salt), milk chocolate (sugar, cocoa butter, milk, chocolate, soy lecithin, vanillin artificial flavoring), peanuts, corn sweetener, sugar, milk, butter, salt, carrageenan, vanillin artificial flavoring.

Shakes

Shake mix: milk, sucrose syrup, corn sweetener, cream, whey, nonfat milk solids, mono- and diglycerides, cellulose gum, guar gum, dipotassium phosphate, sodium citrate, carrageenan, salt, sodium carbonate.

Shake syrups added to mix:

Chocolate: also contains cocoa, and in some areas also corn sweeteners, water, dextrose, whey, salt, xanthan gum, potassium sorbate, citric acid.

Strawberry: also contains corn syrup, high-fructose corn syrup, water, citric acid, artificial flavor, sodium benzoate, red #40.

Vanilla: also contains artificial flavor, annatto and caramel colors.

Fats

Vegetable oil (for entrée products): partially hydrogenated soybean oil, TBHQ, citric acid, dimethylpolysiloxane.

Vegetable oil (for French fries): partially hydrogenated soybean and cottonseed oils, TBHQ, citric acid, dimethylpolysiloxane.

CARL'S JR.

COMPONENTS
Sandwiches

All-Star Chili Dog: hot dog, hot dog bun, chili, chili pepper, onion, mustard.

All-Star Hot Dog: hot dog, hot dog bun, onion, mustard, sweet relish.

Carl's Catch (fillet of fish): breaded cod, kaiser bun, American cheese, lettuce, tartar sauce, tomato.

Charbroiler BBQ Chicken: BBQ chicken breast (charbroiled), honeywheat bun, BBQ sauce, lettuce, tomato.

Charbroiler Chicken Club: BBQ chicken breast (charbroiled), honeywheat bun, lettuce, tomato, mayonnaise, alfalfa sprouts, Swiss cheese, bacon.

Country Fried Steak: breaded steak patty, kaiser bun, tomato, lettuce, mayonnaise.

Double Western Bacon Cheeseburger: beef patties, kaiser bun, American cheese, barbecue sauce, bacon, breaded onion rings.

Famous Star (Famous Star with Cheese): beef patty, kaiser bun, lettuce, dill pickle, mayonnaise, onion, special sauce, tomato. (Add American cheese to Famous Star with Cheese.)

Happy Star Hamburger (Cheeseburger): beef patty, plain bun, ketchup, dill pickle, mustard. (American cheese in cheeseburger.)

Old Time Star Hamburger (Cheeseburger): beef patty, kaiser bun, dill pickle, mustard, onion. (American cheese in cheeseburger.)

Roast Beef Club: roast beef, French roll, au jus, Swiss cheese, bacon, mayonnaise, tomato, lettuce, alfalfa sprouts.

Roast Beef Deluxe: roast beef, French roll, au jus, Swiss cheese, green chilies, mayonnaise, lettuce, tomato.

Super Star (Super Star with Cheese): beef patty, kaiser bun, lettuce, dill pickle, mayonnaise, onion, special sauce, tomato. (Add American cheese in Super Star with Cheese.)

Western Bacon Cheeseburger: beef patty, kaiser bun, American cheese, BBQ sauce, bacon, breaded onion rings.

Promotional Sandwiches

Double Deluxe Cheeseburger: beef patties, plain bun, American cheese, lettuce, dill pickle, mayonnaise, onion, special sauce, tomato.

Guacamole Bacon Cheeseburger: beef patty, kaiser bun, American cheese, bacon, lettuce, guacamole, Mexican chili sauce, tomato.

Guacamole Chicken Sandwich: chicken breast, honeywheat bun, American cheese, lettuce, green chilies, guacamole, tomato.

Baked Potatoes

Bacon and cheese: potato, bacon bits, cheese sauce, whipped margarine, sour cream.

Broccoli and cheese: potato, broccoli florets, cheese sauce, whipped margarine.

Cheese: potato, cheese sauce, grated cheese, whipped margarine.

Fiesta: potato, cheese sauce, whipped margarine, salsa, beef taco filling, sour cream.

Lite: potato, whipped margarine (upon request).

Sour cream and chives: potato, chives, whipped margarine, sour cream.

Salads-To-Go

Charbroiler chicken: BBQ chicken breast (charbroiled), croutons, cucumber, grated cheese, salad mix, cherry tomatoes, salad dressing.

Garden: croutons, cucumber, grated cheese, salad mix, cherry tomatoes.

Breakfasts

Breakfast burrito: flour tortilla, scrambled eggs, bacon, grated cheese.

French toast dips: French toast, pancake syrup, powdered sugar.

Hot cakes: hot cake mix, margarine, pancake syrup.

Scrambled eggs: English muffin, egg, margarine, grape jelly, strawberry jelly.

Super scrambled eggs: English muffin, egg, hash brown nuggets, margarine, sausage patty or bacon, grape jelly, strawberry jelly.

Sunrise sandwich: English muffin, egg, American cheese, bacon or sausage patties, margarine.

Super hot cakes: hot cakes, margarine, pancake syrup, sausage patty or bacon.

INGREDIENTS
Sandwiches and Buns

American cheese, regular size: Cheddar cheese (milk, cheese culture, salt, enzymes), enzyme modified Cheddar cheese (Cheddar cheese, water, sodium phosphate, lactic acid, enzymes), water, milkfat, sodium citrate, kasal, salt, sorbic acid, oleoresin paprika, artificial color.

American cheese, small size: cheese [Cheddar cheese (milk, cheese culture, salt, enzymes), granular], water, milkfat, sodium citrate, kasal, salt, sorbic acid, sodium phosphate, annatto color, oleoresin paprika.

Bacon, precooked: cured with water, salt, sugar, hickory smoke flavor, dextrose, sodium erythorbate, sodium nitrite, BHA, BHT.

Beef patty: ground beef.

Chicken breast, BBQ: containing up to 16% of a solution of water, brown sugar, soy sauce, tomato paste, Worcestershire sauce, cider vinegar toner, salt, MSG, spices, granulated garlic, xanthan gum. *OR* containing up to 19% of a solution of water, tomato paste, seasoning (brown sugar, MSG, salt, spices, dehydrated garlic, xanthan gum), soy sauce, cider vinegar toner, Worcestershire sauce.

Chicken strips: chicken, water, natural flavor, salt, sodium tripolyphosphate, sugar; breaded with wheat flour, bleached wheat flour, salt, MSG, spices, dextrose, yeast, extractives of paprika; battered with water, wheat flour, whey, salt, modified corn starch, spices, MSG, dextrose, garlic powder, xanthan gum, onion powder; "predusted" with wheat flour, rice flour, corn flour, salt, MSG.

Chili: beef, tomatoes, water, seasoning (spices, dehydrated onion, corn flour, maltodextrin, salt, garlic powder, caramel color), isolated soy protein.

Cod, breaded: cod, bleached flour, water, corn flour, leavening (sodium aluminum phosphate, sodium bicarbonate), salt, sugar, hydrogenated soybean oil, yeast, nonfat dry milk, whey, dextrose, egg, guar gum, sodium tripolyphosphate.

Country-fried steak patty: beef, salt, sodium phosphate; breaded with enriched bleached flour, leavening (bicarbonate of soda, sodium phosphate, monocalcium phosphate, tricalcium phosphate), salt, malted barley flour; battered with water, enriched flour, modified starch, rice flour, salt, white corn flour, flavoring, guar gum.

Flour tortilla: enriched flour, water, partially hydrogenated soybean and cottonseed oils, salt, baking powder (sodium bicarbonate, sodium aluminum sulfate, corn starch, calcium sulfate, monocalcium phosphate), nonfat dry milk solids, dough conditioners (one or more of sodium stearoyl lactylate, calcium sulfate, L-cysteine hydrochloride, whey), potassium sorbate, calcium propionate.

French roll: *Southern California:* enriched flour (contains barley malt), water; contains 2% or less of skim milk, yeast, corn syrup, salt, wheat gluten, canola, corn, cottonseed, and/or soybean oils, calcium sulfate, dough conditioners (contains one or more of sodium stearoyl lactylate, mono- and diglycerides, ethoxylated mono- and diglycerides, mono- or dicalcium phosphate), potassium bromate, calcium propionate preservative. *Northern California:* enriched bleached flour, water, corn syrup, yeast, wheat gluten, vegetable shortening, salt, dough conditioners (sodium stearoyl lactylate, mono- and diglycerides), yeast nutrients, enrichment vitamins.

Honeywheat bun: *Southern California:* enriched flour (contains barley malt), water, whole-wheat flour, corn syrup, cracked wheat, canola, corn, cottonseed, and/or soybean oils; contains 2% or less of honey, yeast, salt, wheat gluten, sugar, calcium sulfate, dough conditioners (contains one or more of sodium stearoyl lactylate, mono- and diglycerides, ethoxylated mono- and diglycerides, mono- or dicalcium phosphate), potassium bromate, caramel color, calcium propionate. *Northern California:* whole-wheat flour, water, rolled whole wheat berries, honey, corn syrup, rye meal, yeast, wheat gluten; contains 2% or less of each of the following: wheat bran, rolled oats, rolled barley, canola, corn, cottonseed, and/or soybean oils, molasses, raisin juice concentrate, salt, whey solids, malted barley flour, corn starch, yeast nutrients

(contains one or more of monocalcium phosphate, calcium sulfate, and/or ammonium sulfate), dough conditioners (potassium bromate and/or ascorbic acid), vitamins B1, B2, B3, iron.

Hot dog: beef, water, salt, spice, paprika, hydrolyzed vegetable protein flavor, garlic powder, sodium erythorbate, sodium nitrite, natural flavors.

Hot dog bun: *Southern California:* enriched flour (contains barley malt), water, corn syrup, partially hydrogenated canola, corn, cottonseed, and/or soybean oils; contains 2% or less of yeast, salt, corn flour, wheat gluten, calcium sulfate, dough conditioners (contains one or more of sodium stearoyl lactylate, mono- and diglycerides, mono- or dicalcium phosphate), potassium bromate, soy flour, natural flavor, spice, and color, calcium propionate. If seeded, also contains sesame seeds. *Northern California:* enriched bleached flour (contains malted barley flour), water, corn syrup, whey, vegetable shortening (may contain canola, corn, cottonseed, and/or soybean oils), yeast, wheat gluten, salt, dough conditioners (may contain one or more of ethoxylated mono- and diglycerides, calcium and/or sodium stearoyl-2-lactylate, polysorbate 60, succinylated monoglycerides, potassium bromate), soy flour, mono- and diglycerides, yeast nutrients (calcium sulfate, ammonium sulfate), calcium propionate. If seeded, also contains sesame seeds.

Kaiser bun: *Southern California:* enriched flour (contains barley malt), water, corn syrup, soy, corn, cottonseed, and/or canola oils, yeast, salt, wheat gluten, soy flour, calcium sulfate, dough conditioners (contains one or more of calcium stearoyl lactylate, mono- and diglycerides, ethoxylated mono- and diglycerides, mono- or dicalcium phosphate, potassium bromate), calcium propionate. If seeded, also contains sesame seeds. *Northern California:* enriched flour (contains barley flour), water, corn syrup, yeast, wheat gluten, canola, corn, cottonseed, and/or soybean oils, salt, may contain sesame seeds or poppy seeds, dough conditioners (may contain one or more of ethoxylated mono- and diglycerides, calcium and/or sodium stearoyl-2-lactylate, polysorbate 60, succinylated monoglycerides, potassium bromate), mono- and diglycerides, yeast nutrients (calcium sulfate, ammonium sulfate), calcium propionate.

Plain bun: Southern California: enriched flour (contains barley malt), water, corn syrup, soybean, corn, cottonseed, and/or canola oils, yeast, salt, wheat gluten, soy flour, calcium sulfate, dough conditioners (contains one or more of calcium stearoyl lactylate, mono- and diglycerides, ethoxylated mono- and diglycerides, mono- or dicalcium phosphate, potassium bromate), calcium propionate. If seeded, also contains sesame seeds. *Northern California:* enriched bleached flour (contains malted barley flour), water, corn syrup, yeast, wheat gluten, canola, corn, cottonseed, and/or soybean oils, salt, may contain sesame seeds or poppy seeds, dough conditioners (contains one or more of ethoxylated mono- and diglycerides, calcium and/or sodium stearoyl-2-lactylate, polysorbate 60, succinylated monoglycerides, potassium bromate), mono- and diglycerides, yeast nutrients (calcium sulfate, ammonium sulfate), calcium propionate.

Roast beef, 95% lean: injected up to 10% with a solution of water, salt, sodium tripolyphosphate, hydrolyzed plant protein, spice extractives, oil of onion, oil of garlic.

Swiss cheese: Swiss cheese (milk, cheese culture, salt, enzymes), water, milkfat, sodium phosphate, salt, sodium citrate, sorbic acid.

Condiments

Au jus: lactose, flavoring (hydrolyzed plant protein, corn syrup solids, vitamin B1, partially hydrogenated soybean and cottonseed oils, disodium guanylate and inosinate), hydrolyzed plant protein, modified corn starch, MSG, autolyzed yeast, caramel color, meat powder (meat stock, natural flavor, MSG), salt, beef fat, sugar, disodium guanylate and inosinate, oil of onion.

BBQ sauce: high-fructose corn syrup, water, tomato paste, prepared mustard, natural flavor, salt, vinegar, onions, garlic, spices, hydrolyzed vegetable protein, xanthan gum, sodium benzoate, propylene glycol alginate, lemon juice, potassium sorbate, vitamin B1, disodium inosinate and guanylate, dextrose.

Chili peppers: chilies, distilled vinegar, salt, turmeric, yellow #5, sodium phosphate, sodium bicarbonate, citric acid, erythorbic acid.

Classic sauce: soybean oil, corn syrup, distilled vinegar, pickles, tomato paste, salt, spices, sugar, xanthan gum, sodium benzoate, propylene glycol alginate, caramel color, natural flavor, artificial color (including yellow #5).

Dill Pickles: Southern California: cucumbers, water, vinegar, salt, natural flavors, sodium benzoate, yellow #5. *Northern California:* cured cucumbers, water, distilled vinegar, salt, natural flavors, turmeric, sodium benzoate.

Green chilies: green chilies, water, salt, calcium chloride, citric acid.

Honey sauce: corn sweeteners, honey, sugar, water, sodium benzoate.

Ketchup: tomato paste, water, corn sweeteners, vinegar, salt, onion, natural flavor, garlic.

Mayonnaise: soybean oil, water, egg yolks, vinegar, salt, sugar, spices, lemon juice, dextrose, natural flavor, disodium EDTA.

Mustard: water, vinegar, mustard seeds, salt, turmeric, paprika, spices.

Special sauce: sweet pickle relish, water, tomato paste, high-fructose corn syrup, vinegar, salt, sodium benzoate, onion, natural flavor, garlic.

Sweet & sour sauce: distilled vinegar, water, corn syrup, brown sugar, sugar, modified starch, soybean oil, red bell peppers, salt, citric acid, xanthan gum, spices, natural flavor, dehydrated garlic, molasses, caramel color, anchovies, tamarind, artificial colors (including yellow #6).

Sweet relish: pickles, corn syrup, sugar, distilled vinegar, salt, xanthan gum, locust bean and guar gums, spices, sodium benzoate, natural flavor, red bell pepper, polysorbate 80, extractive of turmeric.

Taco sauce: water, tomato paste, vinegar, salt, corn syrup solids, spices, modified starch, garlic, citric acid, sodium benzoate, caramel color.

Tartar sauce: soybean oil, water, pickle relish, sugar, egg yolks, vinegar, lemon juice, onion, modified starch, salt, red peppers, spices, potassium sorbate, sodium benzoate.

Promotional Toppings

Guacamole: avocado, water, lemon juice, corn starch, sugar, salt, spices, dehydrated onion, sodium acid pyrophosphate, natural flavor, citric acid, sodium alginate, dehydrated garlic, xanthan gum, erythorbic acid.

Mexican chili sauce: chilies (green chili peppers, citric acid), water, tomato paste, high-fructose corn syrup, molasses, jalapeños, brown sugar, vinegar, prepared mustard, onion, salt, natural flavor, spices, garlic, sodium benzoate, hydrolyzed vegetable protein, xanthan gum, propylene glycol alginate, lemon juice, potassium sorbate, vitamin B1, disodium inosinate and guanylate, dextrose.

Potato Toppings

Bacon bits: cured with water, salt, sugar, sodium phosphates, sodium erythorbate, sodium nitrite.

Cheese sauce: water, Cheddar cheese (milk, cheese culture, salt, enzymes), soybean or cottonseed oil, modified starch, granular cheese, sodium phosphate, vinegar, salt, propylene glycol monostearate with monoglycerides, sodium citrate, cellulose gum, artificial flavor (including yellow #5).

Salsa: tomatoes, water, green chilies, tomato paste, jalapeños, vinegar, onion, spices, xanthan gum, garlic, sodium benzoate, potassium sorbate, dextrose, natural flavor.

Sour cream: cultured pasteurized cream, nonfat milk, modified starch, locust bean gum, disodium phosphate, cellulose gum, carrageenan, monoglycerides, lecithin.

Whipped margarine: liquid and partially hydrogenated soybean oil, water, salt, whey, vegetable lecithin, vegetable mono- and diglycerides, sodium benzoate, artificial flavor, carotene color, Vitamin A palmitate, Vitamin D2.

Side Orders

CrissCut Fries: potatoes, partially hydrogenated soybean oil, enriched bleached flour, salt, spices and coloring, modified starch, cornmeal, garlic powder, onion powder, leavening (disodium dihydrogen pyrophosphate, sodium bicarbonate), natural flavor.

Jr. Crisp Burrito: flour tortilla, beef, green chili, pasteurized process Cheddar cheese (Cheddar cheese, water, sodium phosphate, milkfat, sodium hexametaphosphate, salt, artificial color), process Monterey Jack cheese (Monterey Jack cheese, water, milkfat, sodium phosphate, salt, sodium hexametaphosphate), water, seasoning (salt, soy protein concentrate, tomato powder, spices, hydrolyzed plant protein, corn starch, dehydrated onion, garlic powder, caramel color, citric acid), oats, textured vegetable protein (soy flour, caramel color, zinc oxide, ferrous sulfate, vitamins B1, B2, B3, B6, B12, calcium pantothenate, vitamin A palmitate), Cheddar cheese flavor (natural flavors, maltodextrin, sodium caseinate).

French fries: potatoes, partially hydrogenated soybean oil, disodium dihydrogen pyrophosphate, dextrose.

Onion rings, breaded: onions, bleached flour, water, yellow corn flour, salt, leavening (sodium acid pyrophosphate, sodium bicarbonate, monocalcium phosphate), sugar, whey, dextrose, soybean oil, corn starch, nonfat milk, cellulose gum.

Zucchini, breaded: Southern California: zucchini, bleached flour, water, flour, modified starch, partially hydrogenated soy, cottonseed, and/or palm oils, whey, seasoning (salt, onion powder, spices, garlic powder, dehydrated parsley), lemon juice powder, yeast, sugar, leavening (sodium bicarbonate, sodium aluminum phosphate), dextrose, soybean oil, MSG. *Northern California:* zucchini, bleached flour, water, wheat flour, modified starch, seasoning (salt, onion powder, spices, garlic powder, dehydrated parsley), salt, whey, partially hydrogenated soybean oil, yeast, sugar, MSG, maltodextrin, leavening (sodium bicarbonate, sodium aluminum phosphate), dextrose, calcium silicate, lemon juice solids, citric acid, natural flavor.

Salads

BBQ chicken breast: containing up to 16% of a solution of water, brown sugar, soy sauce, tomato paste, Worcestershire sauce, cider vinegar toner, salt, MSG, spices, granulated garlic, xanthan gum.

Beef taco filling: ground beef, water, taco seasoning mix (spices, whey, salt, MSG, potato starch, paprika, garlic powder, extractive of spice, paprika, turmeric, citric acid, autolyzed yeast), modified starch, enriched flour.

Croutons: toasted bread (enriched flour containing malted barley flour and potassium bromate, water, soybean oil, yeast, dextrose, salt, ascorbic acid), partially hydrogenated soybean oil, cheese powder (Parmesan cheese, Romano cheese solid, nonfat milk, whey solid, yeast solid), salt, yellow degermed corn flour, dextrose, hydrolyzed plant protein, onion powder, dehydrated parsley, garlic powder, spices, natural flavors.

Grated cheese: pasteurized milk, cheese culture, microcrystalline cellulose, salt, enzymes, artificial color.

Olives, black: ripe olives, water, salt, ferrous gluconate.

Salad mix: lettuce, romaine, carrots, red cabbage.

Salsa: tomatoes, green chilies, jalapeños, onion, vinegar, spices, water, garlic, sodium benzoate, dextrose, natural flavor.

Tortilla chips: processed corn, water, trace of lime, palm, soybean, or canola oil, salt.

SALAD DRESSINGS

Blue cheese: soybean oil, blue cheese, water, distilled vinegar, egg yolks, sugar, salt, dehydrated garlic, spices, xanthan gum, potassium sorbate, sodium benzoate, cultured buttermilk solids.

House: water, soybean oil, buttermilk powder, egg yolks, vinegar, salt, maltodextrin, MSG, lactic acid, garlic, sugar, onion, spices, xanthan gum, sodium benzoate and potassium preservatives, propylene glycol alginate, cottonseed oil.

Italian: soybean oil, water, white wine vinegar, sugar, salt, garlic, maltodextrin, onion, MSG, spices, paprika, propylene glycol alginate, corn syrup solids, xanthan gum, dextrose, bell peppers, lemon juice solids, natural flavor.

Reduced-calorie French: vinegar, corn syrup, water, soybean oil, tomato paste, salt, propylene glycol alginate, xanthan gum, potassium sorbate, oleoresin paprika, calcium disodium EDTA, natural flavor.

Thousand island: soybean oil, water, pickle relish, tomato paste, egg yolks, vinegar, high-fructose corn syrup, sugar, salt, modified starch, green peppers, onion, spices, Worcestershire sauce, sodium benzoate, natural flavor, garlic.

Salad Bar

Bacon bits: cured with water, salt, sugar, sodium phosphates, smoke flavor, sodium erythorbate, sodium nitrite.

Beets: beets, water, salt.

Blueberry yogurt, low-fat: cultured pasteurized lowfat milk, nonfat milk, sugar, blueberries, whey protein concentrate, kosher gelatin, natural flavors.

Bread sticks: enriched flour (contains malted barley flour), sesame seeds, wheat gluten, malt, hydrogenated soybean and/or cottonseed oils, salt, yeast.

Cottage cheese, 2% milkfat: pasteurized cultured skim milk, milk, cream, salt, guar gum, sodium carboxymethyl cellulose, mono- and diglycerides, dipotassium phosphate, carrageenan.

Macaroni de verano: rotini pasta, diced tomato, mayonnaise, light Italian dressing, green bell peppers, green onions, salt, white pepper, sodium benzoate, potassium sorbate.

Macaroni salad, elbow: macaroni, salad dressing, mayonnaise, celery, sweet relish, sugar, onions, seasoning, cheese powder, red peppers, mustard, sodium benzoate, potassium sorbate.

Macaroni salad, sour cream and Cheddar: cooked macaroni, mayonnaise (soybean oil, egg yolk, water, sugar, vinegar, salt, mustard flour), sour cream (skim milk, cream, gelatin, starch, guar gum, potassium sorbate, carrageenan, enzymes), bell peppers, water, processed Cheddar cheese [Cheddar cheese (milk, culture, salt, enzymes), milkfat, sodium phosphate, sodium citrate, annatto], high-fructose corn syrup, celery, carrots, scallions, salt, vinegar, xanthan gum, mustard flour, spices, potassium sorbate, sodium benzoate.

Pasta Italiano: white rotini, spinach rotini, tomatoes, chili sauce, D'oro dressing (water, vinegar, high-fructose corn syrup, salt, soybean oil, dehydrated garlic, dextrose, lemon juice concentrate, xanthan gum, propylene glycol alginate, dehydrated red peppers, sodium benzoate, lactic acid, spices, dehydrated onion, natural flavors, calcium disodium EDTA, yellow #5), turkey ham, onions, olives, soy sauce, soybean oil, herbs and spices, xanthan gum, cilantro, hickory smoke flavor, sodium benzoate, potassium sorbate.

Peaches in light syrup: peaches, water, corn syrup, sugar.

Pineapple chunks: pineapple chunks, pineapple juice from concentrate.

Potato salad: potatoes, salad dressing, celery, pepper, onions, parsley, red peppers, seasoning, potassium sorbate, sodium benzoate.

Salad mix: lettuce, romaine, carrots, red cabbage.

Three bean salad: green beans, kidney beans, garbanzo beans, onions, red peppers, green peppers, vinegar, water, sugar, salt, spices, xanthan gum, garlic powder.

Breakfast Items

Bacon, precooked: cured with water, salt, sugar, hickory smoke flavor, dextrose, sodium erythorbate, sodium nitrite, BHA, BHT.

Blueberry muffin: enriched bleached flour (contains malted barley flour), blueberries, sugar, skim milk, partially hydrogenated soybean and/or cottonseed oils, whole eggs, corn syrup, egg whites, baking powder, sodium aluminum phosphate, baking soda, corn starch, monocalcium phosphate, modified starch, natural flavors, salt, propylene glycol esters, mono- and diglycerides, xanthan gum, lactylic stearate.

Cinnamon roll: enriched bleached flour (contains malted barley flour, potassium bromate), water, sugar, partially hydrogenated soybean oil, whole eggs, compressed yeast, raisins, cinnamon, artificial flavor, salt, starch, whey, mono- and diglycerides, soy flour, soda, sodium acid pyrophosphate, gluten, sodium caseinate, yeast food, sodium stearoyl lactylate, modified starch, dextrose, calcium carbonate, agar, algin, locust bean gum, polysorbate 60, calcium sulfate, beta-carotene, artificial color (including yellow #5), sodium benzoate, propyl paraben.

English muffin: enriched flour, water, yeast, sugar, salt, vegetable shortening, wheat gluten, corn flour, calcium peroxide, mold inhibitor, white sour (*sic*), fumaric acid.

French toast: *Southern California:* enriched bread, bleached and enriched bleached flour, water, partially hydrogenated soybean oil, sugar, egg, salt, natural and artificial flavors, nonfat milk, gelatinized wheat starch, leavening (sodium bicarbonate, sodium acid pyrophosphate, sodium aluminum phosphate), dextrose, modified starch, whey, oleoresin paprika. *Northern California:* bleached flour (contains malted barley), water, sugar, soybean oil, yeast, wheat gluten; contains 2% or less of salt, soy flour, dextrose, whole egg solids, dough conditioners (sodium and/or sodium stearoyl-2-lactylate), lecithin, turmeric, paprika, yeast nutrients (monocalcium phosphate, calcium sulfate, ammonium sulfate, and/or ammonium chloride), beta-carotene, potassium bromate; battered and breaded with water, flour, soybean oil, sugar, eggs, yellow corn flour, modified corn starch, nonfat dry milk, salt, soy flour, gum arabic, dextrose, leavening (monocalcium phosphate, sodium bicarbonate), natural and artificial flavors, yeast, polysorbate 80.

Fruit danish: bleached enriched flour (contains malted barley flour, potassium bromate), water, sugar, partially hydrogenated soybean oil, whole eggs, compressed yeast, raspberry, apple, high-fructose corn syrup, artificial flavor, salt, starch, whey, mono- and diglycerides, soy flour, soda, sodium acid pyrophosphate, modified corn starch, gluten, sodium caseinate, yeast food, sodium stearoyl lactylate, dextrose, calcium carbonate, agar, algin, locust bean gum, spice, polysorbate 60, calcium sulfate, beta-carotene, artificial color including yellow #5, sodium benzoate, propyl paraben, potassium sorbate, sodium propionate.

Hash brown nuggets: potatoes, partially hydrogenated soybean oil, dehydrated potato flakes, salt, corn flour, enriched flour, natural flavor, disodium dihydrogen pyrophosphate, dextrose.

Hot cake mix: enriched bleached flour, sugar, soy flour, yellow corn flour, leavening (sodium aluminum phosphate, sodium bicarbonate, monocalcium phosphate), dextrose, partially hydrogenated soybean and/or cottonseed oil, whey, egg, lecithin, salt.

Grape jelly: corn syrup, grape juice, high-fructose corn syrup, pectin, citric acid.

Strawberry jelly: strawberries, corn syrup, sugar, pectin, citric acid.

Pancake syrup: corn sweeteners, water, sugar, natural and artificial flavors, caramel color, potassium sorbate, citric acid.

Raisin bran muffin: raisins, buttermilk, enriched flour, sugar, cottonseed oil, honey, whole eggs, bran flour, nonfat milk, baking soda, salt, vanilla, cinnamon, sodium acid pyrophosphate, disodium phosphate.

Sausage patties: pork, water, salt, corn syrup solids, spices, sugar, dextrose, flavorings, BHT, propyl gallate, citric acid.

Desserts

Birthday cake: sugar, buttermilk, enriched flour, soybean oil, eggs, cream cheese, cocoa, margarine (liquid soybean oil, water, whey, lecithin, mono- and diglycerides, beta-carotene), invert sugar, modified starch, salt, baking soda, natural and artificial flavors.

Chocolate chip cookies: enriched flour, chocolate chips (semisweet chocolate with lecithin emulsifier), sugar, cake shortening (partially hydrogenated soybean, palm, and cottonseed oils with mono- and diglycerides), eggs, butter, margarine (liquid soybean oil, whey, lecithin, mono- and diglycerides, beta-carotene), invert sugar, dextrose, salt, baking soda, artificial flavor, lecithin.

Fudge brownie: sugar, enriched bleached bromated flour (contains malted barley flour), partially hydrogenated soybean oil, walnuts, water, egg whites, cocoa processed with alkali, salt, artificial flavor, baking soda.

Fudge brownie mousse cake: water, sugar, partially hydrogenated coconut, cottonseed, palm kernel, and/or soybean oils, walnuts, enriched flour, corn syrup, cocoa, cocoa processed with alkali, dried egg whites, modified starch, whole milk powder, chocolate liquor, chocolate, carob powder, nonfat milk solids, mono- and diglycerides, salt, sodium caseinate, artificial and natural flavors, cocoa butter, artificial color, baking powder, baking soda, lecithin, sodium stearoyl lactylate, polysorbate 60, sorbitan monostearate, carrageenan.

Raspberry swirl cheesecake: cream cheese, sour cream, sugar, graham cracker meal, corn syrup, nonfat milk solids, water, margarine, whole eggs, raspberries, modified starch, gelatin, lemon juice, artificial and natural flavors, citric acid, spice, artificial color, sodium benzoate, potassium sorbate.

Beverages

Coffee creamer: water, partially hydrogenated soybean oil, corn syrup, sugar, mono- and diglycerides, soy protein, dipotassium phosphate, polysorbate 60, sodium stearoyl lactylate, salt.

Hot cocoa mix: sugar, whey, corn syrup, cocoa processed with alkali, partially hydrogenated coconut, cottonseed, palm, palm kernel, safflower, and/or soybean oils, nonfat milk, cellulose gum, salt, sodium caseinate, artificial vanilla flavor.

Chocolate shake base: corn sweeteners, water, cocoa, caramel color, modified starch, artificial color (including yellow #5), salt, cellulose gum, potassium sorbate, natural flavor.

Strawberry shake base: corn sweetener, water, citric acid, artificial strawberry flavor, artificial color, sodium benzoate.

Vanilla ice milk mix: condensed skim milk, sugar, cream, corn sweetener, whey solids, guar gum, calcium sulfate, carrageenan, polysorbate 80, dextrose, whey, salt, artificial

flavor, annatto color.

Fats

Margarine: liquid sunflower oil, partially hydrogenated soybean and cottonseed oils, water, salt, vegetable mono- and diglycerides, soy lecithin, potassium sorbate, citric acid, calcium disodium EDTA, artificial flavor, beta-carotene, vitamin A (palmitate), vitamin D (calciferol).

Grill oil primer: blend of two or more of the following oils: corn, soy, coconut with lecithin; beta-carotene, artificial butter flavor, dimethylpolysiloxane.

Vegetable shortening: partially hydrogenated corn and soybean oils, TBHQ, citric acid, dimethylpolysiloxane.

DAIRY QUEEN/ BRAZIER

INGREDIENTS
Frozen Novelties

Buster Bar: layers of peanuts, fudge, and artificially flavored vanilla ice milk, coated in chocolate. *Filling:* peanuts, water, skim milk, sugar, corn sweeteners, hydrogenated coconut and soybean oils, cocoa processed with alkali, sodium caseinate, sodium alginate, salt, mono- and diglycerides, potassium sorbate, disodium phosphate, chocolate liquor, lecithin, natural and artificial flavors. *Coating:* partially hydrogenated coconut and soybean oils, sugar, cocoa, skim milk, lecithin, salt, artificial flavors.

Dilly Bar: artificially flavored vanilla ice milk with coating. *Coating:* ice milk, partially hydrogenated coconut and soybean oils, sugar, lecithin, salt, artificial flavor. *If chocolate,* coating also contains skim milk and cocoa. *If butterscotch,* coating also contains whey, cocoa processed with alkali, artificial colors. *If cherry*, coating also contains whey, artificial colors, citric acid.

DQ Sandwich: artificially flavored vanilla ice milk between chocolate wafers. *Wafers:* flour, sugar, partially hydrogenated soybean, palm, and cottonseed oils, hydrogenated palm oil, corn sweeteners, artificial color, baking soda, salt, lecithin, cocoa, artificial color.

Ice milk, vanilla: milkfat and nonfat milk, sugar, corn sweeteners, whey, mono- and diglycerides, artificial flavor, guar gum, polysorbate 80, carrageenan, yellow #5.

Starkiss: water, sugar, corn sweeteners, methylcellulose, locust bean gum, guar gum, karaya gum, pectin. *If lemon-lime*, also contains citric acid, natural and artificial flavors, yellow #5 artificial color. *If orange*, also contains citric acid, natural and artificial flavors, yellow #6. *If cherry or grape*, also contains citric acid, artificial flavor, artificial color. *If banana*, also contains artificial flavors, yellow #5 and #6.

DOMINO'S PIZZA

Crust ingredients vary from store to store.

Crusts

#1 Enriched unbleached high-protein flour, water, sugar, salt, partially hydrogenated soybean, canola, and/or cottonseed oils, mono- and diglycerides, yeast, Cheddar cheese solids (Cheddar cheese, pasteurized milk, cheese cultures, salt, enzymes), malted barley, sodium stearoyl-2-lactylate, whey, butter, buttermilk, artificial color, cheese flavor (natural cheese flavor, milk solids, salt, sodium citrate, yeast, disodium guanylate and inosinate, onion powder, garlic powder, spices).

#2 Enriched unbleached high-protein flour, water, sugar, salt, partially hydrogenated soybean, canola, and/or cottonseed oils, yeast.

#3 Enriched unbleached high-protein flour, water, rice flour, wheat bran, sugar, salt, partially hydrogenated soybean, canola, and/or cottonseed oils, mono- and diglycerides, whole egg, soy flour, whey, soybean oil, yeast, sodium caseinate, monocalcium phosphate, ascorbic acid, potassium bromate, natural and artificial butter flavors, lecithin, TBHQ.

#4 Enriched unbleached high-protein flour, water, soybean, canola, and/or cottonseed oils, mono- and diglycerides, salt, whey powder, sugar, yeast, seasoning (onion, garlic, cumin, and/or rosemary), sodium stearoyl lactylate, citric acid.

Toppings

Anchovies: anchovies, salt, olive and/or soy oil.

Beef for pizza: beef, water, seasonings (salt, sodium tripolyphosphate, spices, disodium guanylate and inosinate, BHA, BHT, citric acid).

Black olives: ripe black olives, water, salt, ferrous gluconate.

Bold 'n' Spicy Italian sausage: pork, seasonings (spices, salt, corn syrup solids, sodium tripolyphosphate, garlic powder, caramel color, disodium guanylate and inosinate, BHA, BHT, citric acid), water.

Breakfast sausage: pork, seasonings (spices, salt, MSG, BHA, BHT, citric acid), water.

Canadian-style bacon: pork, water, salt, brown sugar, sodium phosphate, sodium nitrite.

Canned mushrooms: mushrooms, water, salt, may contain citric acid.

Cheese: low-moisture part-skim mozzarella cheese, part-skim mozzarella cheese, semisoft part-skim cheese, pasteurized part-skim milk, cultures, rennet (vegetable or animal origin), salt, enzymes, calcium chloride, microcrystalline cellulose (less than 1%), sodium citrate, natural flavor, sodium propionate.

Chunked tomatoes: tomatoes, tomato juice, salt, citric acid, calcium chloride.

Green chilies: green chilies, water, salt, citric acid, calcium chloride.

Green olives: green olives, water, salt, lactic acid.

Green peppers: green peppers; citric acid, erythorbic acid, calcium sulfate, and/or calcium phosphate.

Ham: pork, water, salt, sodium phosphate, sugar, hydrolyzed plant protein, sodium ascorbate, sodium nitrite.

Hot banana peppers: banana peppers, vinegar, salt, calcium chloride, yellow #5, sodium benzoate, polysorbate 80, turmeric, calcium disodium EDTA.

Italian sausage: pork, seasonings (spices, salt, corn syrup solids, sodium tripolyphosphate, garlic powder, caramel color, disodium guanylate and inosinate, BHA, BHT, citric acid), water.

Jalapeño peppers: jalapeños, water, vinegar, salt, garlic powder, spices, calcium chloride, may contain citric acid.

Pepperoni: pork and beef, salt, water, dextrose, spices, lactic acid starter culture, oleoresin of paprika, hydrolyzed plant protein, sodium nitrite, BHA, BHT, citric acid.

Pineapple: pineapple, pineapple juice, sugar, water.

Raw ground beef: ground beef.

Raw mushrooms: mushrooms.

Sauce: Tomatoes, salt, sugar, spices and herbs, garlic powder (may contain citric acid and an anti-caking agent).

Sauerkraut: sauerkraut/cabbage, water, salt.

DUNKIN' DONUTS

INGREDIENTS

Oat-bran muffin: sugar, enriched bromated flour (contains malted barley flour), oat bran, partially hydrogenated soy oil with mono- and diglycerides, whole-wheat flour, salt, leavening (baking soda, sodium aluminum phosphate, sodium acid pyrophosphate), dried egg albumen, starch, cinnamon, dried honey, dried molasses, natural and artificial flavors. *Raisin and nut* also contains the respective fruit and nuts.

Plain bagel: enriched bromated high-gluten flour (contains malted barley), water, high-fructose corn syrup, salt, yeast, cornmeal. *If Egg,* also contains corn flour, dried egg yolk solids, beta-carotene. *If Onion,* also contains chopped toasted onions. *If Cinnamon 'N' Raisin,* also contains raisins, soybean oil, honey, cinnamon.

HARDEE'S

COMPONENTS
Sandwiches

All Beef Hot Dog: hot dog, bun, mustard, chili, onion.

Bacon Cheeseburger: meat patty, bun, mayonnaise, lettuce, tomato, bacon, cheese.

Big Deluxe Burger: meat patty, bun, mayonnaise, dill pickle, lettuce, tomato, onion, cheese.

Big Twin: meat patty, bun, Big Twin sauce, lettuce, cheese.

Chicken Fillet: breaded chicken fillet, bun, mayonnaise, lettuce.

Fisherman's Fillet: breaded fish fillet, natural grain bun, tartar sauce, lettuce, cheese.

Grilled Chicken Sandwich: grilled chicken, natural grain bun, mayonnaise, tomato, lettuce.

Hamburger: meat patty, bun, mustard, ketchup, dill pickle. Also with cheese.

Hot Ham 'N' Cheese: Ham, bun, Swiss cheese.

Mushroom 'N' Swiss Burger: meat patty, bun, mushroom sauce, Swiss cheese.

Quarter Pound Cheeseburger: meat patty, bun, mustard, ketchup, dill pickle, cheese.

Regular and Big Roast Beef: roast beef, bun.

Roy Rogers Roast Beef Sandwiches (selected stores only):

   Roast beef (small): sesame roll, roast beef, au jus (option with cheese).

   Roast beef (large): kaiser roll, roast beef, au jus (option with cheese).

   Roast beef: pure beef top rounds, salt, pepper.

Turkey Club: Sliced turkey, natural grain bun, mayonnaise, lettuce, tomato, bacon.

Salads

Chef Salad: Garden Salad with ham and turkey.

Chicken and Pasta Salad: lettuce, carrots, red cabbage, chicken, mixture of pasta, beans, and broccoli.

Chicken Fiesta (or Grilled Chicken) Salad: Garden Salad with chicken.

Garden Salad: lettuce, carrots, red cabbage, egg, tomato, Cheddar and Monterey Jack cheeses.

Side salad: lettuce, carrots, red cabbage, tomato.

Breakfasts

Big Country Breakfast: biscuit; bacon, ham, or sausage; eggs; hash rounds.

INGREDIENTS
Sandwiches and Buns

FILLINGS

All meat frank: beef and pork, water, salt, corn syrup, dextrose, spice, sodium erythorbate, paprika, sodium nitrite, natural flavors, oleoresin paprika.

Sliced American cheese: American cheese, milk, salt, cheese culture, enzymes, water, cream, sodium citrate, salt, natural flavor, sodium phosphate, sorbic acid.

Bacon: bacon cured with water, salt, sugar, sodium phosphate, sodium erythorbate, sodium nitrite, may contain one or more of: hickory smoke and/or hickory smoke flavor, sugar, dextrose.

Beef frank (All Beef Hot Dogs): beef, water, salt, dextrose, flavorings, sodium erythorbate, sodium nitrite.

Chicken fillet: chicken with rib meat, water, salt, sodium phosphate, enriched unbleached self-rising flour, flavorings, salt, cereal, corn starch, buttermilk solids, guar gum. *OR* chicken breast fillet with rib meat, water, salt, sodium phosphate, bleached flour, salt, spices, natural flavors, corn flour, dextrose, monocalcium phosphate, sodium bicarbonate, soy flour, may contain one or more of: corn syrup, MSG, sodium alginate, whey, nonfat milk, eggs, corn flour, tapioca flour, gluten, egg whites, yeast.

Chicken sticks: chicken breast fillet, water, salt, sodium phosphate, bleached flour, spices, garlic powder, MSG, natural flavor, dextrose, yeast, corn flour, monocalcium phosphate, sodium bicarbonate, sodium alginate, whey, soy flour, nonfat milk, natural flavors, eggs.

Cheddar cheese: cultured pasteurized milk, salt, calcium chloride, enzymes, microcrystalline cellulose, artificial color, may contain powdered cellulose.

Cheese slices: American cheese, cultured milk, salt, enzymes, artificial color, water, cream, enzyme modified cheese, sodium citrate, salt, sodium phosphates, citric acid, acetic acid, sorbic acid, lecithin, artificial color.

Chili (for hot dogs): beef, water, chili seasoning (chili pepper, salt, dehydrated onion, garlic, spices, caramel color), spices, dehydrated onion, salt, tomato paste, textured vegetable protein, soy flour, caramel color, may contain one or more of: yellow corn flour, structured soy flour, dextrose, hydrolyzed plant protein, spice extractive encapsulated in modified corn starch, sugar, cumin, modified starch, flour.

Cod (for Fisherman's Fillet): cod, bleached flour, water, corn flour, modified starch, potatoes, salt, leavening, sodium acid pyrophosphate, sodium bicarbonate, partially hydrogenated soybean and cottonseed oils, sugar, yeast, spice, may contain bread crumbs, whey, nonfat milk. *OR* cod, bleached flour, water, modified corn starch, yellow corn flour, whey, salt, dextrose, yeast, natural flavor, soybean oil.

Grilled chicken sandwich breast fillets: containing up to 28% of a solution of water, margarine, seasoning (black pepper, salt, citric acid, corn starch, lemon peel, sugar, garlic, onion, natural flavor, disodium inosinate and guanylate), charbroil salt [salt, seasoning (maltodextrin, natural flavor, gum arabic), natural flavor, caramel color, spice], sodium phosphates, seasoned salt (salt, sugar, garlic and onion powder, spices, extractive of paprika and other spices, carotenal coloring), garlic salt (salt, rice flour, garlic powder).

Ham: pork cured with water, salt, dextrose, corn syrup, sodium phosphate, sodium erythorbate, sodium nitrite.

Hamburger: beef, salt, pepper.

Hot dog buns: flour, water, corn syrup, yeast, partially hydrogenated soybean oil, salt, gluten, lactose, soy flour, mono- and diglycerides, ethoxylated mono- and diglycerides, calcium and sodium stearoyl lactylates, potassium bromate, calcium peroxide, corn starch, calcium sulfate, ammonium sulfate, calcium propionate, liquid sugar, buffer, gluten, brewloid, guard,

top mate, azodicarbonamite bromate, whey, soy flour, gluten, succinylated monoglycerides, azodicarbonamide, BHA.

Monterey Jack cheese: cultured pasteurized milk, salt, calcium chloride, enzymes, powdered cellulose, microcrystalline cellulose.

Mushrooms 'N' Sauce: mushrooms, water, modified starch, hydrolyzed vegetable protein, salt, maltodextrin, partially hydrogenated cottonseed and soybean oils, MSG, caramel color, beef-extract powder, beef-fat powder, onion powder, artificial and natural flavors, garlic powder, spice, silicon dioxide, tomato paste, shortening, beef fat, cottonseed oil, may contain seasoning.

Natural grain bun: enriched flour, water, whole-wheat flour, corn syrup, soybean oil, calcium sulfate, yeast, salt, monoglycerides, peroxide, calcium iodate, potassium bromate, ammonium sulfate, calcium sulfate, calcium propionate, liquid sugar, buffer, gluten, brewloid, guar, condition plus, azodicarbonamite bromate, base mix, dicalcium phosphate, diammonium phosphate, lactose, corn starch, shortening, granular sugar, powdered egg, vinegar, potato flour, high-fructose corn syrup, mold inhibitor, fructose, honey, polysorbate 60, bran, oat, rye meal, vitamin B1, oats, BHA, partially hydrogenated soybean and/or cottonseed oils, whey.

Roast beef: beef, water, salt, sodium phosphate or tripolyphosphate.

Roast beef rub: dextrose, caramel color, natural flavor, MSG, onion powder, garlic powder, spice, soybean oil, calcium stearate.

Sea Nuggets (for Crispy Fish Dippers): cod, bleached flour, water, corn flour, sodium aluminum phosphate, sodium bicarbonate, salt, sugar, hydrogenated soybean and/or palm oils, yeast, nonfat dry milk, whey, dextrose, egg, guar gum, sodium tripolyphosphate.

Seeded bun: enriched flour, water, corn syrup, soybean oil, calcium sulfate, yeast, salt, monoglycerides, sodium stearoyl lactylate, barley malt, calcium peroxide, calcium iodate, potassium bromate, ammonium sulfate, calcium sulfate, calcium propionate, ascorbic acid, whey, sesame seeds, vegetable shortening, gluten, ethoxylated mono- and diglycerides, soya flour, corn flour, liquid sugar, buffer, protease, azodicarbonamide, guard, lactose, corn starch, brewloid, top mate, bromate, powdered egg, margarine, vinegar, potato flour, high-fructose corn syrup, syrup, emulsilac, mak-soft, CTC improver, mold inhibitor, fructose corn syrup, ferrous sulfate, polysorbate 60, azodicarbonamide, partially hydrogenated soybean and/or cottonseed oils, BHA, fermaloid, PD321, starplex, wytase.

Sliced Swiss cheese: Swiss cheese, milk, salt, cheese culture, enzymes, water, cream, sodium citrate, salt, sorbic acid, lactic acid.

Swiss cheese: cultured milk, salt, enzymes, water, cream, sodium citrate, salt, sodium phosphates, ascorbic acid, citric acid, lecithin. *OR* cultured milk, salt, enzymes, microcrystalline cellulose.

Condiments

Barbecue sauce: water, vinegar, tomato paste, salt, modified food or corn starch, spices and/or spice extractives, onion powder, paprika, garlic powder, may contain one or more of: molasses, hydrolyzed vegetable protein, anchovies, tamarind, sugar, vegetable gums, citric acid, dextrose, corn syrup, Worcestershire sauce concentrate, hickory smoke flavor, sodium benzoate, brown sugar, soy oil, xanthan gum, MSG, potassium sorbate, caramel color, natural flavor. *OR* tomato purée, vinegar, water, sugar, pepper pulp, Worcestershire sauce, salt, spices, vegetable stabilizer, flavorings.

Club crackers: enriched flour, partially hydrogenated soybean and cottonseed oils, TBHQ, sugar, salt, corn syrup, leavening, sodium bicarbonate, sodium acid pyrophosphate, monocalcium phosphate, whey.

Cocktail sauce: water, tomato paste, sugar, vinegar, horseradish, salt, modified starch, soybean oil, lemon juice, sorbic acid, lactic acid, canthaxanthin color, natural flavor, lecithin.

Croutons: enriched flour, partially hydrogenated soybean and/or cottonseed oils, salt, whey, corn syrup, yeast, spices, Romano cheese made from cow's milk, onion, MSG, distilled vinegar, garlic powder, beta-carotene.

Half-and-half: cream, skim milk, sodium citrate, disodium phosphate, carrageenan, whole milk, sodium tetraphosphates.

Horseradish sauce: soy oil, vinegar, corn syrup, water, horseradish, modified food or corn starch, egg yolk, salt, spice and spice extractives, paprika, may contain one or more of: potassium sorbate, sodium benzoate, calcium sodium EDTA, oil of mustard, oleoresin paprika, xanthan gum.

Ketchup: tomatoes, natural sweeteners, water, distilled vinegar, salt, onion powder, natural flavors. *OR* tomato paste, distilled vinegar, corn sweetener, salt, onion powder, spice, natural flavor.

Lemon juice: lemon juice, sodium benzoate.

Mayonnaise: soybean oil, distilled vinegar, egg yolks, corn syrup, whole eggs, salt, water, calcium disodium EDTA.

Meadowgold for coffee: sodium stearoyl-2-lactylate, salt, disodium phosphate, sodium citrate, artificial flavor, yellow #5 and #6.

Mustard: water, distilled vinegar, mustard seeds, salt, turmeric, paprika, may contain one or more of: spice extractives, mustard bran, natural flavors.

Non-dairy creamer: water, partially hydrogenated soybean oil, corn syrup solids, sodium caseinate, mono- and diglycerides, dipotassium phosphate, salt, stearoyl lactylate, natural color. *OR* water, corn sweetener, partially hydrogenated coconut, soybean, cottonseed, palm, and/or palm kernel oils.

Pickles: pickles, water, distilled vinegar, salt, alum, sodium benzoate, natural spice flavors, turmeric, polysorbate 80, may contain calcium chloride and/or HD256. *OR* cucumbers, water, vinegar, salt, spice flavors, alum, turmeric, sodium benzoate, polysorbate 80.

Reduced-calorie mayonnaise: soybean oil, water, vinegar, modified starch, egg yolks, corn syrup, salt, xanthan gum, spices, sodium benzoate, potassium sorbate, calcium disodium EDTA, beta-carotene, may contain whole eggs and/or natural flavor.

Seasoned salt: salt, sugar, garlic and onion powder, MSG, spices, extractives of paprika and other spices, carotenal, tricalcium phosphate.

Sweet mustard sauce: corn syrup, water, vinegar, modified starch, salt, soybean oil, xanthan gum, turmeric, paprika, spice, oleoresin paprika, oleoresin turmeric, may contain mustard flour.

Sweet 'n' sour sauce: corn syrup, water, vinegar, modified starch, salt, soybean oil, dehydrated red and green bell peppers, natural flavor, spice, xanthan gum, caramel color, oleoresin paprika.

Tartar sauce: soybean oil, onion, water, pickle, vinegar, egg yolks, sugar, salt, artificial and natural flavor, mustard or ground mustard seeds, calcium disodium EDTA, spice, may contain one or more of corn syrup, potassium bromate, sodium benzoate, extractives of paprika, xanthan gum, olives, propylene glycol alginate.

Side Orders

Crispy Curls: potatoes, partially hydrogenated soybean oil, enriched bleached flour (contains malted barley flour), salt, spices and coloring, starch, modified starch, cornmeal, garlic powder, onion powder, whey, leavening (sodium aluminum phosphate, sodium bicarbonate), natural flavor, disodium dihydrogen pyrophosphate.

French fries: potatoes, partially hydrogenated soybean oil, dextrose, may contain sodium acid pyrophosphate and/or disodium dihydrogen pyrophosphate.

Salads

Chicken strips (for Chicken Stix): chicken breast, water, bleached flour, salt, spices, garlic powder, MSG, natural flavors, flour, bread crumbs, corn flour, salt, spice, dextrose, leavening, monocalcium phosphate, sodium bicarbonate, sodium alginate, whey, soy flour, nonfat milk, natural flavors, eggs.

Ham strips (for chef salad): ham, water, salt, sodium phosphates, sodium erythorbate, sodium nitrite, may contain dextrose.

Hard-cooked eggs (for chef salad): eggs, citric acid, sodium benzoate.

Turkey strips (for chef salad): white turkey meat, water, salt, sodium phosphates, may contain modified starch, dextrose.

SALAD DRESSINGS

Blue cheese: soybean oil, water, corn sweeteners, maltodextrin, vinegar, egg yolks, natural blue cheese flavor, blue cheese, salt, xanthan gum, artificial color, propylene glycol alginate, fumaric acid, natural flavors, potassium sorbate EDTA, BHA, BHT, spice, disodium inosinate and guanylate.

House: soybean oil, buttermilk, water, maltodextrin, vinegar, egg yolks, salt, sugar, dried yeast, MSG, dried onion, natural flavors, xanthan gum, fumaric acid, dried garlic, potassium sorbate, disodium EDTA, BHA, BHT, polysorbate 60, spices, propylene glycol alginate.

Parmesan ranch: soybean oil, buttermilk, maltodextrin, water, vinegar, egg yolks, Parmesan cheese, salt, sugar, dried yeast, MSG, dried onion, artificial color, natural flavors, xanthan gum, fumaric acid, dried garlic, polysorbate 60, potassium sorbate, disodium EDTA, BHA, BHT, spices, propylene glycol alginate.

Reduced-calorie French: corn sweeteners, water, reconstituted nonfat yogurt, vinegars, soybean oil, sugar, maltodextrin, tomato paste, modified starch, salt, natural flavors, dried onion, xanthan gum, potassium sorbate, disodium EDTA, artificial colors, paprika, dried garlic, citric acid.

Reduced-calorie Italian: water, vinegar, soybean oil, sugar, natural flavors, modified starch, dried garlic, rehydrated red bell pepper, spices, xanthan gum, dried onion, disodium EDTA, oleoresin paprika.

Thousand island: soybean oil, water, sugar, vinegars, sweet pickles, egg yolks, tomato paste, salt, spices, dried onion, natural flavors, modified starch, xanthan gum, potassium sorbate, disodium EDTA, BHA, BHT, paprika, citric acid, oleoresin paprika, artificial color.

Breakfast Items

Beef steaks (used in all company stores and by some franchisees for breakfast sandwiches): beef, salt, water, bleached flour, buttermilk, monocalcium phosphate, sodium bicarbonate, spices, soybean oil.

Biscuit flour: enriched flour, salt, sodium bicarbonate, monocalcium phosphate, sodium aluminum phosphate.

Biscuit gravy: modified starch, flour, partially hydrogenated vegetable oil, corn syrup solids, salt, dipotassium phosphate, sugar, black pepper, mono- and diglycerides, may contain one or more of vegetable shortening, sodium citrate, carrageenan, sodium caseinate, natural buttermilk flavor, sodium stearoyl lactylate, sodium alginate, artificial flavor, corn starch, lecithin, yellow #5, sodium silicoaluminate, natural flavor, buttermilk solids, artificial color, water, non-dairy creamer, potassium phosphate, sodium phosphate, sodium chloride, sodium calcium alginate, dry vegetable oil base, lactose, tricalcium phosphate, milk, MSG, spice extractive, sodium benzoate, titanium dioxide, butter flavor, butter, corn syrup, milk, dextrose, yeast, hydrolyzed plant protein, whole hog sausage.

Biscuit kit: leavening, sodium bicarbonate, corn starch, sodium aluminum sulfate, monocalcium phosphate, calcium carbonate, sugar.

Biscuit shortening: partially hydrogenated soybean and cottonseed oils.

Boddie Noell biscuit recipe: enriched flour, vegetable shortening, partially hydrogenated soybean and cottonseed oils, baking powder, salt, sugar, whey, maltodextrin.

Breakfast ham: cured with water, sugar, salt, sodium tripolyphosphate, sodium erythorbate, sodium nitrite.

Canadian bacon: cured with water, salt, sugar, sodium tripolyphosphate, sodium erythorbate, sodium nitrite.

Cinnamon chips (for cinnamon raisin biscuit): partially hydrogenated soybean oil, sugar, cinnamon.

Cinnamon 'n' raisin icing: sugar, water, corn syrup, stearic acid, titanium dioxide, salt, agar, potassium sorbate, citric acid, natural and artificial flavors, guar gum, pectin, dextrose, sodium hexametaphosphate.

Cinnamon raisin biscuit: enriched flour, partially hydrogenated soybean and/or cottonseed oils, raisins, leavening, baking soda, sodium aluminum phosphate, monocalcium phosphate, sugar, cinnamon, salt, cellulose gum.

Cinnamon sugar blend: sugar, cinnamon, soybean oil.

Country ham: cured with salt, white sugar, brown sugar, sodium nitrate, pepper, sodium nitrite.

Grape jam: grapes, corn syrup, sugar, pectin, citric acid.

Hash rounds: potatoes, partially hydrogenated soybean oil, salt, natural flavor, dextrose; may contain one or more of disodium dihydrogen pyrophosphate, dehydrated potato flakes, corn flour, flour, partially hydrogenated vegetable shortening, pepper, cream.

Liquid eggs: eggs, citric acid, water.

Sausage: pork, salt, sugar, red and/or black peppers, sage.

Steak fritter (may be used by some franchisees for breakfast sandwiches): beef, enriched unbleached self-rising flour, water, enriched unbleached flour, salt, flavoring, cereal, corn starch, buttermilk solids, guar gum, set in shortening.

Strawberry jam: sugar, strawberries, pectin, citric acid.

Desserts

Apple turnover: flour, lard or shortening, partially hydrogenated soybean and palm oils, water, crust improver, sugar, flour, salt, butter flavor, soda, sodium propionate, dough conditioner, L-cysteine, high-fructose corn syrup, salt, artificial butter flavor, flour, cane sugar, salt, hydrolyzed butter fat, lecithin, vanilla, water, fructose corn syrup, evaporated apples, sulfites, modified starch, spices, salt, butter flavoring, propylene glycol, xanthan gum, dextrose, polysorbate 80, vegetable color. *OR* apples, evaporated apples (sulfite treated), water, high-fructose corn sweetener, enriched flour, modified starch, animal/vegetable shortening, dough improvers, salt, cinnamon, nutmeg, allspice, stabilizer, citric acid, butter flavor, whey derivative, sodium metabisulfite.

Big Cookie: flour, sugar, semisweet chocolate with lecithin, salt, partially hydrogenated soybean and cottonseed oils, margarine with mono- and diglycerides, eggs, molasses, baking soda, natural and artificial flavors, sodium aluminum phosphate, may also contain vanillin, high-fructose corn syrup, enriched flour, invert syrup, palm oil.

Caramel: corn syrup, sweetened condensed skim milk, high-fructose corn syrup, sugar, cream, butter, water, salt, mono- and diglycerides, xanthan gum, sodium bicarbonate, vanilla.

Chocolate ice milk soft-serve mix: whole milk, cream, milk solids, sugar, cocoa processed with alkali, mono- and diglycerides, cellulose gum, guar gum, carrageenan, polysorbate 65, salt, vanilla and vanillin, disodium phosphate, artificial flavor, dextrose, corn syrup, sucrose, corn sweeteners, milk fat, nonfat milk, polysorbate 80, whey, buttermilk, locust bean gum.

Chocolate yogurt: milk, sugar, corn syrup solids, skim milk, cream, cocoa processed with alkali, guar gum, carrageenan, locust bean gum, xanthan gum.

Hot fudge: high-fructose corn syrup, sweetened condensed skim milk, partially hydrogenated palm kernel, coconut, cottonseed, soybean, and/or palm oils, water, cocoa processed with alkali, corn syrup solids, disodium phosphate, salt, sodium alginate, artificial flavor, potassium sorbate.

Ice cream cones: flour, tapioca flour, sugar, partially hydrogenated soybean and/or cottonseed oils, baking soda, salt, artificial flavor, annatto. *OR* unbleached flour, tapioca flour, sugar, partially hydrogenated soybean or cottonseed oils, baking soda, vanillin, salt, yellow #5 and #6.

Lollipops (for children's birthday parties): corn syrup, sugar, citric acid, artificial flavors, yellow #5, may contain yellow #6 and natural flavors.

Low-calorie yogurts contain aspartame (artificial sweetener) instead of some or all of the sugar.

Oatmeal and raisin cookie: enriched flour, flour, white sugar, quick oats, high-fructose corn syrup, midget raisins, partially hydrogenated soybean, cottonseed, and palm oils, margarine, mono- and diglycerides, water, baking soda, whole eggs, molasses, salt, vanilla and vanillin, cinnamon, cloves, nutmeg, sodium aluminum phosphate.

Strawberry topping: strawberries, corn syrup, sugar, water, pectin, citric acid, artificial flavor, sodium benzoate, calcium chloride, artificial color.

Vanilla ice milk soft-serve mix: whole milk, sugar, cream, nonfat milk solids, corn syrup solids, mono- and diglycerides, guar gum, carrageenan, locust bean gum, artificial flavor, dextrose, vanillin, cellulose gum, natural flavor, vanilla, polysorbate 80, sucrose, disodium phosphate, sodium citrate, corn sweeteners, buttermilk solids, whey, calcium sulfate.

Vanilla yogurt: milk, sugar, corn syrup solids, skim milk, cream, active yogurt culture, guar gum, artificial flavor, carrageenan, locust bean gum, xanthan gum, annatto color.

Beverages

Chocolate-flavored shake base: corn sweeteners, water, cocoa processed with alkali, sugar, salt, potassium sorbate, xanthan gum, vanillin, artificial flavor. *OR* corn syrup, water, cocoa, salt, citric acid, potassium sorbate, sodium benzoate, carboxymethylcellulose, xanthan gum, artificial chocolate flavor, yellow #5 and #6, vanillin, ethyl vanillin. *OR* high-fructose corn syrup, corn syrup, water, cocoa processed with alkali, salt, potassium sorbate, artificial flavors, amylase enzyme.

Chocolate milk: milk, corn sweeteners, sugar, cocoa processed with alkali, starch, salt, carrageenan, vanillin, artificial flavors.

Chocolate shake: milk, sugar, nonfat milk solids, corn syrup solids, carrageenan and cocoa processed with alkali, cream, stabilizer, guar gum, mono- and diglycerides, cellulose gum, disodium phosphate, artificial flavor, sodium citrate, dextrose, sucrose, whey, chocolate liquor, dipotassium phosphate, sodium chloride, sodium carbonate.

Hot chocolate: sugar, whey, cocoa (may be processed with alkali), partially hydrogenated coconut, cottonseed, soybean, palm kernel, and/or palm oils, corn syrup solids, sodium caseinate, salt, artificial flavor, may contain one or more of: guar gum, carboxymethylcellulose, sodium aluminosilicate, sodium silicoaluminate, nonfat milk, water, sweetened condensed skim milk, disodium phosphate, natural flavors, soy lecithin.

Low-fat chocolate milk: low-fat milk, corn sweetener, cocoa processed with alkali, modified starch, whey powder, salt, carrageenan, vanillin, artificial flavor, vitamin A palmitate, vitamin B3, sugar, yellow #5.

Strawberry shake base: corn sweeteners, strawberry purée, water, propylene glycol, sodium benzoate, locust bean gum, artificial flavor and color, may contain one or more of: strawberry purée, sugar, malic acid, propylene glycol, locust bean gum, yellow #5, red #40, citric acid, caramel color, natural flavors.

Tea: sugar, citric acid, instant tea, sodium citrate, natural lemon flavor, yellow #5 and #6, tricalcium phosphate. *OR* 100% pure tea.

Vanilla shake: whole milk, pasteurized milk, sugar, nonfat dry milk, cream or butter, corn syrup solids, guar gum, mono- and diglycerides, cellulose gum, carrageenan, disodium phosphate, artificial flavor, sodium citrate, dextrose, vanillin, yellow #5, sucrose, whey, calcium phosphate, artificial color.

Vanilla syrup: corn sweeteners, water, caramel color, sodium benzoate, citric acid, propylene glycol, vanilla extract, ethyl vanillin and vanillin, artificial flavor.

Fats

All vegetable frying oil: partially hydrogenated soybean oil, peanut oil, TBHQ, citric acid, dimethylpolysiloxane.

Margarine solids: Liquid soybean oil, partially hydrogenated soybean oil, water, salt, nonfat dry milk, lecithin, mono- and diglycerides, sodium benzoate, artificial flavor, vitamin A palmitate, beta-carotene.

Liquid shortening: partially hydrogenated soybean oil, lecithin, artificial flavor and color, methyl silicone.

Vegetable frying shortening: soybean oil, partially hydrogenated soybean oil, water, salt, nonfat dry milk, lecithin, sodium benzoate, beta-carotene, vitamin A palmitate, artificial flavor.

Vegetable margarine: partially hydrogenated soybean oil, water, salt, whey, lecithin, mono- and diglycerides, sodium benzoate, artificial color and flavor.

Whipped margarine: Liquid soybean oil, partially hydrogenated soybean oil, water, salt, lecithin, vegetable mono- and diglycerides, sodium benzoate, citric acid, artificial flavor, beta-carotene, vitamin A palmitate, may contain one or more of: whey, soy, nonfat dry milk, artificial color, vegetable oil, vegetable mono- and diglycerides, calcium disodium EDTA.

Jack in the Box

INGREDIENTS
Sandwich Meats

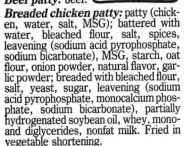

Beef patty: beef.

Breaded chicken patty: patty (chicken, water, salt, MSG); battered with water, bleached flour, salt, spices, leavening (sodium acid pyrophosphate, sodium bicarbonate), MSG, starch, oat flour, onion powder, natural flavor, garlic powder; breaded with bleached flour, salt, yeast, sugar, leavening (sodium acid pyrophosphate, monocalcium phosphate, sodium bicarbonate), partially hydrogenated soybean oil, whey, mono- and diglycerides, nonfat milk. Fried in vegetable shortening.

Breaded fish portion: cod, corn flour, water, bleached flour, modified starch, potato flour, salt, yeast, sugar, cellulose gum, malt syrup, maltodextrin, onion, garlic, natural flavor, calcium propionate. Fried in vegetable shortening.

Chicken fajita meat: chicken, water, seasoning [salt, lactose, natural flavor, spices, onion powder, garlic powder, maltodextrin, hydrolyzed vegetable protein, soy sauce solids (soy beans, wheat, salt), silicon dioxide, chicken fat, modified corn starch, citric acid, dextrin, autolyzed yeast, dextrose, BHA, propyl gallate], soybean oil, salt, concentrated lime.

Chicken fillet: Boneless, skinless chicken breast marinated with water, seasoning (salt, maltodextrin, sugar, natural flavor, hydrolyzed vegetable protein, onion, autolyzed yeast, dextrose, spices, garlic, paprika, chicken fat, silicon dioxide, turmeric, mesquite smoke flavor, BHA, propyl gallate, citric acid), sodium tripolyphosphate, partially hydrogenated soybean and cotton-seed oils, chicken fat, lecithin, BHA, propyl gallate.

Chicken strips: chicken breast tenderloins; breaded with bleached flour, salt, yeast, sugar, leavening (sodium acid pyrophosphate, monocalcium phosphate, sodium bicarbonate), partially hydrogenated soybean oil, dried whey, mono- and diglycerides, nonfat milk; battered with water, bleached flour, salt, spices, leavening (sodium acid pyrophosphate, sodium bicarbonate); MSG, starch, oat flour, onion powder, natural flavor, garlic powder. Fried in vegetable shortening.

Taco salad meat: beef, water, seasonings (corn flour, salt, maltodextrin, dehydrated onion, paprika, dehydrated chili, spices, modified starch, MSG, sugar, dehydrated garlic, caramel color, citric acid), soy protein isolate.

Tortilla shell and meat filling: *Tortilla:* ground corn, water, calcium carbonate. *Filling:* beef, water, textured vegetable protein (soy flour, caramel color), soy grits, salt, chili powder (chili peppers, spices, salt, garlic), tomato paste, Worcestershire sauce, flour, taco mix (chili pepper, garlic, salt, spices, MSG, dextrose, hydrogenated soybean and cottonseed oils, disodium inosinate and guanylate), white pepper. Fried in vegetable shortening.

Side Orders and
Sandwich Ingredients

American cheese: American cheese (cultured milk, salt, enzymes, artificial color), water, sodium phosphate, sodium citrate, enzyme-modified cheese, cream, salt, lactic acid, acetic acid, sorbic acid, lecithin, artificial color.

Bacon: bacon, water, salt, sugar, sodium phosphate, sodium ascorbate, sodium nitrite.

Breaded onion rings: onions, flour, water, corn flour, salt, modified corn starch, soybean oil, dried whey, skim milk powder, dextrose, baking powder, egg yolk solids, MSG, xanthan gum, spices. Fried in vegetable shortening.

Breakfast ham: ham, water, salt, sodium lactate, sugar, corn syrup, sodium phosphate, soy sauce, natural flavor, sodium erythorbate or sodium ascorbate, sodium nitrite.

Cheddar cheese: cultured pasteurized milk, salt, enzymes, artificial color.

Croissant roll: enriched flour (contains malted barley flour), partially hydrogenated soybean oil, water, sugar, yeast, nonfat dry milk, soy flour, salt, butter, natural and artificial flavors, yeast nutrients (calcium sulfate, ammonium chloride, potassium chloride).

Diced egg: egg whites, egg yolks, modified starch.

Egg: fresh eggs.

Egg rolls: *Crust:* flour, water, eggs, partially hydrogenated vegetable oil, salt, calcium propionate. *Filling:* cabbage, pork, celery, carrot, onion, vegetable protein product (soy flour, zinc oxide, vitamin A palmitate, vitamins B1, B2, B3, B6, B12, ferrous sulfate, copper gluconate, calcium pantothenate), salt, sugar, MSG, water, black pepper, anchovy fish extract.

English muffin: enriched flour, water, yeast, corn syrup, gluten, partially hydrogenated soybean oil, salt, vinegar, dough conditioners (calcium sulfate, monocalcium phosphate, potassium bromate), citric acid, yeast nutrients (calcium carbonate, ammonium sulfate, calcium sulfate), corn meal, calcium propionate.

French fries: potatoes, partially hydrogenated soybean oil, dextrose or invert sugar, sodium acid pyrophosphate. Fried in vegetable shortening.

Ham: ham, water, salt, sugar or dextrose, sodium tripolyphosphate, sodium erythorbate, sodium nitrite.

Hamburger bun: enriched flour, water, sugar, partially hydrogenated soybean oil, vital gluten, calcium propionate, yeast, yeast nutrients (high-fructose corn syrup, ammonium salts), salt.

Hash browns: potatoes, partially hydrogenated soybean oil, salt, enriched flour, natural flavor, dextrose, sodium acid pyrophosphate. Fried in vegetable shortening.

Pancake mix: enriched flour, buttermilk, water, high-fructose corn syrup, eggs, corn flour, leavening (sodium aluminum phosphate, sodium bicarbonate), partially hydrogenated soybean oil, salt, polysorbate 65, sodium erythorbate.

Pita bread: enriched unbleached flour (contains malted barley flour), water, salt, yeast.

Pork sausage: pork, seasonings (salt, mustard, spices, sugar, MSG, spice extracts).

Processed Monterey Jack cheese: Monterey Jack cheese, American cheese, water, sodium citrate, cream, salt, sodium phosphate, sorbic acid, lactic acid, lecithin.

Processed Swiss cheese: Swiss cheese, cheese with water, cream, sodium citrate, sodium phosphate, enzyme-modified cheese, salt, lactic acid, acetic acid, sorbic acid, lecithin.

Rye bread: enriched unbleached flour (contains malted barley flour), water, rye flour, imitation or natural rye sour, salt, caraway seeds, partially hydrogenated vegetable oil, yeast, cornmeal.

Salad mix: lettuce, carrots, red cabbage.

Sesame bread sticks: enriched flour, sesame seeds, gluten, malt, partially hydrogenated soybean and/or cottonseed oils, salt, yeast.

Sesame bun: enriched flour, sugar, partially hydrogenated soybean oil, sesame seeds, vital gluten, calcium propionate, yeast, yeast nutrients (high-fructose corn syrup, ammonium salts), salt, water.

Sourdough bread: enriched flour, water, yeast, liquid sugar, soybean oil, salt, ammonia sulfate, bromate, dough conditioners, artificial sour flavor.

Swiss cheese: cultured milk, salt, enzymes, powdered cellulose.

Taquitos: *Filling:* cooked beef (with beef broth and salt), water, onions, tomato paste, seasoning (spices, garlic, onion, hydrolyzed vegetable protein), hydrated isolated soy protein (water, isolated soy protein, caramel color), soybean oil, starch, beef base [beef (roasted beef and concentrated beef stock), salt, hydrolyzed plant protein, MSG, sugar, corn oil, chicken fat, flavoring, dried whey], caramel color. *Tortilla:* ground corn, water, salt, lime, guar gum, mono- and diglycerides, cellulose gum. Fried in vegetable shortening.

Tortilla bowl: enriched flour, partially hydrogenated soybean oil, water, baking soda, calcium propionate.

Tortilla chips: stone-ground corn flour, soybean oil, water, salt, lime.

Turkey: turkey breast meat, water, salt, dextrose, sodium phosphate.

Wheat buns: enriched flour, water, corn sweetener, whole wheat flour, cracked wheat, partially hydrogenated soybean oil, vital gluten, bran, mono- and diglycerides, calcium propionate, yeast, yeast nutrients (high-fructose corn syrup, ammonium salts), salt, caramel color, molasses.

Condiments

Blue cheese dressing: soybean oil, water, corn sweeteners, maltodextrin, blue cheese, vinegar, egg yolks, salt, xanthan gum, artificial color, propylene glycol alginate, fumaric acid, natural flavors, potassium sorbate, disodium EDTA, BHA, BHT, spice, disodium inosinate and guanylate.

Buttermilk house dressing: mayonnaise (soybean oil, vinegar, egg yolks, sugar, salt), buttermilk, maltodextrin, soybean oil, vinegar, water, salt, dried yeast, MSG, sugar, dried onion, artificial color, natural flavors, xanthan gum, fumaric acid, dried garlic, potassium sorbate, disodium EDTA, BHA, BHT, polysorbate 60, spices, propylene glycol alginate.

Guacamole: avocado, water, lemon juice, sugar, salt, spices, dehydrated onion, sodium acid pyrophosphate, citric acid, sodium alginate, natural flavor, dehydrated garlic, xanthan gum, erythorbic acid.

Honey-mustard sauce: soybean oil, honey, Dijon mustard (water, mustard seeds, distilled vinegar, salt, white wine, citric acid, tartaric acid, spices), water, egg yolk, vinegar, spices, salt, potassium sorbate, xanthan gum, calcium disodium EDTA.

Ketchup: tomatoes, corn syrup, vinegar, salt, dehydrated onion, garlic powder, natural flavor.

Mayo-mustard sauce: soybean oil, Dijon mustard, water, egg yolk, distilled vinegar, high-fructose corn syrup, salt, spices, polysorbate 60, xanthan gum, calcium disodium EDTA.

Mayo-onion sauce: soybean oil, egg yolks, vinegar, sugar, salt, dehydrated onion, spices, potassium sorbate, disodium EDTA.

Mustard: water, vinegar, ground mustard seeds, salt, turmeric, dehydrated onion, spice oleoresins, paprika.

Pancake syrup: corn syrup, sugar, water, potassium sorbate, artificial flavor, caramel color, citric acid.

Pickle: cucumbers, water, distilled vinegar, salt, natural flavors, turmeric, sodium benzoate.

Reduced-calorie French dressing: corn sweeteners, water, reconstituted nonfat yogurt, vinegars, soybean oil, sugar, maltodextrin, tomato paste, modified starch, salt, natural flavors, dried onion, xanthan gum, potassium sorbate, disodium EDTA, artificial colors, paprika, dried garlic, citric acid.

Salsa: tomatoes, water, green chilies, distilled vinegar, dehydrated onions, salt, modified starch, sugar, calcium chloride, sodium benzoate, citric acid, spices, chili powder.

Secret sauce: water, soybean oil, tomato paste, mustard, vinegar, high-fructose corn syrup, dehydrated onion, chili sauce, egg yolk, salt, Worcestershire sauce, A-1 sauce, Tabasco sauce, spices, potassium sorbate, sodium benzoate, xanthan gum, extractives of paprika.

Taco sauce: water, tomatoes, distilled vinegar, salt, chilies, modified starch, dextrose, maltodextrin, spices, xanthan gum, dehydrated garlic, sodium benzoate, citric acid, MSG, dehydrated onions, beef powder.

Tartar sauce: soybean oil, dill relish, water, distilled vinegar, high-fructose corn syrup, egg yolk, whole egg, salt, xanthan gum, onion powder, spices, lemon juice concentrate, potassium sorbate, sodium benzoate.

Thousand island dressing: mayonnaise (soybean oil, vinegar, egg yolks, sugar, salt), water, sugar, pickles, vinegar, tomato paste, egg yolks, salt, spices, dried onion, natural flavors, modified starch, xanthan gum, potassium sorbate, disodium EDTA, BHA, BHT, paprika, citric acid, artificial color.

Desserts

Apple turnover: apples, evaporated apples (sulfite treated), water, high-fructose corn sweetener, enriched flour, modified starch, animal/vegetable shortening, salt, cinnamon, citric acid, potassium sorbate, nutmeg, dextrose. Fried in vegetable shortening.

Cheesecake: cream cheese (pasteurized whole milk, cream cheese culture, salt, carob bean gum), sour cream, sugar, nonfat milk, water, graham cracker meal, margarine, eggs, lemon juice, gelatin, natural and artificial flavors.

Beverages

Hot chocolate base mix: sugar and/or corn syrup, water, sweetened condensed milk, partially hydrogenated soybean, coconut, palm, palm kernel, and/or cottonseed oils, cocoa processed with alkali, whey solids, salt, disodium phosphate, natural and artificial flavors, chocolate liquor, soy lecithin, cream.

Ice milk shake mix: milk fat, sugar, nonfat milk, whey, dextrose, carrageenan, guar gum, carob bean gum, or locust bean gum.

Shake syrups added to mix:

Chocolate: high-fructose corn syrup, water, cocoa processed with alkali, sugar, salt, xanthan gum, ammoniated glycyrrhizin, potassium sorbate, ethyl vanillin, citric acid.

Strawberry: high-fructose corn syrup, water, citric acid, artificial strawberry flavor (tragacanth gum, sugar, lactic acid, water, propylene glycol, ethyl alcohol, artificial flavors), red #40, sodium benzoate.

Vanilla: high-fructose corn syrup, water, vanilla concentrate (ethyl vanillin, vanilla bean extract matter, alcohol, sugar, caramel color, water), sodium benzoate.

Ramblin' root beer: carbonated water, sugar, water, caramel color, acacia, natural and artificial flavors, sodium benzoate, quillaia, citric acid.

Fats

Butter: lightly salted butter.

Vegetable shortening: partially hydrogenated soybean and cottonseed oils, TBHQ, citric acid, dimethylpolysiloxane.

KFC

INGREDIENTS

Ingredients of the breading are not disclosed; may contain MSG.

Coleslaw: cabbage, salad dressing (Miracle Whip—soybean oil, water, vinegar, sugar, egg yolks, starch, modified starch, salt, mustard flour, spice, paprika, natural flavor—sugar, tarragon vinegar, soybean oil, salt), carrots, onions.

Little Bucket Parfait, Strawberry Shortcake: strawberries, water, sugar, corn syrup, bleached flour, eggs, partially hydrogenated coconut, palm kernel, and/or palm oils, nonfat milk, modified starch, soybean oil, invert sugar, whey, leavening (baking soda, sodium acid pyrophosphate, monocalcium phosphate), salt, natural and artificial flavors, polysorbate 60, mono- and diglycerides, sodium stearoyl lactylate, cellulose gum, polyglycerol esters of fatty acids, sodium citrate, lecithin, disodium phosphate, xanthan gum, artificial color, citric acid, potassium sorbate.

McDONALD'S

COMPONENTS

Sandwiches

Big Mac: beef patties, sesame-seed bun, cheese, lettuce, pickle, onion, salt, pepper.

Chicken Fajita: tortilla, chicken, Cheddar cheese, green peppers, onions, tomato.

Hamburger (Cheeseburger): beef patty, bun, ketchup, mustard, pickle, onion, salt, pepper. (Cheese added in cheeseburger.)

Filet-O-Fish: breaded fish fillet, bun, tartar sauce, cheese, salt.

McChicken: breaded chicken patty, sesame-seed bun, reduced-calorie mayonnaise, lettuce.

McLean Deluxe (McLean Deluxe with Cheese): lean beef patty, sesame-seed bun, lettuce, tomato, ketchup, mustard, pickle, onion. (Cheese added in McLean Deluxe with Cheese.)

Quarter Pounder (Quarter Pounder with Cheese): beef patty, sesame-seed bun, ketchup, mustard, pickle, onion, salt, pepper. (Cheese added in Quarter Pounder with Cheese.)

Salads

Chef: lettuce, tomatoes, ham, turkey,

egg, cucumber, cheese, carrots, red cabbage, radishes.

Chicken: lettuce, chicken, tomatoes, carrots, green pepper, red cabbage.

Garden: lettuce, tomatoes, egg, cucumber, carrots, red cabbage, radishes.

Side: lettuce, tomatoes, carrots, egg, cucumber, red cabbage, radishes.

Breakfasts

Bacon, egg, cheese biscuit: biscuit, bacon, egg, cheese.

Breakfast burrito: flour tortilla, egg, sausage, cheese, tomato, green chilies, onion, black pepper.

Egg McMuffin: English muffin, egg, cheese, Canadian-style bacon.

Sausage biscuit: biscuit, sausage patty.

Sausage biscuit with egg: biscuit, sausage patty, egg.

Sausage McMuffin: English muffin, sausage patty, cheese.

Sausage McMuffin with egg: English muffin, sausage patty, egg, cheese.

INGREDIENTS

Main Items

Bacon: circular bacon cured with water, salt, sugar, hickory smoke flavor, sodium phosphate, sodium erythorbate, sodium nitrite.

Beef patty: Beef.

Buns: enriched bleached flour, water, high-fructose corn syrup, partially hydrogenated soybean, corn, canola, and/or cottonseed oils, yeast, contains 2% or less of soy flour, corn flour, salt, milk powder, yeast food (calcium sulfate, ammonium chloride, ascorbic acid, azodicarbonamide), emulsifier, dough conditioner (mono- and diglycerides, diacetyltartaric acid esters and/or sodium stearoyl-2-lactylate, calcium peroxide), fungal enzymes, calcium propionate, potassium bromate.

Buttermilk biscuits: enriched bleached flour, buttermilk, partially hydrogenated soybean and/or cottonseed oils, leavening (baking powder, sodium bicarbonate, sodium aluminum phosphate, monocalcium phosphate), sugar, salt. *Biscuit spread* (used to prepare buttermilk biscuits: partially hydrogenated soybean oil, lecithin, artificial flavor, TBHQ, artificial color, methylsilicone.

Canadian-style bacon: sliced bacon with natural juices, fully cooked, smoked, cured with water, salt, dextrose, corn syrup, sodium phosphate, sodium erythorbate, sodium nitrite.

Cheese, slices: Cheddar cheese, enzyme-modified Cheddar cheese (milk, cheese culture, salt, enzymes), water, lactic acid, enzymes, water, milkfat, sodium citrate, kasal, salt, sodium phosphate, sorbic acid, apocarotenal.

Cheerios: oat flour (includes oat bran), wheat starch, sugar, salt, calcium carbonate, trisodium phosphate, vitamins A, B1, B2, B3, B6, C, D, folic acid, iron, BHT.

Chicken McNuggets: chicken (white meat, dark meat, salt, sodium phosphate), marinated up to 8% with water and chicken seasoning; breaded with water, enriched flour, corn and wheat flour, modified starch, salt, leavening (bicarbonate of soda, sodium acid pyrophosphate, sodium aluminum phosphate, monocalcium phosphate, calcium lactate), spices, whey, wheat starch, cornstarch. Fried in vegetable oil.

Chicken Patty: chicken breast with rib meat marinated up to 8% with a solution of water, chicken seasoning (ground chicken meat, natural flavor, salt, maltodextrin, spices, onion powder, wheat starch, vegetable oil, calcium phosphate tribasic, garlic powder, alpha-tocopherol), salt; battered and breaded with water, modified corn starch, yellow corn flour, enriched, bleached, and bromated flour, spices, salt, hydroxypropyl methylcellulose, wheat starch, leavening (disodium pyrophosphate, baking soda, monocalcium phosphate, calcium lactate), guar gum, dextrose, natural flavor. Fried in vegetable shortening.

Danish, apple: enriched flour (with malted barley flour), apples, sugar, partially hydrogenated soybean and/or cottonseed oils, corn syrup, whole eggs, skim milk, modified food starch, water, mono- and diglycerides, yeast, salt, egg whites, cinnamon, dextrin, lemon juice, vanillin, dried apples, agar, propylene glycol alginate, xanthan gum, gelatin, citric acid, annatto extract, ascorbic acid, carob bean gum, carrageenan.

Danish, cinnamon raisin: enriched flour (with malted barley flour), sugar, partially hydrogenated soybean and/or cottonseed oils, corn syrup, whole eggs, water, skim milk, raisins, cinnamon, mono- and diglycerides, yeast, salt, modified starch, egg whites, baking powder (sodium acid pyrophosphate, baking soda, cornstarch, monocalcium phosphate, calcium sulfate), dextrin, vanillin, agar, annatto extract, gelatin.

Danish, iced cheese: enriched flour (with malted barley flour), Neufchatel cheese, sugar, partially hydrogenated soybean and/or cottonseed oils, whole eggs, corn syrup, skim milk, water, modified starch, mono- and diglycerides, yeast, salt, egg whites, lemon juice, vanillin, dextrin, natural flavors, agar, xanthan gum, gelatin, annatto and turmeric extracts.

Danish, raspberry: enriched flour (with malted barley flour), corn syrup, sugar, partially hydrogenated soybean and/or cottonseed oils, raspberries, whole eggs, modified starch, dried apples, skim milk, water, mono- and diglycerides, yeast, salt, egg whites, dextrin, citric acid, vanillin, agar, propylene glycol alginate, xanthan gum, gelatin, citric acid, annatto extract, artificial color, ascorbic acid, carob bean gum, carrageenan.

Egg: fresh egg.

English muffins: enriched unbleached flour (contains malted barley), water, yeast, corn syrup, salt, shortening (contains partially hydrogenated soybean, canola, corn, and/or cottonseed oils), vinegar, calcium propionate, soy flour, farina and/or corn meal, whey solids, dough conditioners (mono- and diglycerides, sodium stearoyl-2-lactylate, polysorbate), yeast nutrients (ammonium chloride, ammonia sulfate, calcium sulfate, potassium bromate).

Fat-free apple bran muffin: nonfat milk, enriched bleached flour, egg whites, apples, sugar, honey, wheat bran, oat bran, modified cornstarch, natural and artificial flavors, baking powder (baking soda, sodium aluminum phosphate), spices, dextrose, salt, potassium sorbate, water, corn syrup, sodium stearoyl lactylate, polysorbate 60, maltodextrin, xanthan and guar gums, modified starch, malic acid, citric acid, carob bean gum.

Fat-free blueberry muffin: nonfat milk, enriched bleached flour, sugar, blueberries, dextrose, egg whites, 2% or less of modified cornstarch, oat bran, oat fiber, wheat bran, leavening (baking soda, sodium aluminum phosphate), malted barley flour, lactic acid, malic acid, dicalcium phosphate, sodium stearoyl lactylate, emulsifiers (sorbitan monostearate, polysorbate 60, mono- and diglycerides), xanthan gum, maltodextrin, guar gum, calcium chloride, potassium sorbate, artificial color, natural and artificial flavors, salt.

Fish patty: white fish; breaded with bleached flour, water, modified corn starch, yellow corn flour, salt, whey, dextrose, cellulose gum, paprika, and turmeric extract colorings, sodium tripolyphosphate, natural flavor. Fried in vegetable shortening.

French fries: potatoes, dextrose, sodium acid pyrophosphate, natural flavor. Fried in vegetable shortening.

Hash browns: potatoes, salt, corn flour, flour, dextrose, sodium acid pyrophosphate, spice. Fried in vegetable shortening.

Hot cakes: enriched flour, whey, yellow corn flour, sugar, soybean oil, leavening (sodium bicarbonate, sodium aluminum phosphate, monocalcium phosphate), cultured buttermilk solids, dextrose, eggs, salt, glyceryl monooleate, natural and artificial flavors.

McLean Deluxe beef patty: beef, water, encapsulated salt, carrageenan, natural beef flavor.

Sausage: pork, water, salt, spices, corn syrup, dextrose, MSG, flavorings, BHA, propyl gallate, citric acid.

Wheaties: whole wheat, sugar, salt, cereal malt syrup, calcium carbonate, calcium chloride, trisodium phosphate, vitamins A, B1, B2, B3, B6, D, folic acid, iron, BHT.

Condiments

Bacon bits: bacon cured with water, salt, smoke flavor, sodium nitrite, may also contain sugar, sodium phosphate, sodium erythorbate, flavorings, dextrose, sodium ascorbate, potassium chloride, brown sugar, citric acid.

Barbecue sauce: high-fructose corn syrup, water, tomato paste, red wine vinegar, vinegar, salt, modified starch, soy sauce (water, wheat, soybeans, salt), mustard flour, dextrose, soybean oil, xanthan gum, sugar, hickory smoke flavor, sodium benzoate, caramel color, garlic powder, spices, cellulose gum, malic acid, onion powder, natural flavors, succinic acid.

Big Mac sauce: water, soybean oil, pickles, vinegar, sugar, modified starch, egg yolk, high-fructose corn syrup, mustard flour, salt, xanthan gum, potassium sorbate, extractives of onion and paprika, spice extractives, garlic, hydrolyzed vegetable protein.

Grape jam: concord grapes, corn syrup, sugar, red grapes, malic acid, pectin.

Hot cake syrup: corn syrup, sugar, water, natural flavor, artificial maple flavor, potassium sorbate, caramel color.

Hot mustard sauce: water, high-fructose corn syrup, vinegar, soybean oil, mustard seeds, sugar, mustard flour, salt, egg yolks, spices, modified starch, xanthan gum, turmeric, sodium benzoate, annatto extract, caramel color, paprika, poppy seeds, natural flavors.

Ketchup: tomatoes, high-fructose

corn syrup, distilled vinegar, salt, natural flavors.

Mustard: water, vinegar, mustard seeds, salt, turmeric, paprika, spices.

Pickles: cucumbers, water, salt, vinegar, calcium chloride, sodium benzoate or potassium sorbate, natural flavor, alum, polysorbate 80, turmeric extract.

Reduced-calorie mayonnaise: water, soybean oil, modified starch, distilled vinegar, salt, egg whites, corn syrup solids, sugar, mustard flour, natural flavor, xanthan gum, lactic acid, potassium sorbate, lemon juice concentrate, calcium disodium EDTA, beta-carotene, turmeric.

Strawberry preserves: sugar, strawberries, pectin, citric acid.

Sweet & sour sauce: high-fructose corn syrup, water, apricot concentrate, vinegar, salt, modified starch, soy sauce (water, wheat, soybeans, salt), dextrose, soybean oil, xanthan gum, mustard flour, sugar, sodium benzoate, dehydrated garlic, spices, natural flavors, cellulose gum, malic acid, dehydrated onion, caramel color, succinic acid.

Tartar sauce: water, pickles, soybean oil, distilled vinegar, onions, modified starch, egg yolks, corn syrup solids, sugar, salt, mustard flour, xanthan gum, capers, potassium sorbate, dehydrated parsley, spice extractives.

Salad Ingredients

Cheddar cheese, shredded: milk, cheese culture, microcrystalline cellulose, salt, enzymes, annatto color.

Chicken, diced: chicken meat, water, chicken seasoning (salt, chicken fat, dextrose, chicken meat, corn starch, onion powder, natural flavor, wheat starch, alpha-tocopherol), salt, sodium phosphate.

Croutons: enriched flour, partially hydrogenated soybean oil, water, corn syrup or dextrose, salt, yeast, yeast nutrients, dough conditioners, calcium propionate, dehydrated parsley, cornstarch, dehydrated Romano cheese, spices, garlic powder.

Ham: ham, water added, cured with a solution of water, salt, sugar and/or dextrose, sodium phosphate, sodium ascorbate, sodium nitrite.

Salad mix: lettuce, carrots, red cabbage.

Turkey, julienne slices: white turkey, turkey broth, modified starch, salt, sodium phosphate.

SALAD DRESSINGS

Blue cheese: water, soybean oil, blue cheese (milk, cheese culture, salt, enzymes, artificial color), whey, egg yolks, sugar, maltodextrin, modified starch, salt, lactic acid, xanthan gum, mustard flour, spice, propylene glycol alginate, potassium sorbate, natural flavor, dehydrated onion powder, calcium disodium EDTA.

Lite vinaigrette: water, vinegar, sugar, soybean oil, dehydrated garlic powder, dehydrated onion powder, ground mustard seeds, salt, propylene glycol alginate, poppy seeds, dehydrated red bell pepper, xanthan gum, paprika, potassium sorbate, beta-carotene, calcium disodium EDTA.

Reduced-calorie red French: water, sugar, soybean oil, dehydrated corn syrup, vinegar, tomato paste, salt, Worcestershire sauce concentrate [vinegar, soy sauce (molasses, artificial color, natural flavor, salt), water, dehydrated corn syrup, salt, garlic, sugar, anchovies, tamarind, spice, natural flavor], paprika, xanthan gum, dehydrated onion, dehydrated garlic, oleoresin paprika, sorbic acid, spice.

Ranch: water, soybean oil, whey, vinegars, sugar, egg yolks, maltodextrin, salt, modified starch, lactic acid, dehydrated garlic, dehydrated onion, spice extractives, buttermilk flavor with other natural flavors, xanthan gum, dehydrated parsley, sodium benzoate, calcium disodium EDTA.

Thousand island: soybean oil, sweet pickle relish (chopped pickles, corn syrup, vinegar, salt), water, sugar, tomato paste, egg yolks, vinegar, salt, dehydrated sour cream (sour cream solids, cultured nonfat milk solids, citric acid, BHA), lactic acid, dehydrated onion powder, paprika, xanthan gum, spices and natural flavors, calcium disodium EDTA.

Desserts

Apple pie filling: apple, water, brown sugar, corn syrup, sugar, modified cornstarch, margarine, spices, natural and artificial flavors, lemon juice, salt, citric acid. *Crust and glaze:* flour, partially hydrogenated soybean oil, water, contains less than 1% of salt, dextrose, natural and artificial flavors, modified cornstarch, lecithin, sodium bicarbonate, guar gum, L-cysteine.

Birthday cake, chocolate: sugar, enriched bleached flour, skim milk, whole eggs, partially hydrogenated soybean and/or cottonseed oils, cocoa, corn syrup, egg whites, baking powder (sodium acid pyrophosphate, baking soda, cornstarch, monocalcium phosphate, calcium sulfate), salt, vanillin, propylene glycol esters, mono- and diglycerides, lactylic stearate.

Birthday cake, yellow: sugar, enriched bleached flour, whole eggs, whey, corn syrup, partially hydrogenated soybean and/or cottonseed oils, mono- and diglycerides, egg whites, baking powder (sodium acid pyrophosphate, baking soda, cornstarch, monocalcium phosphate, calcium sulfate), salt, vanillin, annatto extract.

Chocolate chip cookies: enriched flour, sugar, partially hydrogenated soybean oil, chocolate (with lecithin and vanillin), butter, whey, molasses, cocoa, salt, leavening (sodium bicarbonate, sodium acid pyrophosphate, monocalcium phosphate), emulsifiers (lecithin, sorbitan monostearate, polysorbate 60), cocoa processed with alkali, dextrose, artificial flavor.

Frozen yogurt cone: enriched flour, tapioca starch, sugar, partially hydrogenated soybean oil, leavening (sodium bicarbonate, dicalcium phosphate), annatto extract, salt, vanillin, lecithin.

Hot caramel topping: sweetened condensed whole milk, corn sweeteners, butter, water, sucrose, salt, disodium phosphate, pectin, potassium sorbate, vanillin.

Hot fudge topping: sweetened condensed skim milk, sugar, water, partially hydrogenated palm, soybean, and/or cottonseed oils, cocoa processed with alkali, dextrin, salt, sodium alginate, vanillin, potassium sorbate, disodium phosphate.

Low-fat vanilla frozen yogurt: cultured skim milk (with active yogurt cultures), milk, sugar, corn syrup solids, microcrystalline cellulose, mono- and diglycerides, guar gum, cellulose gum, artificial flavor, carrageenan, annatto extract. *Lowfat chocolate frozen yogurt* also contains cocoa.

McDonaldland cookies: enriched flour, sugar, partially hydrogenated soybean oil, corn syrup, salt, leavening (sodium bicarbonate, sodium acid pyrophosphate, monocalcium phosphate), lecithin, natural flavor.

Orange sorbet ice: water, orange juice from concentrate, sugar, corn syrup solids, natural orange flavor, citric acid, pectin, microcrystalline cellulose, cellulose gum, ascorbic acid, yellow #5, red #40.

Strawberry topping: strawberries, sugar, high-fructose corn syrup, water, natural flavors, corn syrup, citric acid, pectin, sodium benzoate or potassium sorbate, locust bean gum, artificial color, calcium chloride.

Beverages

Hot chocolate drink: water, sugar, sweetened condensed skim milk, partially hydrogenated vegetable shortening, cocoa processed with alkali, salt, chocolate liquor, disodium phosphate, artificial flavors, potassium sorbate, lecithin, cream.

Low-fat milk shake mix: skim milk, milk, corn syrup solids, maltodextrin, microcrystalline cellulose, guar gum, cellulose gum, carrageenan, artificial flavor.

Shake syrups added to mix:

Chocolate-flavored syrup: corn syrup, high-fructose corn syrup, water, cocoa processed with alkali, sugar, malt extract, salt, potassium sorbate, citric acid, vanillin.

Strawberry-flavored syrup: corn sweetener, sugar, water, concentrated strawberry juice, citric acid, sodium benzoate, propylene glycol, artificial colors and flavors.

Vanilla-flavored syrup: corn sweetener, sugar, water, caramel color, propylene glycol, vanillin, tartaric acid, sodium benzoate.

Orange drink: high-fructose corn syrup, invert syrup, water, citric acid, sodium benzoate, gum arabic, orange juice concentrate, natural and artificial flavors, glycerol abietate, artificial color, processed vegetable oil.

Fats

Frying shortening: partially hydrogenated corn and cottonseed oils, monoglyceride citrate, propyl gallate. *For Chicken McNuggets, fish patties, chicken patties, hot pies:* partially hydrogenated corn oil, cottonseed oil, monoglyceride citrate, propyl gallate.

SUBWAY

INGREDIENTS

Honey-wheat roll: White flour, water, whole-wheat flour, yeast, partially hydrogenated soybean, cottonseed, and/or palm oils, salt, dextrose, sugar, vegetable mono- and diglycerides, soy flour, caramel color, honey, diammonium phosphate, potassium bromate, potassium iodate, ascorbic acid, L-cysteine, azodicarbonamide, enzyme. (A roll made with totally refined flour is also sold.)

TACO BELL

COMPONENTS
Burritos

Bean Burrito: flour tortilla, water, red sauce, pinto beans, onion, Cheddar cheese, corn oil, salt.

Beef Burrito: ground beef, flour tortilla, red sauce, meat seasoning, onion, Cheddar cheese.

Burrito Supreme: flour tortilla, ground beef, tomatoes, red sauce, sour cream, water, lettuce, pinto beans, Cheddar cheese, olives, onion, meat seasoning, corn oil, salt.

Chicken Burrito: heat-pressed tortilla, chicken, pizza sauce, lettuce, Cheddar cheese, onion.

Fiesta Bean Burrito: pinto beans, heat-pressed tortilla, red or green sauce, Cheddar cheese.

Tacos/Tostadas

Chicken Taco: chicken, corn tortilla, lettuce, Cheddar cheese.

Chicken Tostada: lettuce, red or green sauce, pinto beans, chicken, sour cream, Cheddar cheese, tomatoes.

Fiesta Taco: seasoned beef, corn tortilla, lettuce, Cheddar cheese.

Fiesta Soft Taco: seasoned beef, heat-pressed tortilla, lettuce, Cheddar cheese.

Fiesta Tostada: Pinto beans, corn tortilla, lettuce, red or green sauce, Cheddar cheese.

Soft Chicken Taco: chicken, heat-pressed tortilla, lettuce, Cheddar cheese.

Soft Taco: ground beef, heat-pressed tortilla, lettuce, Cheddar cheese, meat seasoning. *Soft Taco Supreme* also has sour cream and tomato.

Super Combo Taco: ground beef, corn tortilla, lettuce, pinto beans, pizza

sauce, Cheddar cheese, meat seasoning, corn oil, salt.

Taco: ground beef, corn tortilla, Cheddar cheese, meat seasoning. *Taco
Bellgrande* also has lettuce and tomato.

Taco Light: ground beef, flour tortilla,
sour cream, lettuce, Cheddar cheese,
tomato, meat seasoning.

Tostada: water, lettuce, red sauce,
pinto beans, corn tortilla, Cheddar
cheese, corn oil, salt.

Other Items

Beef Meximelt: heat-pressed tortilla,
ground beef, pico de gallo, Cheddar
cheese, Monterey Jack cheese, meat
seasoning.

Chicken Meximelt: heat-pressed tortilla, chicken, pico sauce, Cheddar
cheese, pepper Jack cheese.

Enchirito: red sauce, ground beef,
water, enchirito tortilla, Cheddar
cheese, pinto beans, onion, olives, meat
seasonings, corn oil, salt.

Mexican Pizza: flour tortilla (fried),
pizza sauce, ground beef, water, Cheddar cheese, Monterey Jack cheese, tomatoes, pinto beans, processed American cheese, olives, onion, meat seasoning, corn oil, sodium citrate, dry milk
fat, salt, cellulose powder, red bell pepper, disodium phosphate, lactic acid,
sorbic acid.

Nachos: cheese sauce, corn tortilla.

Nachos Bellgrande: cheese sauce,
water, corn tortilla, ground beef, tomato, pinto beans, sour cream, olives,
meat seasoning, corn oil, salt.

Pintos 'N' Cheese: water, red sauce,
pinto beans, Cheddar cheese, corn oil,
salt.

Taco salad: lettuce, ground beef, flour
tortilla (fried), water, tomatoes, Cheddar cheese, pinto beans, sour cream,
olives, meat seasoning, onion, corn oil,
salt. Also comes with salsa, with ranch
dressing, or without the shell.

Taco salad, no beans: lettuce,
ground beef, flour tortilla (fried), tomatoes, Cheddar cheese, sour cream,
olives, meat seasoning, onion.

INGREDIENTS
Cheeses and Tortillas

Cheddar cheese: pasteurized milk,
cheese culture, salt, enzymes, calcium
chloride, annatto.

Monterey Jack cheese: cultured
milk, salt, enzymes.

Pepper Jack cheese: Monterey Jack
cheese, American cheese (cultured
milk, salt, enzymes), water, cream,
sodium citrate, cellulose, salt, red bell
peppers, sodium phosphate, sorbic
acid, lactic acid.

*Pasteurized process American
cheese:* cultured milk, salt, enzymes.

Corn tortilla: ground corn, corn or
soybean oil, water, trace of lime.

Enchirito tortilla: enriched bleached
flour, water, soybean oil, salt, sodium
stearoyl lactylate, potassium sorbate.

Flour tortilla: enriched bleached
flour, corn oil, water, soybean oil, salt,
sodium stearoyl lactylate, potassium
sorbate.

Flour tortilla, fried: enriched
bleached flour, corn oil, water, soybean
oil, salt, sodium stearoyl lactylate,
potassium bromate.

Heat-pressed tortilla: enriched
bleached flour, shortening, mono- and
diglycerides, baking powder, water,
salt, nonfat dry milk, dough conditioners
(L-cysteine, sodium stearoyl lactylate), calcium propionate,
potassium sorbate.

Condiments and Side Dishes

Taco Bell does not disclose the ingredients of meat seasoning, cheese sauce,
red sauce, green sauce, chilito sauce.

Guacamole: avocado, water, lemon
juice, sour cream, salt, sugar, onions,
sodium alginate, garlic, xanthan gum,
citric acid, erythorbic acid.

Olives: olives, water, salt, ferrous
gluconate.

Pinto beans: pinto beans, water, corn oil, salt, artificial colors.

Ranch dressing: soybean oil, buttermilk, egg yolk, salt, vinegar, starch, MSG, sugar, natural flavors, garlic, onion, egg whites, xanthan gum, spices, citric acid, sorbic acid, calcium disodium EDTA, calcium stearate, artificial flavor.

Salsa: tomatoes, green chili, onions, water, salt, starch, calcium chloride, garlic, oregano, malic acid, capiscom solution.

Sour cream: cultured cream, nonfat milk, starch, gelatin, vegetable gums (carob bean, guar, locust bean, carrageenan), sodium citrate, enzymes, glyceryl monostearate dextrose.

Hot taco sauce: water, tomato paste, jalapeños, vinegar, salt, spices, onion, xanthan gum, sodium benzoate, natural flavor.

Taco sauce: water, tomato paste, vinegar, spices, salt, xanthan gum, sodium benzoate, natural flavor.

Dessert

Cinnamon twists: fried twists (wheat flour, corn flour, rice flour, salt, corn oil), sugar, cinnamon, natural flavor, partially hydrogenated soybean and cottonseed oils.

TCBY

INGREDIENTS

Regular frozen yogurt: milk, nonfat milk, cane sugar, corn sugar, cream, active yogurt culture, natural gums (locust bean, guar, xanthan, carrageenan).

Aloe piña colada also contains aloe juice, natural and artificial flavors, artificial color including yellow #5.

Aloe tropical fruit also contains aloe juice, natural and artificial flavors, artificial color.

Amaretto also contains natural flavor.

Apricot sundae also contains natural flavor.

Banana also contains natural flavor, artificial color.

Blackberry cobbler also contains natural flavor, artificial color.

Blueberry cheesecake also contains cheesecake base (corn syrup, water, natural flavor, cheese and buttermilk powder), natural flavor, artificial color.

Boysenberry also contains natural flavor, artificial color.

Bubble gum also contains natural and artificial flavors, artificial color.

Butter pecan also contains natural flavor, artificial color.

Cappuccino also contains concentrated coffee, cocoa processed with alkali.

Cherries jubilee also contains natural flavor, artificial color.

Chocolate also contains cocoa processed with alkali.

Chocolate mint also contains cocoa processed with alkali, natural flavor.

Chocolate raspberry truffle also contains natural and artificial flavors, fudge powder (cocoa processed with alkali, cocoa powder, whey, sugar, malt extract, dextrose, salt), artificial color.

Coffee also contains concentrated coffee, caramel color.

Dutch apple also contains natural flavor.

Eggnog also contains sugar, corn sweetener, egg yolks, natural and artificial flavors, artificial color.

Golden vanilla also contains natural flavor, annatto color.

Grape also contains natural flavor, artificial color.

Key lime also contains natural flavor, artificial color.

Lemon also contains natural flavor, artificial color including yellow #5.

Old-fashioned root beer also contains caramel color, natural flavor.

Orange chiffon also contains natural flavor, artificial color including yellow #6.

Papaya also contains artificial flavor, artificial color including yellow #5.

Peach also contains natural flavor, artificial color including yellow #6.

Peanut butter also contains peanuts, peanut and/or cottonseed oil, dextrose, mono- and diglycerides.

Pecan praline also contains artificial color, artificial flavor.

Pineapple also contains natural flavor, artificial color including yellow #5.

Pumpkin also contains natural flavor, artificial color.

Raspberry also contains natural flavor, artificial color.

Strawberry also contains natural flavor, artificial color.

Strawberry cheesecake also contains cheesecake base (corn sweetener, water, natural flavor, cheese, buttermilk powder), natural flavor, artificial color.

Tango Mango also contains natural flavor, artificial color.

Watermelon also contains natural and artificial flavors, artificial color.

White chocolate mousse also contains white chocolate (sugar, cocoa butter, dry milk, lecithin, artificial flavor, salt), natural and artificial flavors.

Nonfat frozen yogurt: nonfat milk, cane sugar, corn sweetener, active yogurt culture, natural gums (guar, locust bean, carrageenan).

Aloha peach also contains natural and artificial flavors, artificial color.

Amaretto cheesecake also contains natural flavors.

Banana macadamia nut also contains natural and artificial flavors, annatto and caramel color.

Boysenberry also contains natural flavor, artificial color.

Butterscotch also contains natural flavor, artificial color.

Cappuccino also contains coffee extract, cocoa processed with alkali.

Carob pecan also contains carob powder, natural and artificial flavors, artificial color.

Dutch chocolate also contains cocoa processed with alkali, natural flavor, artificial color.

Lemon cheesecake also contains natural flavor, yellow #5.

Melon Delite also contains natural flavor, artificial color.

Old-fashioned vanilla also contains natural flavor, artificial color.

Pistachio also contains natural flavor, yellow #5.

Strawberries and Creme also contains natural flavor, artificial color.

Nonfat sugar-free frozen yogurt: nonfat milk, maltodextrin, polydextrose, sorbitol, active yogurt culture, stabilizer (cellulose gel, guar gum, mono- and diglycerides, cellulose gum, xanthan gum, carrageenan), aspartame.

Chocolate also contains cocoa processed with alkali.

Strawberry also contains natural flavor, artificial color.

Vanilla also contains annatto color, artificial vanilla flavor.

Frozen tofu: water, cane sugar, corn sweetener, partially hydrogenated sunflower oil, isolated soy protein, guar gum, lecithin, locust bean gum.

Almond pecan also contains artificial flavor.

Chocolate Moca also contains cocoa processed with alkali, concentrated coffee.

Strawberry banana also contains natural flavor, artificial color.

WENDY'S

COMPONENTS
Sandwiches

Bacon Cheeseburger: ground beef patty, American cheese, bacon, toasted white bun.

Bacon Swiss Burger: ground beef patty, mayonnaise, barbecue sauce, tomato, lettuce, bacon, Swiss cheese, kaiser bun.

Big Classic (Big Classic with Cheese): ground beef patty, mayonnaise, ketchup, pickle, onion, tomato, lettuce, kaiser bun. (Add American cheese in Big Classic with Cheese.)

Big Classic Double: ground beef patties, mayonnaise, ketchup, pickle, onion, tomato, lettuce, kaiser bun.

Chicken Sandwich: chicken breast fillet, mayonnaise, tomato, lettuce, sandwich bun.

Chicken Club Sandwich: chicken breast fillet, mayonnaise, tomato, lettuce, bacon, sandwich bun.

Double (Double with Cheese) Hamburger: ground beef patties, toasted white bun. (Add American cheese in Double with Cheese.)

Fish Fillet Sandwich: fish fillet, tartar sauce, lettuce, sandwich bun.

Grilled Chicken Sandwich: grilled chicken breast fillet, honey mustard sauce, tomato, lettuce, sandwich bun.

Junior Hamburger (Cheeseburger): ground beef patty, ketchup, mustard, pickle, onion, sandwich bun. (Add American cheese in Junior Cheeseburger.)

Kids' Meal Hamburger (Cheeseburger): ground beef patty, ketchup, mustard, pickle, sandwich bun. (Add American cheese in Cheeseburger.)

Plain Single (Plain Single with Cheese): ground beef patty, toasted white bun. (Add American cheese in the Plain Single with Cheese.)

Single Cheese with Everything: ground beef patty, American cheese, mayonnaise, ketchup, mustard, pickle, onion, tomato, lettuce, toasted white bun.

Baked Potatoes

Plain: potato.

Bacon & Cheese: potato, cheese sauce, bacon bits, liquid margarine.

Broccoli & Cheese: potato, broccoli, cheese sauce, liquid margarine.

Cheese: potato, cheese sauce, grated imitation American cheese, liquid margarine.

Chili & Cheese: potato, chili, cheese sauce, shredded imitation cheese, liquid margarine.

Sour Cream & Chives: potato, sour cream, whipped margarine, chives.

INGREDIENTS
Sandwich Fillings and Buns

American cheese: American cheese (cultured pasteurized milk, salt, enzymes, artificial color), water, cream, sodium citrate, enzyme-modified cheese, salt, sodium phosphate, acetic acid, sorbic acid, lecithin, artificial color.

Bacon: bacon, cured with water, salt, sugar, sodium phosphate, sodium erythorbate, sodium nitrite.

Chicken breast fillet (for chicken club): chicken breast fillet (with rib meat) marinated with approximately 17% of a solution of water, sodium phosphates, salt; breaded with flour, modified corn starch, salt, wheat gluten, spices, egg white powder, garlic powder; battered with water, flour, modified corn starch, salt, spices, corn flour, leavening (sodium acid pyrophosphate, sodium bicarbonate, monocalcium phosphate), garlic powder; breading set in vegetable oil.

Chicken breast fillet (for chicken sandwich): chicken breast fillet (with rib meat) marinated with approximately 6% of a solution of water, sodium phosphates, salt; breaded with flour, modified corn starch, salt, wheat gluten, spices, egg white powder, garlic powder; battered with water, flour, modified corn starch, salt, spices, corn flour, leavening (sodium acid pyrophosphate, sodium bicarbonate, monocalcium phosphate), garlic powder; breading set in vegetable oil.

Fish Fillet: Boneless, skinless fish fillet, flour, water, modified corn starch, shortening (beef fat, lard, partially hydrogenated soybean, cottonseed, and/or palm oils), corn flour, salt, leavening, sodium bicarbonate, yeast), dextrose, lecithin, MSG, onion powder, garlic powder, hydrolyzed plant protein, malt syrup, spices.

Ground beef patty: 100% ground beef.

Kaiser bun: enriched flour, water, high-fructose corn syrup, lard, buttermilk solids, cornmeal, yeast, vital wheat gluten, salt, calcium sulfate, sodium stearoyl lactylate, calcium stearoyl-2-lactylate, turmeric, paprika, whey solids, sodium caseinate, monocalcium phosphate, potassium bromate, azodicarbonamide.

Ketchup: tomato paste, distilled vinegar, corn syrup, salt, onion powder, spices, natural flavors.

Mayonnaise: soybean oil, egg yolks, water, distilled vinegar, corn syrup, mustard flour, salt, lemon juice, calcium disodium EDTA.

Mustard: distilled vinegar, water, mustard seeds, salt, turmeric, spices, natural flavors.

Pickles, dill: pickles, water, distilled vinegar, salt, alum, potassium sorbate, sodium benzoate, natural flavors, tur-

meric, polysorbate 80.

Tartar sauce: soybean oil, pickle relish, eggs, water, vinegar, onion, spices, salt, turmeric.

White bun: enriched flour, water, high-fructose corn syrup, lard, buttermilk solids, yeast, vital wheat gluten, salt, calcium sulfate, sodium stearoyl lactylate, calcium stearoyl-2-lactylate, turmeric, paprika, whey solids, sodium caseinate, monocalcium phosphate, potassium bromate, azodicarbonamide.

OTHER ITEMS

Chicken nuggets: white chicken meat, chicken skin, water, sodium tripolyphosphate, salt, MSG; battered and breaded with water, flour, partially hydrogenated soybean oil, bleached flour, modified starch, salt, corn starch, leavening (sodium bicarbonate, monocalcium phosphate, sodium acid pyrophosphate, sodium aluminum phosphate), spices, buttermilk, corn flour, citric acid, garlic powder, dextrose, yeast, xanthan gum. Fried in vegetable oil.

Chili: ground beef, chili base (tomatoes, salt, citric acid, calcium chloride), seasoning (maltodextrin, tomato powder, spices, modified starch, salt, sugar, onion powder, MSG, garlic powder, silicon dioxide, soybean oil, xanthan gum, citric acid, oleoresin paprika, disodium guanylate and inosinate, artificial flavor), vegetable mix (onions, celery, green pepper), kidney beans (dark red kidney beans, sugar, corn syrup, salt, calcium chloride, calcium disodium EDTA), chili beans (pink beans, water, sugar, corn syrup, salt, natural flavors, onion powder, calcium chloride).

French fries: potatoes, partially hydrogenated soybean oil, dextrose, disodium dihydrogen pyrophosphate. Fried in vegetable oil.

Nugget sauces: *Barbecue sauce:* corn syrup, tomato paste, water, vinegar, molasses, salt, modified starch, torula yeast, smoke flavor, spices, onion and garlic powder, sodium benzoate. *Sweet & sour sauce:* water, sugar, vinegar, pineapple juice concentrate, tomato paste, modified starch, salt, spices, sodium benzoate, artificial color, calcium disodium EDTA. *Sweet mustard:* vinegar, honey, mustard seeds, water, brown sugar, salt, modified starch, spices, soybean oil, xanthan gum, dehydrated onion and garlic, turmeric, paprika.

PREPARED SALADS

Chef: lettuce, tomatoes, eggs (eggs, water, citric acid, potassium sorbate, sodium benzoate), cauliflower, carrots, broccoli, imitation cheese (water, casein, partially hydrogenated soybean oil, natural flavor, sodium aluminum phosphate, whey, skim milk, salt, powdered cellulose, lactic acid, modified starch, sodium citrate, sodium phosphate, sorbic acid, acetic acid, vitamin A palmitate, magnesium oxide, zinc oxide, ferric orthophosphate, vitamins B1, B2, B3, B6, B12, folic acid, artificial color, guar gum), turkey ham (turkey thigh meat, water, salt, flavorings, dextrose, brown sugar, sodium phosphate, sodium erythorbate, sodium nitrite), white turkey (white turkey meat, water, salt, dextrose, sodium phosphates), cucumbers.

Garden: lettuce, tomatoes, cauliflower, carrots, broccoli, imitation cheese (see chef salad), cucumbers.

Taco: lettuce, chili (ingredients listed above), cheese [American cheese (cultured milk, salt, enzymes, artificial color), water, sodium citrate, cream, enzyme-modified cheese, salt, powdered cellulose, acetic acid, artificial color], tomatoes, taco chips (corn, partially hydrogenated soybean oil, salt).

POTATO TOPPINGS

Cheese sauce: cheese solids (aged sharp Cheddar cheese, partially hydrogenated soybean oil, natural cheese culture, buttermilk, blue cheese, artificial color), modified starch, hydrogenated soybean oil, whey, salt, corn syrup solids, sodium caseinate, natural cheese flavor, MSG, dipotassium phosphate, citric acid, spices, sodium silicoalumi-

nate, mono- and diglycerides, carrageenan.

Margarine, liquid: liquid and hydrogenated soybean oils, water, salt, mono- and diglycerides, lecithin, sodium benzoate, citric acid, calcium disodium EDTA, artificial flavor, beta-carotene, vitamin A palmitate.

Margarine whipped: liquid and partially hydrogenated soybean oil, water, salt, whey, lecithin, mono- and diglycerides, sodium benzoate, citric acid, artificial flavor, beta-carotene, vitamin A palmitate.

Sour cream: cultured cream, nonfat milk solids, tapioca starch, locust bean gum, carrageenan, sodium citrate, enzymes.

SALAD/SUPER BAR

Not all salad bar components are at every Wendy's, and some Wendy's have other components.

American cheese, imitation: water, partially hydrogenated soybean or cottonseed oil, sodium and calcium caseinate, sodium and calcium phosphate, American cheese (cultured pasteurized milk, salt, enzymes), sodium citrate, salt, adipic acid, sorbic acid, artificial flavor, artificial colors, powdered cellulose.

American cheese, shredded: American cheese (cultured pasteurized milk, salt, enzymes, artificial color), water, sodium citrate, cream, enzyme-modified cheese, salt, powdered cellulose, acetic acid, artificial color.

Bacon bits: bacon, water, salt, sugar, sodium phosphate, smoke flavor, sodium erythorbate, sodium nitrite, BHA, BHT.

Bread sticks: enriched flour, sesame seeds, partially hydrogenated soybean, palm, and/or cottonseed oils, yeast, salt, corn syrup, malt syrup, torula yeast.

Cheddar cheese, imitation: water, casein, partially hydrogenated soybean oil, natural flavor, sodium aluminum phosphate, whey, skim milk, salt, powdered cellulose, lactic acid, modified starch, sodium citrate, sodium phosphate, sorbic acid, acetic acid, vitamin A palmitate, magnesium oxide, zinc oxide, ferric orthophosphate, vitamins B1, B2, B3, B6, B12, folic acid, artificial color, guar gum.

Cheddar chips: enriched flour, partially hydrogenated soybean, cottonseed, and/or sunflower oils, bulgur wheat, sesame seeds, salt, Cheddar cheese (milk, cheese cultures, enzymes), whey, soy protein isolate, annatto, apo-carotenal, natural flavors, citric acid.

Cheese tortellini in spaghetti sauce: *Cheese tortellini:* durum semolina flour, eggs, water, ricotta cheese, Romano cheese, provolone cheese, mozzarella cheese, fontina cheese, Parmesan cheese, bread crumbs, parsley, pepper. *Spaghetti sauce:* dehydrated vegetables (tomatoes, onions, garlic, bell peppers, parsley, celery), sugar, corn syrup solids, modified starch, salt, dextrose, MSG, spices, lemon juice powder, soybean oil, hydrolyzed vegetable protein, citric acid, disodium inosinate and guanylate, artificial flavor, artificial color.

Cherry peppers, mild: peppers, water, distilled vinegar, salt, calcium chloride, sodium benzoate, yellow #5.

Chicken salad: *Chicken:* cooked diced chicken fryer meat. *Salad base:* soybean oil, water, sugar, corn vinegar, eggs, pickle relish, cider vinegar, modified starch, salt, mustard, distilled vinegar, spice, lemon juice concentrate, potassium sorbate, xanthan gum, dehydrated onion, natural flavor, calcium disodium EDTA. *Vegetables:* onions, celery, green peppers.

Chow mein noodles: enriched flour, partially hydrogenated soybean and/or cottonseed oils, salt, water, yeast.

Coleslaw: cabbage, salad dressing (sugar, oil, vinegar, corn syrup, eggs, salt, spice, natural flavors, arabic, guar, and xanthan gums, EDTA), carrots,

water, oil, sodium erythorbate, vegetable gum, sodium benzoate, potassium sorbate.

Corn relish: corn, vinegar, sugar, green and red peppers, onions, salt, spices, Tabasco sauce, potassium sorbate.

Cottage cheese: cultured skim milk, cream, salt, citric acid, guar gum, locust bean gum, mono- and diglycerides, carrageenan, calcium sulfate, potassium sorbate, enzymes.

Croutons: enriched flour, partially hydrogenated soybean, palm, and/or cottonseed oils, salt, whey, corn syrup, yeast, spices, Romano cheese, onion powder, MSG, distilled vinegar, garlic powder, beta-carotene.

Eggs, hard cooked: frozen chopped eggs. *OR* eggs preserved in water, sodium benzoate, citric acid, potassium sorbate.

Fettucini: egg noodles (enriched durum flour, water, egg solids), soybean oil.

Flour tortilla: flour, water, baking powder, shortening, margarine, salt, sodium propionate.

Garlic toast: *Toast:* enriched flour, water, high-fructose corn syrup, lard, buttermilk solids, cornmeal, yeast, vital wheat gluten, salt, calcium sulfate, sodium stearoyl lactylate, calcium stearoyl-2-lactylate, turmeric, paprika, whey solids, sodium caseinate, monocalcium phosphate, potassium bromate, azodicarbonamide. *Garlic blend:* partially hydrogenated soybean oil, salt, oleoresin of garlic, artificial flavors, vegetable lecithin, dimethyl silicone, beta-carotene, vitamin A palmitate.

Jalapeño peppers: jalapeños packed in water, vinegar, salt, corn and/or safflower oils, sugar, spices.

Parmesan cheese, grated: pasteurized part-skim cultured milk, salt, enzymes, sodium silicoaluminate.

Parmesan cheese, imitation: casein, water, modified starch, partially hydrogenated soybean oil, natural flavors, salt, powdered cellulose, sodium phosphate, citric acid, potassium sorbate, acetic acid, MSG, betacarotene.

Pasta medley: white rotini (semolina flour and water), broccoli, zucchini, red peppers, soybean oil.

Pasta salad: pasta (enriched semolina flour), Italian dressing (soybean oil, vinegar, corn syrup, salt, xanthan gum, garlic, propylene glycol alginate, spice, oleoresin paprika, yellow #5), celery, carrots, red peppers, pepperoni (pork, beef, salt, dextrose, spice, oleoresin paprika, dehydrated garlic, sodium nitrite, BHA, BHT, citric acid), onions, black olives, distilled vinegar, Romano cheese, oregano.

Pasta salad, deli: water, mushrooms, macaroni (enriched durum semolina, water, egg whites, tomato solids, spinach solids), soybean oil, carrots, vinegar, cauliflower, red peppers, black olives, mustard (vinegar, water, mustard seeds, salt, spices, turmeric, natural flavor), sugar, salt, garlic powder, spice, Cheddar cheese, natural flavor.

Peaches: peaches packed in water, corn syrup, sugar.

Pepper rings, mild: peppers, water, distilled vinegar, salt, calcium chloride, sodium benzoate, yellow #5.

Pepperoni, sliced: pork and beef, salt, water, dextrose, spices, lactic acid, starter culture, oleoresin of paprika, flavoring, sodium nitrite, BHA, BHT, citric acid.

Picante sauce: tomato juice, unpeeled tomatoes, jalapeños, onions, citric acid.

Pineapple chunks: pineapple, pineapple juice.

Potato salad: potatoes, mayonnaise, onions, water, vinegar, salt, sugar, spices, sodium benzoate, potassium sorbate, xanthan gum, soybean oil.

Puddings: *Butterscotch:* skim milk, water, sugar, partially hydrogenated soybean oil, modified starch, salt, sodium stearoyl lactylate, artificial flavor, disodium phosphate, artificial color including yellow #5. *Chocolate:* skim milk, water, sugar, partially hydrogenated soybean oil, modified starch, cocoa processed with alkali, salt, sodium stearoyl lactylate, artificial color, artificial flavor.

Refried beans: *Refried beans:* pinto beans, lard, salt, artificial colors. *Taco sauce:* water, tomato powder, modified starch, salt, dehydrated green pepper, dehydrated onion, spices, citric acid, autolyzed yeast.

Rotini: white rotini macaroni (semolina flour and water), soybean oil.

Seafood salad: *Seafood blend:* pollock, water, sorbitol, sugar, wheat starch, salt, calcium carbonate, modified starch, egg whites, natural flavors, crab extract, artificial flavors and colors (annatto, cochineal). *Salad base:* soybean oil, water, sugar, corn vinegar, eggs, pickle relish, cider vinegar, modified starch, salt, mustard, distilled vinegar, spice, lemon juice concentrate, potassium sorbate, xanthan gum, dehydrated onion, natural flavor, calcium disodium EDTA. *Vegetables:* onions, celery, green peppers.

Tuna salad: *Tuna:* tuna, water, salt. *Salad base:* soybean oil, water, sugar, corn vinegar, eggs, pickle relish, cider vinegar, modified starch, salt, mustard, distilled vinegar, spice, lemon juice concentrate, potassium sorbate, xanthan gum, dehydrated onion, natural flavor, calcium disodium EDTA. *Vegetables:* onions, celery, green peppers.

PASTA SAUCES

Alfredo sauce: Parmesan cheese blend (whey, milk, salt, cheese culture, enzymes, lactic acids), modified starch, whey cheese solids (aged Cheddar cheese, milk, salt, cheese culture, enzymes, whey), partially hydrogenated soybean or cottonseed oil, Parmesan cheese solids, corn syrup solids, buttermilk solids, natural flavors, MSG, tricalcium phosphate, red peppers, xanthan gum, mono- and diglycerides, carrageenan, artificial color including yellow #5 and #6.

Cheese sauce: cheese solids (aged sharp Cheddar cheese, partially hydrogenated soybean oil, natural cheese culture, buttermilk, blue cheese, artificial color), modified starch, hydrogenated soybean oil, whey, salt, corn syrup solids, sodium caseinate, natural cheese flavor, MSG, dipotassium phosphate, citric acid, spices, sodium silicoaluminate, mono- and diglycerides, carrageenan.

Spaghetti meat sauce: water, ground beef, dehydrated vegetables (tomatoes, onions, garlic, bell peppers, parsley, celery), sugar, corn syrup solids, modified starch, salt, dextrose, MSG, spices, lemon juice powder, soybean oil, hydrolyzed vegetable protein, citric acid, disodium inosinate and guanylate, artificial flavor and color.

Spaghetti sauce without meat: tomato powder, corn syrup solids, sugar, salt, modified starch, garlic powder, chopped onion, onion powder, MSG, spices, oregano, basil, citric acid, parsley, natural beef flavor, celery, lemon flavor, natural flavors.

Sour topping, imitation: water, coconut and soybean oils, nonfat milk solids, starch, mono- and diglycerides, lactic, citric, and acetic acids, salt, locust bean gum, carrageenan, artificial flavor and color, potassium sorbate.

Spanish rice: enriched parboiled long-grain rice, dehydrated vegetables (tomatoes, onions, garlic, bell peppers, parsley, celery, beets), salt, modified starch, hydrolyzed vegetable protein, spices (including paprika and turmeric), sugar, MSG, natural and artificial flavors, citric acid, partially hydrogenated soybean and cottonseed oils, tricalcium phosphate, extractives of paprika, disodium inosinate and guanylate, liquid margarine (ingredients listed above).

Taco chips: corn, partially hydrogenated soybean oil, salt.

Taco meat: water, ground beef, textured vegetable protein, spices, dehydrated onion, salt, garlic powder, MSG, potato flour, cocoa powder, citric acid, soybean oil, water.

Taco sauce: water, tomato powder, modified starch, salt, dehydrated green pepper and onion, spices, citric acid, autolyzed yeast.

Taco shells: stone-ground corn flour, sunflower, coconut, corn, soybean, and/or palm oils, water, trace of lime.

Three-bean salad: green beans, wax beans, dark red kidney beans, water, sugar, vinegar, salt, soybean oil, dehydrated onion, dehydrated red pepper, natural flavor, spice.

Turkey ham: turkey thigh meat, water, salt, brown sugar, sodium phosphate, smoke flavor, sodium erythorbate, sodium nitrite.

SALAD DRESSINGS

Blue cheese dressing: soybean oil, blue cheese, water, distilled vinegar, egg yolk, salt, sugar, garlic powder, xanthan gum, potassium sorbate, citric acid, natural flavor, calcium disodium EDTA, artificial flavor, oleoresin paprika.

Celery seed dressing: soy oil, corn syrup, water, distilled vinegar, egg yolk, salt, celery seeds, spices, xanthan gum, sodium benzoate, potassium sorbate, caramel color, onion powder, natural flavors, yellow #5, blue #1.

Creamy peppercorn dressing: soybean oil, water, distilled vinegar, whole eggs, dehydrated sour cream, cultured nonfat milk solids, modified starch, whey powder, salt, spices, garlic powder, xanthan gum, sugar, disodium phosphate, yeast, BHA.

French dressing: soybean oil, distilled vinegar, sugar, tomato paste, corn syrup, salt, water, onion powder, xanthan gum, spices, natural flavor, artificial color, calcium disodium EDTA.

Golden Italian dressing: soy oil, water, corn syrup, distilled vinegar, salt, dehydrated garlic, xanthan gum, spices, natural flavors, sodium benzoate, potassium sorbate, dehydrated peppers, caramel color, oleoresin paprika, yellow #5.

Hidden Valley Ranch dressing: soybean oil, water, buttermilk, egg yolk, salt, vinegar, modified starch, MSG, sugar, natural flavors, dehydrated garlic, dehydrated onion, dehydrated egg whites, xanthan gum, spices, citric acid, sorbic acid, calcium stearate, artificial flavor, calcium disodium EDTA.

Italian Caesar dressing: soybean oil, water, wine vinegar, whole egg, Parmesan cheese, salt, lemon juice concentrate, Romano cheese, spices, lactic acid, garlic powder, polysorbate 80, natural flavors, anchovy paste, dehydrated garlic, sorbic acid, MSG, xanthan gum, sodium benzoate, hydrolyzed vegetable protein, citric acid, dehydrated onion.

Reduced-calorie bacon and tomato dressing: water, soybean oil, distilled vinegar, corn syrup, tomato paste, salt, corn syrup solids, "bacon" granules (soy flour, soybean oil, salt, hydrolyzed vegetable protein, natural and artificial flavors, caramel and artificial colors), xanthan gum, propylene glycol alginate, natural flavors, polysorbate 60, sodium benzoate, potassium sorbate, onion powder, garlic powder, MSG, artificial flavors, calcium disodium EDTA.

Reduced-calorie Italian dressing: water, soybean oil, distilled vinegar, corn sweeteners, salt, granulated sugar, dehydrated minced garlic, propylene glycol alginate, granulated red peppers, xanthan gum, black pepper, sodium benzoate, potassium sorbate, whole oregano, yellow #5, beta-carotene, EDTA.

Sweet red French dressing: corn syrup, soybean oil, distilled vinegar, tomato paste, salt, onion powder, beet juice powder, xanthan gum, propylene glycol alginate, potassium sorbate, natural flavors, caramel color, oleoresin paprika, calcium disodium EDTA.

Thousand island dressing: soybean oil, water, sweet pickle relish, tomato paste, granulated sugar, salted egg yolks, white vinegar, salt, onion powder, sorbic acid, xanthan gum, natural flavors, EDTA, ground white pepper.

Salad oil: soybean oil.

Wine vinegar: grape wine vinegar.

Breakfasts

MAIN BREAKFAST ITEMS

Omelet #1: eggs, ham (cured with water, dextrose, salt, corn syrup, sodium phosphate, sodium erythorbate, sodium citrate, sodium nitrite, natural flavors), cheese.

Omelet #2: eggs, ham (see omelet #1), cheese, mushrooms.

Omelet #3: eggs, ham (see omelet #1), cheese, onion, green pepper.

Omelet #4: eggs, mushrooms, green pepper, onion.

Breakfast Sandwich: white bread, egg, cheese.

Egg, fried or scrambled: egg.

French toast: bread seasoned with sugar, dextrose, modified starch, salt, silicon dioxide, soybean oil, xanthan gum, natural and artificial flavors, spices, yellow #5 and #6. *Apple and blueberry toppings:* ingredients not available.

SIDE BREAKFAST ITEMS

Bacon: bacon cured with water, salt, sugar, sodium phosphate, sodium erythorbate, sodium nitrite.

Breakfast potatoes: potatoes, partially hydrogenated soybean, palm, and/or cottonseed oils, beef fat, dehydrated potato flakes, salt, corn flour, enriched flour, natural flavor, sodium acid pyrophosphate, dextrose.

Buttermilk biscuit: bleached white flour, sugar, leavening (sodium bicarbonate, monocalcium phosphate, sodium aluminum phosphate), salt, gelatinized wheat starch, nonfat milk solids, egg whites, partially hydrogenated soybean oil, palm oil.

Danish, apple: bromated flour (contains malted barley flour), apples, water, sugar, partially hydrogenated soybean, cottonseed, and/or palm oils, corn syrup, yeast, dextrose, modified corn starch, emulsifiers (mono- and diglycerides, lecithin, polysorbate 60), nonfat milk solids, salt, egg yolks, whey, soy flour, natural and artificial flavors, corn flour, dextrin, dry malt, cinnamon, baking powder (sodium acid pyrophosphate, baking soda, monocalcium phosphate), lemon, sodium phosphate, sodium stearoyl lactylate, calcium carbonate, maltodextrin, calcium sulfate, ammonium sulfate, spices, sodium caseinate, agar, tragacanth, potassium sorbate, sodium benzoate, artificial colors, citric acid.

Danish, cheese: bromated flour (contains malted barley flour), Neufchatel cheese (pasteurized milk and cream, salt, cheese culture, carob bean gum), sugar (sucrose, dextrose, corn syrup), water, margarine (partially hydrogenated corn, soybean, and/or palm oils, water, salt, mono- and diglycerides, lecithin, beta-carotene, sodium benzoate, artificial flavor), nonfat milk; partially hydrogenated soybean, cottonseed, palm kernel, and/or palm oils with one or more of: glyceryl lactoesters of fatty acids, propylene glycol monoesters, mono- and diglycerides, polysorbate 80; eggs, emulsifiers (mono- and diglycerides, ethoxylated monoglycerides, lecithin, polysorbate 60), yeast, corn starch, salt, whey, natural and artificial flavors, soy flour, corn flour, baking powder (sodium acid pyrophosphate, sodium phosphate, baking soda, monocalcium phosphate), dough conditioners (potassium bromate, malt, sodium stearoyl lactylate), yeast food (ammonium sulfate, calcium sulfate), lemon juice, spice, sodium caseinate, potassium sorbate.

Danish, cinnamon raisin: bromated flour (contains malted barley flour), sugar (sucrose, dextrose), water, margarine (partially hydrogenated corn, soybean, and/or palm oils, water, salt, mono- and diglycerides, lecithin, beta-carotene, sodium benzoate, artificial flavor), nonfat milk; partially hydrogenated soybean, cottonseed, palm kernel, and/or palm oils with one or more of: glyceryl lactoesters of fatty acids, propylene glycol monoesters, mono- and diglycerides; raisins, corn starch, yeast, filberts, emulsifiers (mono- and diglycerides, ethoxylated monoglycer-

ides, lecithin, polysorbate 60), cinnamon, egg whites, whey, salt, soy flour, stabilizer (sugar, calcium sulfate, calcium carbonate, agar, dextrose, salt, mono- and diglycerides, titanium dioxide, carob bean gum, artificial flavor), corn flour, natural and artificial flavors, molasses, spices, dough conditioners (potassium bromate, malt, sodium stearoyl lactylate), yeast food (ammonium sulfate, calcium sulfate), sodium caseinate, potassium sorbate, xanthan gum.

Sausage gravy: enriched flour, nondairy creamer (partially hydrogenated coconut, cottonseed, palm kernel, safflower, and/or soybean oils, corn syrup, sodium caseinate, potassium phosphate, mono- and diglycerides, sucrose, sodium hexametaphosphate, salt, calcium alginate, artificial color and flavor), chicken fat, whey, salt, modified corn starch, MSG, hydrolyzed plant protein, spices, sugar, garlic powder, onion powder, caramel color, silica gel.

Sausage patty: whole boned hog, salt, spices (sage, black pepper, red pepper), sugar, MSG, caramel color.

Syrup: corn syrup, sugar, water, high-fructose corn syrup, artificial flavor, caramel color, potassium sorbate, citric acid.

Wheat toast with margarine: enriched flour, water, whole-wheat flour, corn syrup, molasses, yeast, honey, partially hydrogenated soybean, palm, and/or cottonseed oils, wheat gluten, salt, wheat bran, raisin juice concentrate, rye meal, oatmeal, soy flour, corn flour, barley flakes, rice flour, mono- and diglycerides, dough conditioners (sodium stearoyl lactylate, calcium stearoyl lactylate, potassium bromate), monocalcium phosphate, ammonium sulfate, calcium propionate.

White toast with margarine: enriched bromated flour, water, corn syrup, partially hydrogenated soybean, palm, and/or cottonseed oils, salt, yeast, soy flour, sodium stearoyl lactylate, nonfat milk solids, buttermilk, calcium sulfate, barley malt, monoglycerides, calcium propionate, ammonium sulfate.

Desserts

Chocolate chip cookie: wheat flour, chocolate chips (bittersweet chocolate, lecithin, artificial flavors), sugar, margarine, corn syrup, high fructose corn syrup, whole eggs, molasses, vanilla, salt, modified food starch, leavening (baking soda, aluminum phosphate).

Frosty Dairy Dessert: milk, cream, sugar, nonfat milk solids, corn syrup, cocoa, dextrose, guar gum, carboxymethyl cellulose, mono- and diglycerides, carrageenan, disodium phosphate, natural and artificial flavors.

Beverages

Hot chocolate: sugar, whey, cocoa, corn syrup, nonfat milk solids, partially hydrogenated coconut, cottonseed, palm, palm kernel, safflower, and/or soybean oils, sodium caseinate, cellulose gum, salt, vanillin.

Lemonade: concentrated and pulpy lemon juice, corn syrup, sugar, lemon oil, sugar, citric acid, calcium carbonate, sodium carbonate, potassium carbonate, dextrose, trisodium citrate, natural lemon flavor, maltodextrin, vitamin C, monoglycerides, dimethylpolysiloxane, yellow #5, xanthan gum, propyl gallate, BHA.

The Fast-Food Chains

Arby's
Consumer Affairs
6917 Collins Ave.
Miami Beach, FL 33141
800-223-8473

Baskin-Robbins
Consumer Affairs
31 Baskin Robbins Place
Glendale, CA 91201
818/956-0031

Burger King Corporation
Consumer Information
P.O. Box 520783
General Mail Facility
Miami, FL 33152-0783
800-937-1800

Carl's Jr.
Carl Karcher Enterprises
Consumer Affairs
P.O. Box 4349
Anaheim, CA 92803
714/774-5796

Dairy Queen/Brazier
Consumer Affairs
5701 Green Valley Drive
P.O. Box 35286
Minneapolis, MN 55435
612/830-0200

Domino's Pizza, Inc.
Consumer Affairs
30 Frank Lloyd Wright Dr.
P.O. Box 997
Ann Arbor, MI 48106-0997
313/930-3030

Dunkin' Donuts
Consumer Affairs
14 Pacella Park Drive
Randolph, MA 02368
617/961-4000

Hardee's Food Systems, Inc.
Consumer Affairs
1233 North Church St.
Rocky Mount, NC 27801
800-346-2243

Jack in the Box
Foodmaker, Inc.
Consumer Affairs
P.O. Box 783
San Diego, CA 92112-4126
619/571-2121

KFC (Kentucky Fried Chicken)
Consumer Affairs
P.O. Box 32070
Louisville, KY 40232-2070
502/456-8300

Long John Silver's
Jerrico, Inc.
Consumer Affairs
P.O. Box 11988
Lexington, KY 40579
800-735-5555

McDonald's
Nutrition Information Center
McDonald's Plaza
Oak Brook, IL 60521
708/575-FOOD

Pizza Hut
Consumer Affairs
Department
P.O. Box 428
Wichita, KS 67201
316/681-9000

Popeyes/Church's Fried Chicken
Consumer Affairs
1333 S. Clearview Pkwy.
Jefferson, LA 70121
504/733-4300

Roy Rogers Restaurants
% Hardee's Food Systems, Inc.
Consumer Affairs
1233 North Church St.
Rocky Mount, NC 27801
800-346-2243

Subway
Consumer Affairs
325 Bic Drive
Milford, CT 06460
800-888-4848

Taco Bell Corporation
Public Affairs
17901 Von Karman
Irvine, CA 92714
714/863-4500; 800-225-8226

TCBY Systems, Inc.
Consumer Affairs
1100 TCBY Tower
425 West Capitol Ave.
Little Rock, AK 72201
501/688-8229

Wendy's International, Inc.
Consumer Affairs
P.O. Box 256
Dublin, OH 43017
614/764-3100

Our address is:
Center for Science in the
Public Interest
1875 Connecticut Ave. NW,
3rd floor
Washington, DC 20009

Now That You've Read the Book...

SEE THE POSTER!

The Fast-Food Eating Guide will make a colorful — and educational — addition to your kitchen or office wall. This information-packed wall chart compares the nutritional values of over 200 fast-food items and meals. Comparing the various products on this cheerfully illustrated chart may be the best way to get your child, patient, or self to pass up the junk and select the good stuff at fast-food restaurants.

The 24 × 18-inch wall chart comes in a standard paper version ($4.95) and in a durable, stain-resistant, plastic-laminated version ($8.95) — ideal if you dare to use the chart while eating a greasy meal.

READ THE NEWSLETTER

For the latest news on foods and health, join the Center for Science in the Public Interest (CSPI) and receive *Nutrition Action Healthletter*. This is the newsletter relied upon by consumers and journalists from coast to coast, because the information is original, it's no-holds-barred honest, and it's understandable.

With billion-dollar advertising blitzes telling you how wonderful the all-new tongue teasers are, *Nutrition Action Healthletter* gives you the real poop. Ten times a year, this nationally acclaimed, 16-page, illustrated newsletter provides tasty, nourishing recipes, debunks deceptive ads, and gives you the lowdown on vitamin supplements.

Beyond giving you the information that will improve your diet and health and save you money, CSPI serves as your personal lobbyist and watchdog in Washington. We've stopped numerous deceptive advertising campaigns, gotten restrictions on unsafe food additives, and obtained improved food labeling. We hope you'll join 200,000 other concerned citizens and become part of our membership family.

✔ Please send me

☐ CSPI's latest catalog of publications (no charge)

☐ 1-year membership in CSPI   ____ @ $19.95 = _____

☐ Fast-Food Eating Guide poster (paper)   ____ @ $4.95 = _____

☐ Fast-Food Eating Guide poster (laminated) ____ @ $8.95 = _____

☐ More copies of this book   ____ @ $7.95 = _____

Postage & Handling ____ @ $1.00 = _____

Total Enclosed _____

NAME

STREET	APT

CITY	STATE	ZIP

TOTAL AMOUNT ENCLOSED

Mail the coupon and a check for the total amount to:

Center for Science in the Public Interest
1875 Connecticut Avenue NW, 3rd floor
Washington, D.C. 20009